D1037452

QUALITY PLANNING, CONTROL, AND IMPROVEMENT IN RESEARCH AND DEVELOPMENT

QUALITY AND RELIABILITY

A Series Edited by

1. Designing for Minimal Maintenance Expense: The Practical Application of Reliability and Maintainability, *Marvin A. Moss*
2. Quality Control for Profit: Second Edition, Revised and Expanded, *Ronald H. Lester, Norbert L. Enrick, and Harry E. Mottley, Jr.*
3. QCPAC: Statistical Quality Control on the IBM PC, *Steven M. Zimmerman and Leo M. Conrad*
4. Quality by Experimental Design, *Thomas B. Barker*
5. Applications of Quality Control in the Service Industry, *A. C. Rosander*
6. Integrated Product Testing and Evaluating: A Systems Approach to Improve Reliability and Quality, Revised Edition, *Harold L. Gilmore and Herbert C. Schwartz*
7. Quality Management Handbook, *edited by Loren Walsh, Ralph Wurster, and Raymond J. Kimber*

8. Statistical Process Control: A Guide for Implementation, *Roger W. Berger and Thomas Hart*
9. Quality Circles: Selected Readings, *edited by Roger W. Berger and David L. Shores*
10. Quality and Productivity for Bankers and Financial Managers, *William J. Latzko*
11. Poor-Quality Cost, *H. James Harrington*
12. Human Resources Management, *edited by Jill P. Kern, John J. Riley, and Louis N. Jones*
13. The Good and the Bad News About Quality, *Edward M. Schrock and Henry L. Lefevre*
14. Engineering Design for Producibility and Reliability, *John W. Priest*
15. Statistical Process Control in Automated Manufacturing, *J. Bert Keats and Norma Faris Hubele*
16. Automated Inspection and Quality Assurance, *Stanley L. Robinson and Richard K. Miller*
17. Defect Prevention: Use of Simple Statistical Tools, *Victor E. Kane*
18. Defect Prevention: Use of Simple Statistical Tools, Solutions Manual, *Victor E. Kane*
19. Purchasing and Quality, *Max McRobb*
20. Specification Writing and Management, *Max McRobb*
21. Quality Function Deployment: A Practitioner's Approach, *James L. Bossert*
22. The Quality Promise, *Lester Jay Wollschlaeger*
23. Statistical Process Control in Manufacturing, *edited by J. Bert Keats and Douglas C. Montgomery*
24. Total Manufacturing Assurance, *Douglas C. Brauer and John Cesarone*
25. Deming's 14 Points Applied to Services, *A. C. Rosander*
26. Evaluation and Control of Measurements, *John Mandel*
27. Achieving Excellence in Business: A Practical Guide to the Total Quality Transformation Process, *Kenneth E. Ebel*
28. Statistical Methods for the Process Industries, *William H. McNeese and Robert A. Klein*
29. Quality Engineering Handbook, *edited by Thomas Pyzdek and Roger W. Berger*
30. Managing for World-Class Quality: A Primer for Executives and Managers, *Edwin S. Shecter*
31. A Leader's Journey to Quality, *Dana M. Cound*
32. ISO 9000: Preparing for Registration, *James L. Lamprecht*
33. Statistical Problem Solving, *Wendell E. Carr*
34. Quality Control for Profit: Gaining the Competitive Edge. Third Edition, Revised and Expanded, *Ronald H. Lester, Norbert L. Enrick, and Harry E. Mottley, Jr.*
35. Probability and Its Applications for Engineers, *David H. Evans*
36. An Introduction to Quality Control for the Apparel Industry, *Pradip V. Mehta*

37. Total Engineering Quality Management, *Ronald J. Cottman*
38. Ensuring Software Reliability, *Ann Marie Neufelder*
39. Guidelines for Laboratory Quality Auditing, *Donald C. Singer and Ronald P. Upton*
40. Implementing the ISO 9000 Series, *James L. Lamprecht*
41. Reliability Improvement with Design of Experiments, *Lloyd W. Condra*
42. The Next Phase of Total Quality Management: TQM II and the Focus on Profitability, *Robert E. Stein*
43. Quality by Experimental Design: Second Edition, Revised and Expanded, *Thomas B. Barker*
44. Quality Planning, Control, and Improvement in Research and Development, *edited by George W. Roberts*

ADDITIONAL VOLUMES IN PREPARATION

QUALITY PLANNING, CONTROL, AND IMPROVEMENT IN RESEARCH AND DEVELOPMENT

edited by

GEORGE W. ROBERTS

Babcock & Wilcox

Alliance, Ohio

Marcel Dekker, Inc. New York • Basel • Hong Kong

Library of Congress Cataloging-in-Publication Data

Quality planning, control, and improvement in research and development / edited by George W. Roberts.
p. cm. — (Quality and reliability; 44)
Includes bibliographical references and index.
ISBN 0-8247-9585-7
1. Quality control. I. Roberts, George W. II. Series.
TS156.Q364 1994
658.5'62—dc20 94-39690
CIP

The publisher offers discounts on this book when ordered in bulk quantities. For more information, write to Special Sales/Professional Marketing at the address below.

This book is printed on acid-free paper.

Marcel Dekker, Inc.
270 Madison Avenue, New York, New York 10016

Current printing (last digit):
10 9 8 7 6 5 4 3 2 1

PRINTED IN THE UNITED STATES OF AMERICA

To my wife, Sara Ray

ABOUT THE SERIES

The genesis of modern methods of quality and reliability will be found in a simple memo dated May 16, 1924, in which Walter A. Shewhart proposed the control chart for the analysis of inspection data. This led to a broadening of the concept of inspection from emphasis on detection and correction of defective material to control of quality through analysis and prevention of quality problems. Subsequent concern for product performance in the hands of the user stimulated development of the systems and techniques of reliability. Emphasis on the consumer as the ultimate judge of quality serves as the catalyst to bring about the integration of the methodology of quality with that of reliability. Thus, the innovations that came out of the control chart spawned a philosophy of control of quality and reliability that has come to include not only the methodology of the statistical sciences and engineering, but also the use of appropriate management methods together with various motivational procedures in a concerted effort dedicated to quality improvement.

This series is intended to provide a vehicle to foster interaction of the elements of the modern approach to quality, including statistical applications, quality and reliability engineering, management, and motivational aspects. It is a forum in which the subject matter of these various areas can be brought together to allow for effective integration of appropriate techniques. This will promote the true benefit of each, which can be achieved only through their interaction. In this sense, the whole of quality and reliability is greater than the sum of its parts, as each element augments the others.

The contributors to this series have been encouraged to discuss fundamental concepts as well as methodology, technology, and procedures at the leading edge of the discipline. Thus, new concepts are placed in proper perspective in these evolving disciplines. The series is intended for those in manufacturing, engineering, and marketing and management, as well as the consuming public, all of whom have an interest and stake in the improvement and maintenance of quality and reliability in the products and services that are the lifeblood of the economic system.

The modern approach to quality and reliability concerns excellence: excellence when the product is designed, excellence when the product is made, excellence as the product is used, and excellence throughout its lifetime. But excellence does not result without effort, and products and services of superior quality and reliability require an appropriate combination of statistical, engineering, management, and motivational effort. This effort can be directed for maximum benefit only in light of timely knowledge of approaches and methods that have been developed and are available in these areas of expertise. Within the volumes of this series, the reader will find the means to create, control, correct, and improve quality and reliability in ways that are cost effective, that enhance productivity, and that create a motivational atmosphere that is harmonious and constructive. It is dedicated to that end and to the readers whose study of quality and reliability will lead to greater understanding of their products, their processes, their workplaces, and themselves.

Edward G. Schilling

FOREWORD

The results of research and development are like the foundations of a great building. They may appear slight and represent only a small portion of the total cost of the edifice, but if they are defective, all the subsequent work may be imperiled.

A few years ago, George Roberts' book *Quality Assurance in Research and Development* (Dekker, 1983) helped managers learn practical methods to achieve the high standards of quality in results that are demanded in R & D activity. His first book helped a lot of us in R & D to do things right. In the years since, R & D organizations throughout industry have come to realize that doing things right, while it is necessary, is not sufficient to meet all of today's business requirements. What is needed is timely and efficient integration of the R & D program with the business as a whole. This means doing the right things at the right time, as defined by a creative understanding of both the technical and the business possibilities. It is a process in which teamwork, planning, communication, and continuous improvement play a large part. In short, it is what has come to be called a total quality management process. *Quality Planning, Control, and Improvement in Research and Development* is intended to give practical advice on this process. In keeping with its goal, this book collects the ideas of many people, who have a variety of experiences in the world of innovation and R & D. They have contributed their ideas with the hope that each of us can continue the process of learning to be more effective in the way we manage the process that leads to innovation.

E. Allen Womack, Jr.
McDermott International, Inc.
New Orleans, Louisiana

PREFACE

This is a book about total quality management, but it may not be immediately recognizable as such. Some of the chapters don't talk about the usual tools of TQM. But they do relate to critical aspects of managing quality—identifying the right areas for research, basic concepts of research project management, qualification of software, approaches for technical measurements, and verification beyond the single tool of peer review (with its built-in limitations). The book emphasizes techniques that have proved useful in achieving, verifying, and improving quality. An underlying goal is to provide tools to help researchers reduce risks and minimize costs. For example, Chapter 8 discusses how the use of statistical process control with measurement systems can save the researcher money by predicting when problems are about to begin rather than waiting for a recalibration, often months after the test is complete.

The basic premise of this book is that good research is a result of (among other things) good project management. Too often, the results of excellent technical work are obscured by not attending to details such as communicating with the customer (or knowing who the customer really is), not handling test materials properly, or not caring about the uncertainty of measurements. The work may be good, but we do not know how good it really is. Conversely, very bad work may be conducted very well and still be bad work. But in this instance, the limitations are usually readily apparent because there is an easy trail to follow.

My message is that there is a need to integrate what the casual reader assumes to be "the new management style," with its emphasis on team problem solving, with what is already known and practiced—a synthesis of project management, quality assurance, and quality improvement. AT&T calls them *best practices*. At Babcock & Wilcox, we call them *standard practices*, and others know them as *good engineering and scientific practices*.

The familiar TQM discussions are here as well. Chapter 2 deals with culture, including mission, vision, and empowerment. Chapter 3 talks about organizing for improvement and includes a special appendix on the operation of teams. Chapter 5 provides details about the use of techniques such as quality function deployment and departmental activity analysis. Chapter 11 discusses the problem-solving process with an appendix that gives case studies of research problems. Chapter 12 tells how to measure the results of TQM applied to research.

Juran* has a comprehensive way of visualizing quality. He calls it his "trilogy": quality planning, quality control, and quality improvement. Chapters 2 to 6 deal with planning, Chapters 6 to 10 deal with control, and Chapters 2, 3, and 6 and 10 to 12 deal with improvement and measurement of the overall research process.

The size of research projects varies at any laboratory. At Babcock & Wilcox, projects range from $5000 to $20 million. For labs such as the National Laboratories funded by the Department of Energy, the range of funding is much larger. The reason for emphasizing tailoring in Chapter 5 is that in several chapters, there is considerable discussion of activities affecting research that are only encountered when building a large test apparatus such as a physics detector or reactor flow model. For bench-top projects or literature studies, the reader may select only those tools and techniques pertinent to the task at hand. Another factor to consider is whether the researcher is operating alone or within an institution such as a central laboratory for a large corporation. Institutions provide a supporting infrastructure consisting of facilities and service personnel as well as an overall mission and vision. Chapters 2 to 4 and 10 to 12 speak mostly to the institution, whereas Chapters 5 to 9 are more directly aimed at the individual researcher.

My purpose is integration. I have described concepts that are drawn from experience with many research projects in an attempt to avoid having certain concepts labeled as "that's OK for that kind of work, but we're different." I also avoided being too prescriptive, again to acknowledge the need for tailoring.

A thorough discussion of each area (e.g., design of experiments) is not possible, since each of these subjects warrants a complete text of its own. The level of discussion was also a concern. Statistical designs are used frequently by some organizations, while others, equally competent in their field, do not

*J. M. Juran, "The Quality Trilogy," *Quality Progress*, August 1986.

regularly plan their research using these tools. I have grouped this information with other tools on planning, including Quality Function Deployment; but statistical designs are also used during continuous improvement. It appears to me that some scientists and engineers who are adept at using sophisticated experimental designs for technical data collection may shy away from employing a more basic application of the same tool during business process improvement.

The ideas for this text come from the significant depth of experience gained by the chapter authors throughout their professional association with research work. Much effort was spent talking to researchers and their managers at industrial research centers and government laboratories. These ideas have been discussed and either affirmed or challenged and modified through several forums provided by the American Society for Quality Control, Energy Division Committee for Research and Development, the American Society for Mechanical Engineers' Nuclear Quality Assurance Working Group for Research and Development, several conferences and workshops sponsored by the Department of Energy, the Industrial Research Institute, the Juran Institute Symposiums on Research Quality, and others.

The reader should gain the realization that there are many tools available for achieving and improving research quality. Obviously, if all you own is a hammer, every problem is perceived as a nail. But a variety of tools and concepts are presented here. An abstract artist uses the same basic colors as the house painter, but each adds his perspective and shading to create a unique expression. The tools are available for basic or applied research quality. The scientist, engineer, or manager can apply them as described here or modify them to fit a particular need.

The section on software was developed after more than fifteen years of research by Babcock & Wilcox into what works and by consulting with other industrial and government-funded laboratories. This information was integrated with national standards and then modified to fit a research environment. As always, we stress the need for maintaining a culture conducive to constantly looking for ways to improve our processes, because without such a culture, no real progress can be made.

I have attempted to pull together the many aspects that must be considered to manage research quality. Such a task is too formidable for one person and I am thankful for the wide variety of expertise provided by the chapter authors and other contributors. There was substantial additional contribution by Ernst & Young, including the area activity analysis (originally identified as Department Activity Analysis by Dr. H. J. Harrington of Ernst & Young), quality function deployment, and the seven-step problem-solving process. Chapter 4 was provided in its entirety by Arthur D. Little and is excerpted from their book, *Third Generation R & D* authored by Tamara Erickson, Philip Roussel, and Kamal Saad. I also sincerely appreciate the reviews by David Burrows of the Savannah

River Ecology Laboratory, John Dronkers of Lawrence Livermore National Laboratory, and Ron Geoffrion of Los Alamos National Laboratory. Others who lent a valuable hand were Larry Basar, Bob Casebeer, Tom Davis, Ken Everett, Bob Hoff, Andy Kisik, Tim Kriner, Laura McClung, Dr. Phil Pfund, Barb Reinsel, Maxine Schnabel and Janet Slease, all of B & W, and Natalie Elhardt of PA Consulting Group. My thanks to all of them.

George W. Roberts

CONTENTS

CONTRIBUTORS

David D. Boath PA Consulting Group, Hightstown, New Jersey

Mark Bodnarczuk National Renewable Energy Laboratory, Golden, Colorado

Robert J. Cartwright Research and Development Division, Babcock & Wilcox, Alliance, Ohio

Michael T. Childerson Research and Development Division, Babcock & Wilcox, Alliance, Ohio

Taz Daughtrey Naval Nuclear Fuel Division, Babcock & Wilcox, Lynchburg, Virginia

Tamara J. Erickson North America Management Consulting Directorate, Arthur D. Little, Inc., Cambridge, Massachusetts

David T. Farrell Ernst & Young, San Jose, California

Robin K. Gill Research and Development Division, Babcock & Wilcox, Alliance, Ohio

Joseph J. Gufreda The Kaizen Institute of America, Greenwood, Indiana

William H. Hamilton Consulting Engineer, Ligonier, Pennsylvania

Robert H. Lochner Consultant in Quality Improvement, Milwaukee, Wisconsin

Richard P. Maddams PA Consulting Group, Hightstown, New Jersey

Maureen J. Psaila-Dombrowski Research and Development Division, Babcock & Wilcox, Alliance, Ohio

George W. Roberts Research and Development Division, Babcock & Wilcox, Alliance, Ohio

Philip A. Roussel Arthur D. Little, Inc., Cambridge, Massachusetts

Kamal N. Saad Arthur D. Little, Inc., Brussels, Belgium

Nick G. Sandru Research and Development Division, Babcock & Wilcox, Alliance, Ohio

Rolf B. F. Schumacher COAST Quality Metrology Systems, Inc., San Clemente, California

QUALITY PLANNING, CONTROL, AND IMPROVEMENT IN RESEARCH AND DEVELOPMENT

INTRODUCTION

George W. Roberts
Babcock & Wilcox
Alliance, Ohio

There is an intense competition for research dollars, just as there is for industrial product market share. There is no shortage of problems to be solved, and although it seems axiomatic that problems can be solved with technology and money, there is a limited amount of money as well as increasing demands for its proper use. Regardless of the success that technology may have enjoyed in the past in solving problems, everyone, especially funding agencies and technology managers, continues to want to get the most value for the dollar. The realization is growing that quality—"true" quality—is the key.

But what is "true" quality and how do we get it? Juran (1) describes two types: quality of grade and quality of conformance. Quality of grade has its basis in the technical features of the service or product provided. In the case of research results, the grade of quality is determined by attributes such as novelty, applicability to immediate needs, timeliness, direct usefulness, and the ability to efficiently transfer the technology to practical use—or in the case of basic research, the degree to which the results challenge the prior understandings or paradigms of the scientific community.

Quality of conformance is concerned with reducing costs due to time spent correcting errors. It is generally recognized that research does not progress by doing everything right the first time, since the premise is that the activity is experimental and the results often uncertain. But the concept of reducing costs by improving performance has value not only in support group operations, but in the actual performance of research. The concepts of customer focus, understanding work processes, and employee empowerment are real issues for the industrial researcher, because they affect the way that the researcher responds to a business unit with a problem. When a group of researchers comes together to look at how they do their work and the processes they follow, the stage is set for improvement.

If we look at scientific work as having three phases—proposal, execution, and analysis—we can consider that each of these has certain characteristics that are generally thought of as indicating a level of quality. As we consider these characteristics, we can formulate means toward achieving that level. Since ''quality'' is often defined as meeting or exceeding customer expectations, steps must be taken to determine what those expectations are. This is no easy task, particularly for basic science.

Once the customers and their needs are defined, proposals are generated to obtain agreement on specific work projects. Proposals need to clearly describe the problem. They should indicate the boundaries of the investigation and state the hypothesis in a manner that permits testing for proof or disproof. Included in the proposal must be a statement of need, why the work should be done at all. Is there any novelty involved? The greater the novelty, the greater the possibility for the big success—or failure—and the more risk there is. This points to a need to estimate the probability for success, probability that is often inversely proportional to the novelty of the work. Some government laboratories will not accept work that does not have high risk because their mission is to do work that industry cannot afford to do for itself. Industrial researchers balance incremental improvements that have predictable results and known investment requirements, but generally a low rate of return, against much less certain projects that look for the chance to turn the market upside down. Often, a research proposal's acceptability will hinge on the qualifications of the lead investigator and whether he or she has a proven record for success.

Once the project is complete, the final report is looked upon as the ''proof of the pudding.'' Results are recorded along with their interpretation, and conclusions are presented. What has been accomplished, and is it quality? And how is that determined? If it is ''quality,'' it may meet the customers expectations even if the answer obtained is not what was expected. Sometimes the very highest quality research has rendered totally unexpected results. But anyone who pays millions of dollars for research facilities is not interested in work that gives poor data and conclusions that are suspect. Prudence requires steps to assure that the right work is to be done and, once authorized, done right.

Where do you apply the emphasis on achievement and verification? Current practice seems to concentrate on research proposals and the final reports. These are subjected to formal peer review for work funded by the National Institutes of Health (NIH) and the National Science Foundation (NSF). The apparent expectation is that full verification of research quality will only come when other competing institutions reproduce the experiment. While this has worked well in the past, especially for small experiments, the tremendous cost of big science prohibits construction of duplicate facilities merely to reproduce someone else's results. There is a real possibility for errors to be introduced into the data-taking process that cannot be verified by review of a final report. Even if the data are self-evident, is the peer review process the optimum means self-evident, is the peer review process the optimum means for verifying quality?

It appears that the middle phase—execution—is the missing link to fulfilling the promises of the proposal and in establishing the basis for learned analysis. And it is here that the prudent project leader can reap significant benefits from good project management coupled with good engineering and scientific practices. Small projects are more visible and require less intermediate control. But as complexity increases, so does the need for periodic assessment at various levels of detail. Daughtrey in Chapter 9 describes two ''pathological'' routes often taken during software development and provides guidance in the proper level and timing of verifications.

But what are the criteria to which each of these phases should be held? What are the characteristics that are embodied in good research? How can they be achieved and verified? If peer review has its limitations, what other means are there to determine whether the data are good? Key issues in each phase of a research project are listed below:

Proposal

- The hypothesis is clearly and adequately described.
- The investigation is properly bounded.
- The experiment is designed based on appropriate statistical planning.
- The proposal has sufficient novelty.
- The investigation will solve a need.
- The proposer's qualifications match the needs of the project.
- The work is within the mission of the research organization.
- There is a reasonable chance for success.

Execution

- Background material is relevant and properly applied.
- Team members are appropriately trained and qualified.
- Theoretical/calculational derivations are adequate.
- Experimental models and equipment are suitable.
- Physical/functional configuration of experimental apparatus is known and understood.

- Functional performance of experimental apparatus is known and understood including:
  - Measurement uncertainty
  - Measurement repeatability
  - Process characteristics
  - Environmental influences
  - Influence of operators
- Project documents provide a clear record of the path followed.
- Project was completed on time.
- Project was completed within budget.

Analysis

- Results were interpreted correctly.
- Investigators provided a learned analysis defensible by the results.
- Conclusions were technically appropriate given the analysis.
- The final report is concise, clear, and complete.

We intend to assist the researcher in addressing most of these issues and to show a process a research institution may use to critically examine itself and take action to improve. While recognizing that the quality of inductive and deductive thinking is of paramount importance in setting up and analyzing the results of research work, no attempt is made to address that special area. The strength of peer review is that it provides an opportunity for thorough verification of the quality of those thought processes.

REFERENCE

1. J. M. Juran, *Quality Control Handbook*, McGraw-Hill, New York, 1974

1

THE CHALLENGE FOR RESEARCH QUALITY

George W. Roberts
Babcock & Wilcox
Alliance, Ohio

> The notion of scientists as independent scholars, motivated solely by a thirst for knowledge and unconcerned about the eventual utility of their results has been banished for good (even if it only half-existed in practice). Basic scientists—like company managers—are now clearly expected to keep their eyes on the bottom line; the only difference is of time scale.
>
> *David Dickson* (1)

Company managers judge the success of any activity by its ability to attract customers and by its productivity. Products and services that satisfy a need or enhance the quality of life will attract customers. But the activity, whether it is a profit-oriented enterprise or not, must be able to give customers what they want at a price they are willing to pay. Increased productivity permits industrial corporations to give more value per dollar to their customers and to thereby increase sales and profits. This ability to increase productivity is directly related to advances in science and technology—the direct result of organized research and development.

Once people trusted technology to provide them with the good things of life. But in recent years, the public has realized that our scientific achievements, although impressive, can sometimes produce unpleasant by-products or unwanted results. The tragedies of Three Mile Island, Bhopal, and Chernoble and concerns over acid rain, the greenhouse effect, and toxic wastes have caused such public outcries that these issues and their technological problems can no longer be ignored. Therefore, those that practice technology may be suspected of wanting to "play with a new toy" before they understand the dangers presented by that toy or of rushing into an application in order to beat the competition before adequate steps have been taken to safeguard society and its environment.

Dickson describes the rise and fall of the fortunes of scientific funding (1). Direct government support had its significant beginnings in the years of World War II with establishment of the National Laboratories to help develop nuclear weapons. With the startling realization that the Soviet Union had placed a satellite in orbit and could navigate the length of the United States with impunity, scientific funding grew rapidly through the 1950s and 1960s. A unified goal captured the nation's imagination and the thought of being the first to place a man on the moon was enough to spur engineers, office workers, and technicians through long days and weekends.

As the public mood turned more and more against the Vietnam War, a corresponding disenchantment set in with respect to the technology institutions that provided the means to prosecute the conflict. Social programs were the proper place to spend the nation's resources. The need was for less basic research and more direct application to solve problems. One new goal was a ten-year war on cancer, and it seemed to represent our thinking about technology. After all, if we could put a man on the moon, why couldn't we———?

By the mid-1970s, there was a realization that industry needed help. An automotive executive had his paradigm shaken. He may have derived years of comfort by looking out of his office window over a parking lot saturated with his company's product with a few other Big Three models driven by a few rebels in the work force. Look at all that Detroit iron! Surely, this represented the "real world." Then he took a trip to the West Coast and started noticing swarms of foreign cars and hearing for the first time wonderful tales of low maintenance and attentive sales people from dealers like Datsun, Toyota, and Honda (2). It took a long time for the message to sink in, but gradually the realization spread that U.S. industry was losing its leading position and it wasn't just in automotive and electronic goods. The competition was everywhere. Once again, technology was called to the rescue. Government funding was prioritized to those areas that could help business regain its competitive edge. The national labs were given the mandate to focus on application. There is a belief that there is a positive relationship between science and economic growth. Economist Edwin Mansfield concluded that U.S. industries that spend relatively large amounts of money on R & D are also the leading industries in manufactured exports, foreign direct investment, and licensing (1). With a direct tie of science to industry and the historic relationship of science to the military, it is no surprise that the U.S. government is actively using science as a tool of government policy to reward friendly nations with transfer of knowledge and to achieve its own economic agenda internal to this nation. But the result for laboratories that are the recipients of the government's largess is greater control, more expectations of accountability, and return on investment.

The need to support business is also felt by universities; their motivation is further stimulated by potential financial rewards from patents. Business funds the university, exercises less control over the technical direction of the work, but stands ready to use any promising results for commercialization. Dickson describes two lucrative arrangements, citing the success of stannous fluoride toothpaste discovered by the University of Indiana and licensed to Proctor & Gamble as Crest. He also cites over $100 million received by the University of Wisconsin from licenses (1).

As the universities turn to industry for money and scientific freedom, they must learn the business language. Accountability is the bottom line and one of the industry rules is that good research is facilitated by good project management.

Juran (3) has stated that losses due to poor quality can range from 2 to 25 percent of sales for a manufacturing company depending upon the complexity of its products. The losses can be as high as 35 percent of sales for a service company. Research and development appears to be more of a service than a manufacturing operation, but if a 25 percent estimated loss is applied to a $158 billion level of annual research in the United States (4), the potential loss would be over $39 billion to our economy. Feigenbaum tells us that "traditional industrialization practices have created today what our General Systems engineers call 'hidden plant'—both factory and office—sometimes amounting to as much as 40 percent of total productive capacity. This is the plant that exists because of bad work. It exists to replace products recalled from the field, or to retest and reinspect rejected units, or to maintain unduly high stacks of spare parts" (5). If this concept of the "hidden plant" holds true, most of the waste and losses will occur in rather innocuous areas. It is doubtful that these losses will show up as spectacular failures or significant technical errors—they will be due to inefficient systems and poor project management.

One of the most notorious instances of an apparent technical error (which may have been prevented by good project management) is the experiment reported by Fleischmann and Pons (F-P) where they stated that they had detected the presence of "cold fusion" (6). The scientific world was shocked not only by the fact that someone had claimed to measure this phenomenon but that the results were published in a press release rather than the accepted method of peer review and publishing in a respected scientific journal. The natural result of a discovery of this magnitude was a flurry of action by laboratories around the world to duplicate the results. One by one, the scientists checked in with their results—most found no data to confirm the presence of cold fusion. As the negative results piled up, so did the criticism of how F-P conducted their experiment. One research group complained that F-P had not given enough details to permit an exact duplication of their results (7). An apparent confirmation of cold fusion by researchers at Texas A & M University was later questioned as a possible result of fraudulent spiking of the electrolyte (8). At the least, the Texas A & M problem could have been caused by tritium contamination found in the palladium. Other research teams that had detected tritium had purchased their palladium from the same supplier so it is conceivable that their stock was contaminated as well. Another team reported a problem with consistency in a reference standard purchased from a reputable chemical company (9). The variation in tritium content in two bottles (supposedly from the same lot) bracketed the tritium levels reported by F-P and could mislead researchers to think that the tritium had been created during a cold fusion process.

Throughout all of this, there was argument whether the traditional scientific peer review process is outdated given the speed with which new information is disseminated (i.e., electronic mail), the need to obtain patent protection, and, generally, the desire to be the first to break the news. Others will say that the flap over cold fusion was exactly what a good peer review would have prevented. It is questionable whether a good peer review could have been performed, however, since the specifics of the experimental setup were not given. Was this bad science or bad project management? Could any peer review have detected problems arising from erroneous measurement standards given that peer reviews do not usually include an attempt to actually duplicate experimental results? Experimental confirmation could only come after public disclosure. An analogy in the commercial world would be to distribute completed automobiles to the public and let them perform the first and only inspection—a risk no prudent manufacturer will ever take.

A research project can be thought of as having three phases: proposal, execution, and analysis. The first and third phases can be described in detail in a prospectus or a report. A competent reviewer can then assess the acceptability of the proposed work against target standards and the adequacy of the author's final analysis and conclusions based on the reviewer's own knowledge and experience in dealing with the information the author has provided. These phases are what is subject to the normal peer review process and, where experimental results are involved and the experiment may be replicated at reasonable cost, there may be no need to delve further into the details to determine if the findings are valid. But as the complexity increases, so does the potential for the middle phase, the execution, to become an increasingly significant factor in masking whether the work was high quality or whether it represented an excellent analysis of suspect or inaccurate data. In such instances, the normal tools of peer review are inadequate and the tools of research project management must be employed. While these tools are similar to those employed on any project, there are some subtle differences, particularly when the project manager may also be the primary technical investigator. Although the tools may seem to take precious time away from getting to the "real work" of the research, they have been proven to be valuable savers of time and aggravation to the scientist and engineer.

There is a pressure on commercial industrial establishments (particularly, but not exclusively) to make decisions that are justified by the marketplace. While those decisions may be correct technically and financially as the supplier considers the needs of a majority of customers, those decisions may not be in the best interest of all users, particularly those from a research facility. If a laboratory assumes that its supplier of chemical standards is good only because the supplier is a big company and well known, the laboratory assumes a risk. That supplier may not control its processes as tightly as needed by the laboratory because the controls that are used are "good enough" for most of the supplier's customers. For that reason, the laboratory needs to investigate their significant suppliers thoroughly and, if necessary, formally qualify them. Having to balance features against cost is not limited to the commercial industrial sector, however. Most managers have to operate within tight budget limitations and have to make judgments as to where to best spend their money. One of those judgments is whether to assume that work is performed correctly without obtaining any corroborating objective evidence.

Research quality is not only a matter of doing the job right, it also means doing the right job. And from the perspective of an industrial research and development lab, that means identifying the right products and getting them to market. There is a natural tension between giving researchers enough freedom to exercise their creativity by pursuing their own pet projects and the need to focus their attention on areas that are likely to be useful commercially. The United States has been taken to task for funding too many grand projects such as the Superconducting Super Collider and the Space Station as "engines" for generating technological spin-offs, particularly since many good ideas that were "spun off" from prior government programs never reached the marketplace through U.S. businesses. A litany of missed opportunities include Western Electric licensing transistor technology to Sony, Unimation licensing Kawasaki to make robots, Ampex video recorder technology taken over by the Japanese, and Canon taking away the laser printer from Xerox.

An excellent case history of how a good idea in the lab got blocked from going to market is presented in the book *Fumbling the Future: How Xerox Invented, Then Ignored, the First Personal Computer* (10). Case studies of "lessons learned" are always painful even when used in the relative privacy of the research organization's staff management meetings. While it is unfortunate that any company as fine as Xerox should be held up as an example of what could go wrong, the Xerox experience is of value to all managers.

Perhaps what needs to be stressed is that any research organization has customers. In industry, the customer of the research organization is the business unit that will ultimately be responsible for marketing and producing the product.

Again and again, we read the admonitions to institute research, product development, and manufacturing process design in parallel with each other and get away from the old form of "over-the-wall" product development where each step is taken in series with no concern for the impact of one group's actions on another's. If for no other reason, having the business unit "customer" closely involved in the work will keep the attention of the researchers focused on the needs of that customer and the realities of the commercial world. If you expect support from the business unit managers, they must believe you are going to help them succeed. But a constant business unit customer focus will create tension. If the researcher's work must always be tied to a business unit, the result is a focus on short-term results. There is no time for creative abstraction about "what could be." Researchers do not like to just follow orders and only perform tests assigned to them by the business units. They are always restless to probe the unknown and explore new ideas.

A balance is needed. Xerox changed management at its Palo Alto Research laboratory and, while they are steering the research to areas with a higher prospect for commercialization, abstraction is not forgotten. They have retained 20 percent of their research budget for the blue sky projects (11). General Electric funds 75 percent of its central research lab by individual contracts with the business units. The remaining 25 percent for exploratory work is from assessments. The focus of its central lab is for "game changing" technology, leaving the incremental improvements to the business units. (12) These ratios suggest that, in both cases, decisions were made to allocate 20–30 percent of R & D funds to true innovation and the rest is used to convert technical knowledge for the corporate bottom line.

This predilection for the "game changers" is counter to the idea of continuous improvement. Continuous improvement stresses the need to improve a thousand things by 1 percent, but the home-run hitter wants to improve one thing by 1000 percent. If the rewards accrue to the star player who hits an infrequent home run, no one wants to spend (or waste) their time on small projects with limited payoff. But the Japanese have shown that it is the continuous striving to make small improvements that adds up to major successes.

The advantages for a sound quality program in an R & D activity stem from four areas of concern: product liability, government regulation, sales, and costs. Once into development testing, preproduction data is obtained that characterizes a product's design. Product performance parameters are obtained that form the basis for design trade-offs for reliability assessments, cost optimization studies, and product liability risk assessments.

The results, particularly of industrial R & D, are often part of the justification for a new or changed design. Manufacturers are being held to the level of experts in their field by the courts (13). If R & D provides the confidence that the design is safe, then the R & D will be subject to scrutiny by the courts in deciding to what extent a manufacturer was reasonably prudent in providing protection to the public. Therefore, the records to substantiate the conclusions and recommendations from an R & D project should be both available and understandable.

Certain government specifications and federal codes require extensive testing to prove the capability of a product to function under adverse conditions. This qualification testing, whether done for reliability assessments or for safety assessments, is required to be done under carefully monitored conditions.

Properly qualified R & D provides substantiated data to back up performance claims to customers and can be used as an effective tool in promoting sales. For high-cost items, it provides added confidence to a customer that the item will perform as intended, particularly for first-of-a-kind or custom-built facilities or systems. The data then back up the manufacturer's specification for maintenance and optimum operating conditions for the item. They provide a clear set of documentation to evaluate the reasons for success or failure of the item and can be used as a warning to customers to avoid certain undesirable practices.

Ultimately, the question comes down to a matter of cost. Constructing R & D test facilities costs money for the company. The company needs the confidence that those facilities will provide accurate data and meet performance standards established by the customer or its own research scientists and engineers. Then, by properly qualifying R & D with adequate systems and controls, the overall effort to qualify a product design can be reduced (14). Records from the R & D are capable of justifying specific design concepts without the need for complete qualification testing of production first articles. Certain design assumptions tested during R & D can be supported by the quality verification activity. Qualification testing or full-size prototype testing can then concentrate on proving large-scale production processes or system-level design concepts.

Malcolm Baldrige, U.S. Secretary of Commerce, stated that, "For managers, the challenge is to create an organizational environment that fosters creativity, productivity, and quality consciousness" (15). He pointed out that 40 percent of all costs in getting a product to the marketplace are in the design cycle. He also stated that top management must emphasize prevention, rather than correction. This is a recurring theme with all experts in the field of quality management. The National Advisory Council for Quality identified eight universal quality improvement steps (16). Number 7 states that emphasis for quality must be shifted from error detection to error prevention. This is the same principle used for navigational correction. The earlier you make the change, the greater the effect down the road. But, obviously, you need good information before you make decisions to change. Money spent assuring the quality of research data will pay dividends over again when the product enters the marketplace. Crosby said, "Quality improvement through defect prevention . . . is the foundation of all ITT quality programs" (17).

When designs are rushed to production without proper verification or based on technology derived from hastily conceived and executed experiments, the stage is set for a long and costly process of piecemeal defect detection, analysis, and feedback for design correction actions that could have been prevented. If any company has a research and development organization, that company should heed the advice contained in the National Advisory Council for Quality's Universal Quality Improvement Step Number 4, "Formal quality improvement activity must be launched in every organization" (16).

George Hardigg, vice-president and general manager of Westinghouse's Advanced Power Systems Division, put it very succinctly when he said, "studies and experiments that are not conducted under controlled verifiable conditions and thereby produce data that is useless in supporting subsequent design decisions are a waste of time and money" (18).

SUMMARY

The heyday of unfettered research is over. There is fierce competition for scarce funds and sponsors want to get the most out of their investment. Government is demanding that taxpayer-funded research be aimed at improving our industrial competitiveness. There is little tolerance for spectacular failures and no tolerance for waste. It is generally believed that research leads technology, which leads product innovation and improvement, which leads sales and profits. So, much is expected of our research community. But industry uses other methods for product and process improvement as well. As those methods become familiar, they can be seen to be effective upstream in the research process. We are truly in a time of great change. Researchers must not only be the catalysts for change, they must live in a culture of change.

REFERENCES

1. D. Dickson, *The New Politics of Science*, University of Chicago Press, Chicago, 1988
2. J. Peters and N. Austin, *A Passion for Excellence*, Random House, New York, 1983
3. J. M. Juran, *Quality Control Handbook*, 3rd ed., McGraw-Hill, New York, 1979
4. T. Studt, "$158 Billion and Holding for U.S. R & D in 1991," *R & D Magazine*, January 1991, p. 38
5. A. V. Feigenbaum, "Quality and Business Growth Today," *Quality Progress*, Vol. 15, No. 11, 1982, p. 22
6. M. Fleischmann and S. Pons, "Electrochemically Induced Nuclear Fusion of Deuterium," *J. Electroanal. Chem.*, Vol. p. 261, p. 301 (1989)
7. M. Keddam, "Some Comments on the Calorimetric Aspects of the Electrochemical 'Cold Fusion' by M. Fleischmann and S. Pons," *Electrochemica Acta*, Vol. 34, No. 7, pp. 995-997 (1989)

8. G. Taubes, "Cold Fusion Conundrum at Texas A & M," *News & Comment*, Vol. 248, p. 1299 (June 1990)
9. H.-S. Bosch, G. A. Wurden, J. Gernhardt, F. Karder, and J. Perchermeier, "Electrochemical 'Cold Fusion' Trials at IPP Garching, U.S. Department of Energy, Oak Ridge, TN, 1989
10. D. K. Smith and R. C. Alexander, *Fumbling the Future: How Xerox Invented, Then Ignored, the First Personal Computer*, William Morrow, New York, 1988
11. J. Pitta, "Bean Counters Invade Ivory Tower," *Forbes*, September 18, 1989, p. 198
12. M. Garfinkle, "Quality in R & D," Juran Institute Symposium on Managing for Quality in Research & Development, Juran Institute, Wilton, CT, 1990
13. *Product Liability: The Present Attack*, American Management Association, 1970, p. 24.
14. G. Roberts, Quality Assurance in R & D, *Mechanical Engineering*, Vol. 100, No. 9, p. 41 (1978)
15. M. Baldrige, "Designing for Productivity," *Design News*, Vol. 38, No. 13, p. 11 (July 1982)
16. "News & Trends," *Production Engineering*, Vol. 29, No. 7, p. 8 (July 1982)
17. P. Crosby, *Quality Is Free*, McGraw-Hill, New York, 1979
18. G. Hardigg, Proc. Sixth Annual National Energy Division Conference, ASQC, EP1.17 (1979)

2

ESTABLISHING CULTURE

Joseph J. Gufreda
The Kaizen Institute of America
Greenwood, Indiana

George W. Roberts
Babcock & Wilcox
Alliance, Ohio

I. THE MANDATE FOR CHANGE

"What is impossible now, but, if it were possible, would fundamentally change the way you do business?" Joel Barker challenges everyone with this question in his book, *Future Edge: Discovering the New Paradigms* (1). He points out that, while paradigms are useful for establishing boundaries and helping us identify and solve problems, they can also blind people to new opportunities which can then radically and adversely affect them. As an example, Barker cites the Swiss watchmakers who disregarded the quartz movement innovation (invented by the Swiss in 1967) and who, within 10 years, saw their market share plummet from 65 percent to less than 10 percent as Seiko took over. Barker says that when a paradigm shifts as it did with quartz movements, everyone goes back to zero, adding "Your past success guarantees nothing." He provides a powerful argument for a climate of receptiveness to change.

The research laboratory is expected to be the harbinger of change. Companies such as 3M have stated objectives like requiring that 40 percent of their current business come from products *unknown* four years ago. Such a mandate brings with it tremendous pressure to constantly identify new opportunities. And the burden for leading the search lies squarely on the shoulders of the research laboratory. But, with this responsibility for being the change agents for the rest of the world (or the rest of the company), there must also be a recognition of the need for constant change within the research institution itself. As with the Swiss, anyone can be blinded by their own success and arrogance, unable to foster an openness to creativity because of their own regimen. The young turks of yesterday are the bureaucrats of today. They seek to hold the gains they made on the way up and to protect others from the errors they made on their journey in a vain hope that the path to success can somehow be shortened or made easier. It is an error-driven, risk-averse mentality that avoids change for fear of creating a problem and upsetting the status quo. Tom Peters would have to look hard to find another laboratory manager sitting in his own personal barber chair in his office, asking whether his colleagues and understudies were "having fun" (2). To be creative, one must act creative. The research institution must practice what it preaches and be an agent of change within itself. But they still must be safe, responsible members of society and their community. How to balance the two dictates of change agent and safe practices? Only the world class competitors will succeed.

II. BECOMING WORLD CLASS

World class athletes train in all kinds of weather—in sunny and warm weather as well as in rain, sleet, and snow. They also train on days they don't feel like it, when they are sore, depressed, lonely, or busy. Keeping this world class athlete in mind, each person must ask themselves whether they want their organization to be "world class"? Are they quality focused? Do they want to be the best? Usually, whenever these questions are asked of organizations, most people say "yes," "certainly," "without a doubt. . . ."

What is the difference between most organizations and a world class athlete? The answer is that the athlete is doing something about becoming world class *today*. Many managers however, want to wait for the "right time" or come up with excuses like "there is no money in the budget," "I don't have time," "we have too many things on our plate now," or "we're already doing everything we can."

People are hard pressed to think of when they last *had* extra time, extra money, or extra resources of any kind. That rarely happens in organizations of the 1990s. If an organization is not willing to begin a passionate pursuit of becoming the best *today*, then they can never hope to achieve it.

Consider today's economic and competitive realities. Competition for funds is fierce. Budgets are tight. In most organizations, it is rare to hear "We have extra money and time available so you can focus on continuous improvement." If people wait for things to get better, to be less busy, or to get more money, they will never work on improvement. They need to get started *now*.

III. VISION-BASED IMPROVEMENT

Who are we? What is our business? What is our attitude toward quality, our customers and our own employees? Where do we wish to go from here? Leaders know they have a continuing challenge to inspire their followers. When someone says "I am the way . . ." (3), there had better be some substantial reason why anyone would want to follow in that person's path. There is a great need for people to have meaning in their lives. The search for purpose may very well be the primary motivational force in mankind (4). Those leaders who can harness this force, inherent within all employees, by linking it with a cooperative drive toward goals beneficial to the research organization, will find they have a tremendous advantage in creative energy and the ability to overcome adversity. Victor Frankl, writing from his personal experiences as a prisoner in concentration camps such as Auschwitz, observed that it was this ability to define oneself in terms of some purpose in life that kept the inmates going on in spite of terrible suffering. Those who lost their sense of meaning soon lost their lives. Those who retained it fared much better. As Frankl quoted Nietzsche, "He who has a *why* to live for can bear almost any *how*" (4). Certainly, it was this overall sense of purpose and urgency during the 1960s that kept aerospace workers going around the clock constructing and testing the Apollo launch vehicle. Twelve-hour days, three shifts, and seven-day work weeks were the norm, but everyone knew why. They were going to put a man on the moon before the end of the decade!

It is important that management and employees create this sense of purpose that all can identify with and support. Team builders know how important this idea is. They also know that the benefit of working in teams is at least twofold: they will get done faster because they are all working together; and they will do it better, because, if they all agree, there is a better chance that they are right (5).

But if the research organization is to have a vision, that vision must fit with the business units or funding agencies it supports (6). If the customers don't know where they want to go, technology will not help. Once that hurdle is overcome, a vision for the technology should be developed that is shared by the research organization and its customer(s). But that vision must be shared by others as well—it has to be constantly communicated up and down the organization. The vision must be measurable. Everyone must be able to see for themselves what progress is being made so they can understand what needs to be done. The individual researchers and support organizations will make it all happen. They must understand, help establish, and completely buy into the vision of the research center. People will follow where they want to be led.

There are two main types of organizational visions—external and internal. The external vision is related to the mission of the organization—it defines what the research center is chartered to do and how it will relate to the outside world. External visions for industry often include phrases such as "Number one in our market," "World class," "Best in class," "Benchmark," "Best quality," etc. External visions often serve as rallying points for the organization as was NASA's vision to reach the moon by the end of the decade. They can be seen hanging on the wall. People carry them around on cards in their purses or wallets and they are sometimes quoted by people in the organization. Sometimes, however, these visions are not remembered. People may not know what the vision means or where it came from. The vision has been reduced from a picture of where the organization is heading to just words.

External visions help gain consensus on direction and serve as a way to focus attention, but if the vision is only words on a document, much of the power is lost. Unfortunately and more often than not, this is the case. The organization has to quantify each vision and measure the progress toward achieving it. More specifics on this process are given in Chapter 12.

While many organizations have external visions, few have internal visions. An internal vision is an agreed-upon look at the future culture and operations of the organization. It is helpful to put this vision into words to facilitate consensus building, but the vision that everyone "sees" in their heads is much more important than the words. Rather than focusing on improvement or quality as just good ideas that the organization should pursue, an internal vision creates a friction between the present state and the desired future state.

The internal vision should be devoid of buzzwords and jargon. It should be in plain language and understandable by all. There should be less emphasis on the words and more on the picture in everyone's head; the picture is the vision. One should think of this vision in the same way as when they "see" their favorite vacation spot in their mind's eye when remembering a pleasant holiday. The goal is to let the words help "paint the picture" rather than focusing on the words themselves. For example, individuals could picture a research facility that they are familiar with and compare it with what it could be five years from now after working diligently to improve the lab. Everything is neat and organized, the equipment is state of the art and well maintained. Everyone is well trained and working at a good pace, but no one seems to be in a panic. The atmosphere is cordial, with healthy verbal banter that bespeaks good working relationships. The laboratory is noted for its innovation and ability to bring new products to market ahead of its competition. It is the "supplier of choice" of technology to its business units or funding agencies. How can that be measured?

If there is a difference between what everyone in the organization sees now and what they could see five years from now, it will create a healthy friction every time a person in the organization looks at the current state. Friction means that every time someone looks at the area and its measurements, they want to improve because they believe in the vision and are dissatisfied with the present condition. The vision puts focus into an organization's improvement efforts and is more concrete than just "improving for the customer," "to save money," or because it is "the right thing to do." Most successful improvement efforts are based on the vision of a handful of key people at the top. Getting consensus by all of the key leaders and gaining everyone's input gives the vision power.

It is necessary for the top management team to focus on the question of quality and ask, "What do *we* mean by 'quality'?" This is an excellent topic for an off-site discussion, away from the distractions of jangling phones, visitors, and other meetings that always seem to intrude on any attempts at contemplation. One way to begin the soul searching is to go around the table and ask each manager what they perceive quality to be. While there might be a struggle to concisely define what quality is, there will be no lack of description of what "un-quality" is, with plenty of anecdotes from the manager's sad experiences with that concept. This exercise will bring the realization that quality is complex and defined in various ways by various customers. Sometimes, even defining *who* the customer is may be difficult.

In a laboratory made up of distinctly separate technical disciplines, it is interesting to note how the leader of each department or laboratory defines quality. Scientists and engineers have all had different experiences that have shaped their attitudes on the subject. And all can offer sage advice about avoiding pitfalls. Often, people perceive quality as a lack of errors and they think the best way to avoid errors is to identify the ones they have already made and to prevent them from recurring. This is quality improvement, but it is basically error-driven. And while that type of improvement is necessary, that is not all or even most of the job. Quality is directly related to pleasing customers and must be customer-driven, while at the same time leavened with the experience of the research center's leadership. The customer is not aware of all the issues that may bear on what is received, the "product" as it were. For instance, an airline took a survey of what was important to its customers. Number one on the list was for the baggage to arrive on time. The customers assumed *they* would arrive safely, if not on time. But no airline manager would make that an automatic assumption.

So the task for laboratory management is to understand the dimensions of quality. It involves pleasing the customer, applying the manager's own experience to guide their people toward doing the work right, and it involves learning from past mistakes and working to improve the system. The gathering together of top management to concentrate on this question is an important first step and must be replicated regularly *forever* if any real progress is to be made. Failure to adhere to this first quality discipline will doom any serious attempt to improve quality. And it is generally recognized that failure to continuously improve quality portends the eventual failure of the business enterprise.

IV. PLAN TO ACHIEVE THE VISION

After the vision is established, a plan should be developed for how to achieve it. Included in very general terms should be any activities that should occur, but the focus should first be on the short term—90 days to one year. The plan should define critical success factors for achieving the vision—factors that are measurable. The plan should be specific with target dates, responsibilities, and resources allocated. If no one is responsible, nothing will happen. Three key areas not to forget in the plan are education and training, infrastructure, and improvement projects.

Education and training are critical to the success of any improvement effort. People can't be empowered to improve unless they know what to do. Some organizations focus on awareness training (which is educational but not designed to be put into action immediately). This works in some cases, but in more results-oriented (rather than process-oriented) organizations, it is usually better to pull training in whenever it is needed. With this approach, training is given in smaller doses to people who will immediately apply what they have learned to a work situation or improvement project. While everyone needs to know a little of why we need to improve, they most often lack the knowledge of how to do it. A process for how adults learn is to tell them, have them talk about it, let them see it, and have them *do* it. When the application follows the presentation of a concept or technique, better learning occurs. A valuable adjunct to formal training is the use of case studies from problems identified within the organization. Examples of this technique are provided in Appendix A to Chapter 11.

Infrastructure is everything needed to support the improvement effort. This includes resources such as time, money, training, and facilitation support as well as step-by-step instructions and plans on how people in the organization are to improve. Leaders of the effort need to decide on several issues that determine focus and direction, such as what kinds of teams to have—family/department teams, cross-functional teams, ongoing versus more project-related teams, and how to organize for improvement on an individual basis without being on a team. If people are empowered to improve in ways with which they are not familiar, they will not be successful without help.

Other things to consider in infrastructure are: What organizational systems will need to change as a result of the improvement effort, such as measurements, reward and recognition, work scheduling and organization structure; how to communicate the vision, plan and results to the organization; how to celebrate positive efforts and successes; and how to develop an attitude that it is sometimes OK to fail?

Improvement projects are the fuel that keep the process going. The long-term goal is for continuous improvement to become so much a part of the fabric of the organization that people make improvements without thinking about it and teamwork is the norm. It won't happen without hard work, successes and failures that the organization can learn from. When a group of people who have been provided the training, time, and resources needed to improve, work together, positive things happen. Examples of results of some improvement teams include cutting cost by 65 percent, cutting lead times by 90 percent, and reducing paperwork by 40 percent. People learn best by doing, and teams working on improvement projects are one of the most effective ways for people to learn.

The key to a successful project is to bound it and to work on small low cost/no cost improvements. Don't take on "world hunger," or one of the organization's biggest problems as a first project. People should avoid trying to completely fix the problem or process they are working on. Small incremental improvements are fundamental to the philosophy of Kaizen. It is easy to get caught in this trap. Many have the tendency to look at problems as something to "fix," "put to bed," or "get off my to-do list." In a continuous improvement culture, problems are looked at as treasures which must be identified before improvements can happen. The thought of human perfection implies that one would never be able to improve and is satisfied with the status quo. The focus should be on small rapid improvements with later efforts to improve again and to reach the next level. A real test of the maturity of an improvement effort is the desire and willingness to improve on an area that has recently been improved. When a work group attacks this same area and makes more improvements, their efforts illustrate to the organization that continuous improvement never ends.

V. CHAMPIONS FOR CHANGE

A common characteristic of companies that are involved in continuous improvement is the personal commitment of top management to lead the effort. This is gained because the executives know that this is the right thing to do because of their personality or previous education, or because they personally experienced making improvements and found that this improvement experience was so powerful that they began to fundamentally change the way they manage. Grass roots efforts can get some positive results, but do not remain effective. These initiatives die out because of other priorities. Most people have no desire to fight the system and blatantly pursue goals not supported by the organization's leadership. Leaders don't come out and say they are against quality or improvement, but if they do not give support with their own hard work, money, or resources, the effort will fail. They instead send the message that their focus is on maintaining the work (getting the job done) rather than improving the work. Many measurement systems reinforce this. For example, how many recognition/pay systems reward good firefighters rather than problem preventers?

The one recurring theme of experts is that top management commitment is the key to any significant effort to create a quality-oriented culture. (There is a growing chorus that says turning top management around is the easier part—it's the middle managers who give most of the trouble.) Starting at the top is that necessary first step, without which nothing but frustration will ensue. But where does that elusive initial spark come from? Everyone looks to their next level of management for the clue. In industry, if a division executive is perceived to be the key decision maker, what can lower levels do to convince that person that quality improvement is worth the effort?

Could it be that division executive needs a nod of approval from the next higher level? Where does the buck really stop? Someone has to be willing to finance the movement. A division executive will only take the step if it can be shown that there will be a positive effect to that divisions's bottom line. And the time allowed to show results is measured in months, not years. This is not due to a personal bias against improvement by the executive. But that executive must also answer to a higher authority during periodic financial and operational reviews. The accountability extends to the corporate board of directors, who are in turn accountable to the financial analysts and investors. Could it be that people are the victims of their own impatience and greed? The cycle can only be broken by someone who is willing to be a champion and demonstrate by example or witness that the results can be obtained. The higher this champion is in the corporate structure, the easier it will be to gain acceptance at least for experimentation.

VI. TWO STRUCTURES FOR CHANGE

Ernie Huge, in the book *Total Quality Management: An Executive's Guide for the 1990s* (7) described two basic structures for getting leadership commitment to improvement as a culture in the organization. The first structure is called "go with the flow." Executives in this culture say "Here is a concept. Try it." They introduce an improvement initiative and let their people decide if they want to participate or not. For example, in a company that has over thirty business units and several locations, the executive vice president had a consultant introduce the concepts of Continuous Improvement/Total Quality Management and Just-In-Time manufacturing to several of the locations. Each unit manager was told, "take a look at this and see if you would be interested in trying this at your site." Two locations decided to try it as an experiment.

The plants worked through learning and applying the concepts and achieved improvements in throughput, quality, morale, and communication. Some of the facilities they supplied noticed the improvements in delivery and the quality of the product. Other units started to inquire as to what was going on in these plants. When they explained that the change was because of the improvement efforts, other facilities became interested in trying them themselves. The momentum snowballed until the entire company, including the corporate staff, championed cause.

The executive vice president achieved the results he wanted—the entire organization was on board and the majority of the executives and upper managers were committed to the success of the effort because they personally were involved. They experienced the process first hand and felt it was to their benefit to participate. The executive vice president felt that if he tried to force his people to commit to something they might not understand and didn't experience themselves, they would give it lip service and the improvement process would fail. The idea was that if people personally experience the change, they will have no problem making it happen across the organization.

This approach can also be described as an organic approach with the executive "sprinkling seeds" to see where they will grow or where there is interest and using the places where the seeds have grown as model sites which can be used as examples for the entire organization. People at these sites can be teachers and paradigm busters for the rest of the organization.

Another approach is the "steamroller" approach. In this version, it is as if the top executive is on a steamroller as wide as the room his or her managers are in. There is only one door in the room—in the back. There is no other way to get out! The steamroller is rolling very slowly toward the managers. Regardless of the speed the executive chooses to drive, it is only a matter of time before the steamroller reaches them!

The managers have a few choices. They can go out the door. This occurs when they just can't change or really don't want to change. In this situation, it may be better for them to leave and join another organization. Their goals and those of the organization are in conflict and may be in opposition. Another option is to get on board the steamroller—to behave as if they really believe in the new process. They may or not really believe yet, but they have committed to the organization (and to themselves) that they will try to make it work.

The third choice, of course, is to be squashed by the steamroller. The organization passes them by (in this case, passes over them!). They are forced into a dead-end job, or may even be forced to leave the organization. This option, when presented to most people is the least desirable of the three!

The steamroller approach to culture change can be illustrated by a story of a manufacturing facility that changed plant managers some time ago. The culture toward quality was that it was "OK if you could get it, as long as it did not affect production." When there was a choice between quality and production, production always won. In fact, production was so important that at the end of the month, when the plant had to "hit the numbers," the plant manager was seen tearing "reject—hold for inspection" tags off of dozens of boxes of product so they would be counted in the current month's numbers. He knew that the entire lot would be rejected by an irate customer, but this would not count against his month's production figures. Everyone in the plant knew that the plant manager approved of shipping what they knew was obviously bad product, and they all knew that it was acceptable for them to let bad product go to the customer.

When the plant manager moved on, he was replaced by a person committed to continuous improvement and quality. One day during the new plant manager's first week, a machine setup person came to the office with a part that was visually out of specifications. You could see the defects. He had been having a difficult time with the setup and hoped that someone would approve the quality of the part so he would not have to fight the machine any more. He had almost run out of ideas. He showed the part to the plant manager, said that he was having problems and asked if the parts were acceptable.

The plant manager looked at the parts and said "These parts are not even close to being good. The hole is off and they look bad visually. Fix it, or don't run!" The setup person was disappointed. Most of his earlier requests for deviations had been approved, and he didn't want to fight the machine to make good quality. Production suffered. This sort of activity went on for the plant manager's first week, with every request for a deviation from what the customer wanted denied personally by the new plant manager. After a week, the plant had nearly ground to a halt. The plant manager had let everyone know that the new culture was to be one of quality, one where the customer came first. He had used the steamroller approach at its best. Everyone knew that if they wanted to get along with the new boss, they had better focus on quality.

As a follow up to the story, the plant manager had to loosen his standards a little after the first week. The plant was so used to running poor quality, that they weren't able to turn it around that quickly. The plant manager let his staff know that he would give them some time to turn it around, but that he would not be patient. The pressure was on. He immediately went on a clean-up and organization campaign to instill some pride in the workforce (as well as make them become more efficient). If a worker did not cooperate by cleaning his area, the manager would come to "help" and clean it himself. This certainly got people's attention.

VII. CULTURE DEFINED

Culture is defined in the *American Heritage Dictionary* as "the totality of socially transmitted behavior patterns, arts, beliefs, institutions, and all other products of human work and thought characteristic of a community or population" and "a style of social and artistic expression peculiar to a society or class." Jim Harrington, in his book *The Improvement Process* (8) describes culture as being "molded from its heritage, its background, and the total intellectual and artistic content in its manner, style and thought. Culture also includes the basic beliefs of a company and the foundation on which it was built.''

Masaaki Imai, in the book *Kaizen* (9), uses the term *culture* to mean ''factors of industrial structure and psychology that determine the company's overall strength, productivity, and competitiveness in the long term; such factors include organizational effectiveness, industrial relations, and the capacity to produce quality products economically.'' . . . ''If management is successful in improving the culture of the organization, the company will be more productive, more competitive and more profitable in the long run.''

Perhaps an organization's culture could be defined a little less formally as ''the way we do it around here.'' People new to an organization can quickly learn about the culture of their adopted organization by doing something that is not consistent with the organization's culture. What they do may have been consistent with the culture from their old organization, but may not work with the new. Someone will usually call the person aside and say ''I don't know how you did it where you used to work, but that is not the way we do it here.'' The person will find if they want to fit in, they must conform to the ''way things are done here.'' The culture is then a combination of the history, economic condition, geographic location, leadership style, performance feedback and recognition system and attitudes toward customers and employees of the organization. It is very slow to change without proactive stimulus and can only be managed if the leadership intentionally works on what they want the culture to be and how they will get there.

One can learn about the culture of the organization by observing how people work and listening to what they say. A visitor to Federal Express in the year that they won the Malcolm Baldrige National Quality Award observed that their attitude toward delivery of packages—fast and efficient—was also reflected in their attitude toward improvement: they don't like to wait and want to improve immediately. Some organizations are slow and methodic. In the observer's opinion, Federal Express' culture is to be fast and furious when they improve. Similarly, some organizations' cultures reflect a high value on creativity—it is good to throw out ideas without much prior data gathering or analysis, while others prefer to be more systematic and put much less value on creativity. In the systematic organization, it is more appropriate to have all the data and to have carefully studied the situation before you make suggestions.

A. Examples of Culture

In any organization, the total of everyone's attitudes makes up the culture. The population of the organization can be depicted in a bell-shaped curve (Figure 1).

On the left side of the curve is a small percentage of the people, typically around 5 percent, who are not interested in the least in any new improvement.

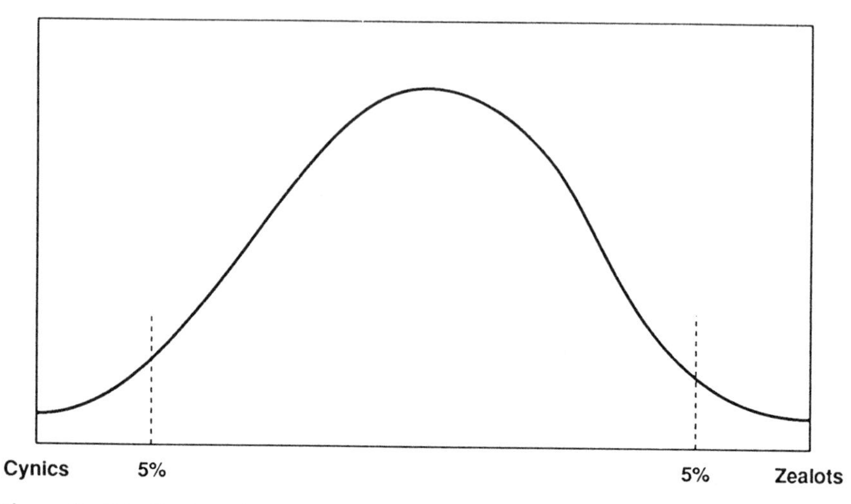

Figure 1 Distribution of attitudes toward TQM in an organization.

They resist. They find a million reasons why not and no reasons why. They have been described by saying that if they were on the road to Damascus and were struck from the horse like Saint Paul, they still wouldn't believe. In short, they are too skeptical or cynical to want to change. It is usually best to identify who these people are to minimize any negative effects they have on anyone else. In short, don't try to change them, but make sure they don't hurt the rest of the organization as it changes to the new culture.

On the right are the zealots. These are the people whose beliefs are already consistent with the desired new culture as individuals, and have been waiting for the organization to catch up to them. The first thing many of them say to outsiders brought in to help the organization change is "What was the holdup? I have been waiting for this for years." All that is needed with these people is sign them up, steer them in the right direction and get out of the way. This group is typically also 5 percent of the organization. Everyone already knows who the cynics and the zealots are in their organization.

The last group, the critical multitude in the middle, has 90 percent of the people and is the group that really needs to change, but takes the longest. This group must be brought over to the new culture little by little, one by one. These people change because they see progress and value in the new way of doing things, and because they eventually believe that the new way is better than the way they had been doing it. It is unreasonable for these people to change just because a new way has been announced. They have seen many "programs of the month" that were popular before they eventually fizzled out. Is it reasonable or fair to expect that a manager who is successful by any measure—financial, perks, office size, influence, position, etc.—should change what he or she has been doing because someone else thinks a different way is better? These people need to see it, feel it, touch it and experience it personally before they can commit to the new way.

This large middle group that is on the fence can come over to the new way slice by slice, group by group, until around 50 percent of the people conform to the new culture. When this happens, the rest of the "on-the-fence" group comes over to the new way, because this "new" way has become the culture of the organization. It can often take as long as three to five years to change the culture of the organization. There is no turnkey implementation of a new culture!

When 95 percent of the people in the organization have moved over to the new culture, it is probably time to deal with the hard-core 5 percent skeptics who have little chance of changing. They will begin to see that their way is not "the way" any more and that the performance management system rewards different behavior than they are exhibiting. It may be time for the Steamroller Effect!

B. R&D Culture

One of the exciting aspects of business is learning and comparing the differences in culture in different organizations and industries. Some things are considered terrific in one organization or industry and the same thing is taboo, irrelevant, or may even upset people when it is mentioned in another organization. Understanding the organization's culture is key to understanding how the organization works.

One of the phrases that some people use to distinguish elementary concepts from subjects that are more complex is "This isn't rocket science!" One particular consultant uses this phrase to make a point with his clients. At one of the first meetings with a group of researchers he described how easy he felt the idea of process improvement was by saying "You know, this isn't rocket science!" Several of the participants peered at him with curious looks on their faces and he realized that they *were* rocket scientists! They were thinking "Yes you're right. We know *exactly* what rocket science is and you aren't even close to describing it!" Realizing this had not worked as well as it had in previous organizations, he tried to salvage as much as he could by adding "and I guess you'd know that." He learned how important it was to understand each organization's unique culture.

A difficulty when examining the desired culture for an R&D organization is how to strike the balance between several conflicting concepts—creativity versus having all the data, small incremental improvements versus big technological innovations and teamwork versus letting highly skilled people go in their own direction. Many R&D organizations establish a goal to create new technologies for their customers' products or processes. Small incremental improvements—

an important aspect of a Kaizen culture—may be inconsistent with the big technological improvements sought by some researchers. Where should the organization put their resources, in embryonic technology, which can be very expensive to develop and may not always work but has potential for radical change and big profits, or in waste reduction and quality improvement, which may be much less expensive but also has much less potential?

What is needed is a blend of continuous improvement and technological breakthroughs. Removing waste from all processes and satisfying customer needs should be key to everyone in the organization. People in research and development can contribute not only by continuing to create and develop new technologies, but by improving their own work processes and by lending their technical and creative skills to the rest of the organization to aid in the overall improvement process.

Researchers and scientists can contribute to the organization's improvement effort by eliminating waste from such R & D processes as report generation, inefficiently designed experiments, or inappropriately planned testing. They can aid the rest of the organization's improvement effort by applying their technical skills and knowledge by helping gather or analyze data, suggesting new ideas of what might be possible technically, or providing an objective viewpoint to another group.

R & D professionals' nature makes them an unusually hard "sell" for TQM. But their technical and professional experience gives them a unique perspective from which to aid the organization's improvement effort. They can provide valuable input in defining the new organizational culture.

VIII. EMPOWERMENT

As the organization begins to implement TQM by forming teams and providing mechanisms through which individuals can contribute, the question of empowerment arises. The term may have been used during training with the natural implication that this was a concept being embraced by the management. Perhaps only a general definition of the term may have been given without specific examples of what was meant. A definition used by one organization was "to provide individuals with the means and opportunity to identify, recommend and implement changes in the workplace." This sounds grand, but what does it mean specifically? The organization should assume that employees have heard about empowerment from the media and other sources. The problem, then, is whether the organization's intentions with respect to empowerment coincide with employees' expectations.

Empowerment is a complex issue that should be thought out at the highest level. It involves issues of leadership and trust and no one can assume that what is appropriate for one organization is just as good for another. In its simplest sense it can be thought of as giving decision making or approval authority to employees where traditionally such authority was only a management preroga- tive. In a research center, one would expect a high degree of empowerment since there may be a multitude of simultaneous projects, some very small scale—one or two individuals—each under the direct control of a principle investigator (PI). The PI may be the only one in the center who understands what is to be done and he or she is expected to make it all happen on time and within budget. Technical section managers may keep an eye on overall budget performance, particularly for larger projects, but do not delve into details until time for a project review. The PI feels responsibility to make technical progress, stay under budget and keep the team functioning smoothly. Does she feel empowered? Ask her after she just finished obtaining six signatures for permission to buy a $5000 computer for a $10 million test program. She's responsible for the success of the $10 million program but can't be trusted for the $5000. Yet every group repre- sented by the signatures will make a strong case for their being kept in the loop because the procurement affects them. The safety officer, the environmental specialist, the quality assurance staff, all need to be kept informed so they can "help" the PI avoid pitfalls. But the process drags on. And when it becomes stifling, the mavericks that learn how to get around the bureaucracy and get the job done quickly, at low cost (but at much higher risk) become heros and are secretly lauded by top management. Were the mavericks empowered? Unoffi- cially, because of top management's clandestine approval—as long as there are no repercussions. Heaven help the offender if the EPA slaps a fine on the research center! But isn't that the essence of high risk, high reward? The problem is, a culture of beat the system will eventually not be able to discriminate which systems are most likely to adversely affect the organization. Whether it is disa- bling a safety valve to get a few more pounds of pressure or shipping a barrel of unknown liquid off to the local landfill, sooner or later the chickens will come home to roost. There is no reward commensurate with the ultimate risk.

The challenge is to empower *everyone* to be efficient and still be responsive to regulatory and business requirements. It involves setting expectations and establishing accountability, agreeing on the measurements and providing for interim checkpoints. Some characteristics of a good philosophy of empowerment follow:

1. *A customer focus*. Johnson and Mollen (10) state that "the primary purpose of empowered workteams in the value adding work process is to focus on internal and external customer requirements. That focus is necessarily a short-term one, and is the sine qua non of customer satisfac- tion. Once that responsibility is given to those that add the value, manag- ers should refocus their attention from now time to next quarter time, or next year time or even further out."
2. *Specific, measurable goals*. Semco S/A, Brazil's largest marine and food processing machinery manufacturer, runs a completely transparent sys-

tem. All employees know exactly what the company's financial status is. They've cut the monthly reports to three with seventy line items that tell them how to run the company, tell the managers how well they know their units, and tell the employees if there is going to be a profit. And the employees know that 23 percent of the after-tax profit is theirs (11). Although specific internal goals may be set that stretch employee assumptions of their own capabilities, ultimately the goals of the organization and the goals of the employee are understood to be in sync.

3. *Enabling, not delegating*. The empowerment process acts as a facilitating device that creates conditions for worker success and enhanced self-esteem. If management believes in the workers they are empowering, the workers will believe in themselves (12). Employees need to have a common knowledge of what is happening and what the goals are. They need to have the skills to tackle problems, in teams and individually. There needs to be open communication within and between all levels.

What specific decisions and authority to be delegated will vary, because the current culture varies widely from one organization to another. There will be tentative steps involving minute decisions at first. Fewer approvals of certain purchases, encouraging teams to pilot their solutions and then report back on the results. This may be followed by a feeling of loss of control, particularly if a mistake was made. Then there will be a tendency to tighten up as the more conservative managers press their "I told you so's." Everyone has to feel reasonably comfortable with the process, but, as the employees are challenged to improve their performance, so will the managers be likewise challenged to maintain the movement to greater autonomy.

But how much autonomy? Again, it will vary. In many instances, specific actions are dictated by law. Semco S/A does not require expense reports. President Semler feels that if he can't trust his people with the company's money or their judgment, he shouldn't be sending them overseas to do business in the company's name in the first place. It's not likely that U.S. federal law would look favorably on the lack of such documents, particularly for government contracts, no matter how noble the cause or fantastic the results. But a general set of examples are:

Maintaining a group's own annual budget
Keeping the team's own time records
Recording statistics on problems or failures
Making work assignments within the group
Training team members
Redesigning or modifying operational processes
Setting team goals
Resolving internal conflicts

Advanced cultures also include giving the teams their own hire/fire authority, using skill-based pay rather than job descriptions, freely available information on everyone's salary, and much more flexible procurement authority.

The top management of the research center must first understand and agree on what empowerment means to it before attempting any modification or clarification. This issue will arise and must be dealt with. There *will* be a cultural interpretation of this issue. The only question is whether it can be guided.

IX. SUMMARY

If one is to avoid the painful lessons of the Swiss watchmakers, one must be open to new ideas. The research arm of the industry or government has the responsibility to identify new opportunities and threats and to foster and facilitate change to capitalize on them. While assuming this role for their customers, they must also be open to constant change within their own structure and methodology, all the while maintaining safe working environments and being a good citizen to the local and global community. Only the world class laboratories will succeed and that is because they will work at it endlessly. It requires vision, directed externally and internally and understandable by all. Top management must lead the development of the visions and plan to achieve them. The infrastructure must be set up, people trained and projects authorized to tackle problems that have been broken down to reasonable size, one by one, forever. Management must create and maintain the culture for change, such that no one can dodge the issue. Either be part of the solution or be identified as the problem. If anyone wants to know what the culture is, he should get out in the work area and listen—not talk, listen and watch. One should support the culture with one's words, but people watch what one does more than what one says. R & D people are trained skeptics, but they have a powerful talent that, properly harnessed, can motivate, demonstrate and facilitate constructive change throughout the organization.

The term *empowerment* means many things to many people. Management should not leave the definition of this concept to chance. There is a general policy to be developed, but the real definition comes in a thousand small details. The philosophy for empowerment must have a customer focus, have specific measurable goals, and be enabling, not merely delegating.

All in all, the stage will be set for immense change. New roles will be tried, perhaps discarded and tried anew. But there will be a dynamic environment created that will shift with changing needs. How to keep on top of it all? How to keep it under control? In all probability, there will be a significant loss of control, felt more so at certain times than others. But the organization will move toward the "loose, tight" type of structure described by Peters (13)—loose in the way it reacts to specific challenges, but tight in adhering to basic organizational values. The actual structure of such an organization will fluctuate to meet the needs, and some approaches for managing the chaos is described in Chapter 3.

REFERENCES

1. J. A. Barker, *Future Edge: Discovering the New Paradigms of Success*, William Morrow, New York, 1992
2. T. Peters and N. Austin, *A Passion for Excellence: The Leadership Difference*, Random House, New York, 1985
3. Holy Bible, New Revised Standard Version, John 14:6, 1989
4. V. E. Frankl, *Man's Search for Meaning*, Beacon Press, Boston, MA, 1963.
5. L. Cain and M. Cotter, *Real People, Real Work*, SPC Press, Knoxville, TN, 1991
6. Roland W. Schmitt, "The Strategic Measure of R & D," *Research/Technology Management*, November–December 1991, p. 13
7. E. C. Huge, *Total Quality Management, An Executive's Guide to the 1990s, Business One*, Irwin, Homewood, IL, 1990
8. H. J. Harrington, *The Improvement Process: How America's Leading Companies Improve Quality*, McGraw-Hill, New York, 1987
9. M. Imai, "Kaizen: The Key to Japan's Competitive Success," McGraw-Hill, New York, 1986
10. J. Johnson and J. Mollen, "Ten Tasks for Managers in the Empowered Workplace," *Journal for Quality and Participation*, Vol. 15, No. 7, p. 18 (1992)
11. R. Semler, "Managing Without Managers," *Harvard Business Review*, September–October, p. 76 (1989)
12. R. Ripley and M. Ripley, "Empowerment, the Cornerstone of Quality: Empowering Management in Innovative Organizations in the 1990s," *Management Decision*, Vol 30., No. 4, p. 20 (1992)
13. T. J. Peters and R. H. Waterman, Jr., *In Search of Excellence*, Warner Books, New York, 1982

3

ORGANIZATION

Robin K. Gill, George W. Roberts, and Robert J. Cartwright
Babcock & Wilcox
Alliance, Ohio

I. ORGANIZING FOR R&D

A. Organizational Structures

Organization. Ask researchers and they will tell you that the less there is the better. For researchers, freedom and independence are the cornerstone of their profession. The fewer the rules and controls, the more conducive the environment for the free flow of ideas and creativity. Controls are seen as antagonistic, imposed by those who have no idea what research truly is or how researchers work. How often has this lament been heard? However, press researchers hard enough and they will admit that a certain level of organization is necessary. Without it, there would be chaos. Researchers are a highly educated, highly motivated group of individuals. If they can be shown the value added in doing something they are asked to do, whether it's a technical improvement or a better way of organizing their work, they will adopt it and even improve upon it. But, if it does not add value and hinders their work in any way, the more militant of them will fight the system openly; others will just ignore it and proceed as they choose. Blake states that for organization to be successful "control must be based on mutual trust, agreed upon and rational rules, qualitative as well as quantitative judgement, and above all, personal communication both vertically and horizontally" (1).

Organizing a company, division, group, or project to perform research and development is no small undertaking. The issues to be considered include:

- The purpose of the research or development
- Deciding on the intended products or services
- Customer needs
- Market influences
- Corporate structure
- The goals and strategies of the organization
- Whether the intent is to organize to control or to empower
- Whether there are projects requiring different organizational structures within an overall organizational structure
- Whether the purpose is to do research only, development only, or a combination of both
- Whether the intent is to organize to communicate and what is to be communicated
- The characteristics of the personnel needed to fill various organizational positions

After searching for the answers to these types of questions, an organization is put into place. If the homework is done well, the organization may serve well for years to come. More often than not, reorganization will be required every so often. This does not mean that the original organizational design was a failure. Most reorganizations are the result of outside influences, changes in the market the organization serves, or a shift in customer's needs. To stay viable and to serve customers effectively and efficiently, periodic reorganization is a must. For some entities, reorganization has become a way of life. The organization that works today may not, and probably will not, work tomorrow. And an organization that works well for one research and development organization will not work for another. Kerzner (2) put it this way:

> Management has come to realize that organizations must be dynamic in nature; that is, they must be capable of rapid restructuring should environmental conditions so dictate. These environmental factors evolved from the increasing competitiveness of the market, changes in technology, and a requirement for better control of resources for multiproduct firms.
>
> Organizational structures are dictated by such factors as technology and its rate of change, complexity, resource availability, products and/or services, competition, and the decision-making requirements. There is no such thing as a good or bad organizational structure; there are only appropriate or inappropriate ones.

This section of this chapter does not attempt to cover all the various aspects of organizing a research and development effort. It is limited to looking at the various types of organizations and discussing their advantages and disadvantages.

In addition, the decision to use project management will be discussed along with its advantages and disadvantages. Throughout the remainder of this chapter "researcher" will be used as an all-inclusive term including scientist and engineer.

The common organizational structures are functional/line, project, and matrix.

1. Functional/Line Organization

Figure 1 demonstrates a functional organization. It is also known as a line organization, function/line organization, or classical or traditional organization. Functional organizations are aligned along disciplines, technology, or technical specialties. Within this organization, researchers are grouped by specialty. This allows them to share the knowledge they have and allows them to keep up with the various projects being conducted by those in their group as well as with the latest developments being made in their field. In fact, it seems that the ability to share knowledge is actually enhanced in this type of structure and this ability provides for a technical continuity. All projects personnel are participating in or conducting can benefit from the knowledge of the group as a whole through their communications. As a result, personnel can be easily switched between projects giving the manager great flexibility in the allocation of technical resources. The reporting structure is vertical and well established with each person reporting to

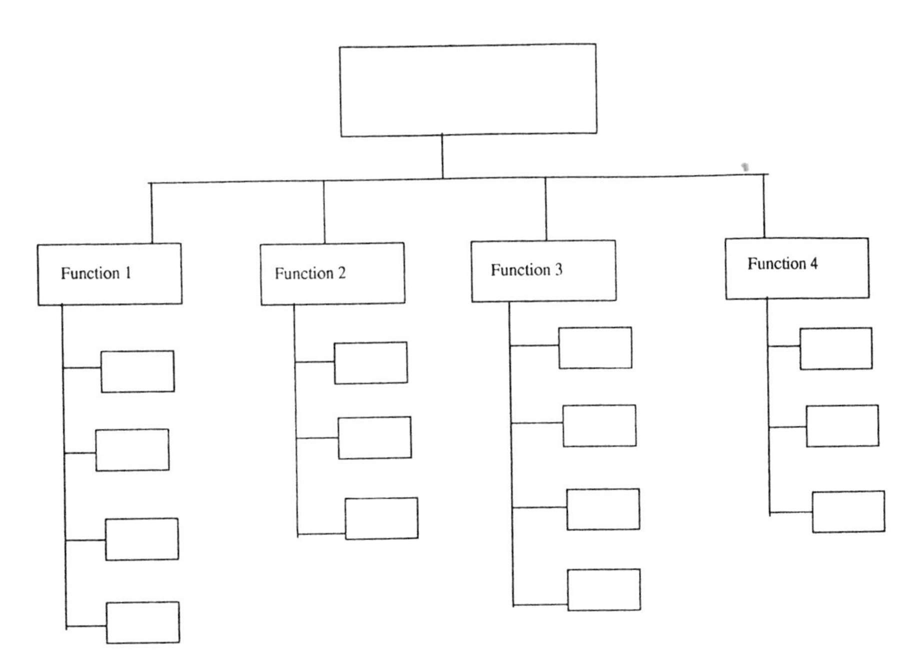

Figure 1 Functional/line organization.

only one person. The structure also allows for uniformity and continuity of policies and procedures. All of this provides for good technical control. Personal advancement is through technical project success.

Unfortunately, the list of disadvantages for a functional organization is longer than the list of advantages. Functional organizations are not customer focused, which leads to priority conflicts, that, in turn, can lead to slow response to a customer's needs. Success within the function is not necessarily project oriented. Functional priorities may outweigh project priorities. There is no project emphasis. There is no one individual responsible for the success of a project. As a result, there is no real project planning or authority.

Input to any research and development organization is technical and scientific knowledge, and a traditional organization tends to be "input" oriented. Output is generally a new design, process, or product. But effective product or process designs usually require cross-functional coordination between several technical disciplines. Allen (3) notes: "This coordination problem can be severe in a functional organization. The specialized functional departments or groups present a barrier to coordination that can become very difficult to manage." Coordinating the tasks required to complete a project can become a monumental undertaking. Little wonder that timeliness is not something the functional organizational structure is known for. The conflict between the needs of the customer and the demands of the functional organization can lead to demotivation of the professionals involved.

2. *Project Organization*

A project organization is one in which everything needed to complete the project—personnel, materials, and hardware—is contained within the organization. Figure 2 shows an example. The project manager has total authority for completion of the project. Personnel are taken out of their function and assigned to specific projects. They work directly for the project manager and are not shared with other projects.

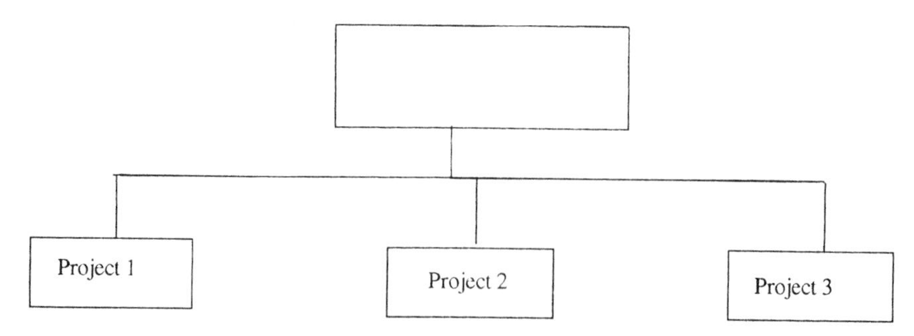

Figure 2 Project organization.

The project organization tends to be "output" oriented. It has all the disciplines needed to produce a finished product, process, or design goal for the project. The structure is simple, yet flexible. Personnel tend to be highly motivated. Loyalty to the project and team spirit abound. Everyone has one assignment, one goal to obtain, and they can be focused. Communication lines are short and well defined. The project focuses on meeting the customer's needs on time. With a customer focus comes the flexibility to determine cost and performance trade-offs. Unlike the functional organization, upper-level management is freed to perform executive-level tasks rather than running the daily activities of the project. Once in place, a project organization can be readily used for successive or similar projects, thus reducing project start-up times.

On the other hand, multiple projects cause duplication of people, facilities, and effort. Each project may have chemists, manufacturing personnel, calibration technicians, and a host of other essential personnel. Various projects may have duplicated the same types of equipment and test facilities. One of the greatest criticisms of a project structure is that, over time, the technical depth of project personnel will diminish. They do not have the contact with others in their discipline to keep them abreast of the latest advances. They become too focused on one particular aspect of their field, thus limiting their ability to work in other areas of their technology. Often, there is very little or no technical interchange even between projects. Policies and procedures used by various projects are not the same. This leads to inconsistencies in the way similar tasks are accomplished.

Projects can take on a life of their own. Personnel may be kept on long after their usefulness to the project is done. Competition between project groups may arise. A little competition can be healthy, even fun; however, competition can rise to the level where it becomes destructive to the people involved with the projects and the projects themselves. Project personnel can also develop an "us versus them" attitude which may be directed toward other projects or against laboratory management. The strong bonds formed by project personnel working for a similar goal can also spawn destructive competitiveness and attitudes. Fights for funding and high-quality personnel, facilities, and equipment can lead to great frustration, and anger.

The costs of operating such an organization are high because of all the duplication. This is not a very resource-efficient approach. Finally, when the project is complete or when the skills of certain groups within the project are no longer needed, the future of its personnel is uncertain. They may be picked up by another project requiring their skills, if they have not become too specialized. Or they may become unemployed.

3. Matrix Organization

The third common organizational structure combines the functional organization and the project organization. This is a matrix organization and is shown in Figure

3. One of the strengths of this structure is that emphasis is placed on the project while at the same time not separating the project's specialists from their functional technological base. Within a matrix structure, the functional organizations exist to support the project. A single person, the project manager, has control over all resources, including cost and personnel, through the functional managers. He has the authority to commit these resources, as needed, to his project provided that scheduling has been worked out with other projects. The matrix organization is project and output focused leading to timely solutions of customers' needs. There can be some flexibility regarding policies and procedures. There can be consistency between the policies and procedures used on each project, or policies and procedures can be set up specific to each project.

The project manager knows the status of the project at all times. Management has one contact point for the project. Project communications move in two directions, across the project and within the technical discipline. Problems and conflicts can be identified and resolved quickly. The majority of the conflicts are resolved by the project manager. When required, problem resolution can be easily obtained from a higher level of management.

As with the functional organization, specialists share their knowledge freely. When problems arise there is a greater talent pool to call on for solutions. This

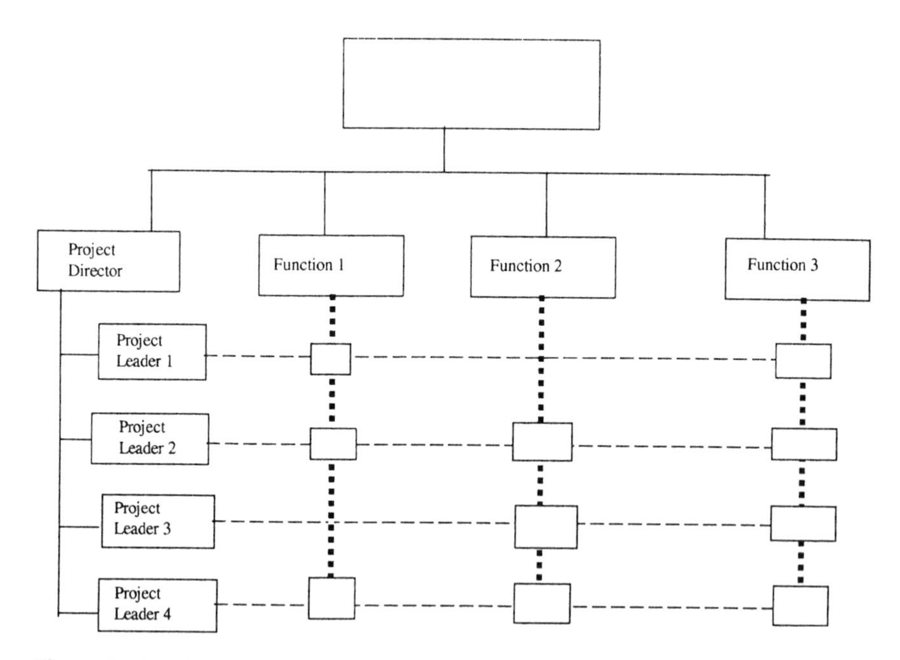

Figure 3 Matrix organization.

talent pool is available to all projects equally. Lessons learned on one project are shared and used on other projects. Functional managers have flexibility in assigning personnel to projects. Personnel can work on a variety of problems at one time. This allows for better control of people resources, and personnel do not risk boredom. It has been found that the matrix organizational structure can develop specialists and generalists in a short period of time because of the variety of work assignments. Too, because resources are shared among many projects, the organization operates more efficiently and overall project costs can be minimized.

When personnel are through working on a project they move to another. Job stability is greatly enhanced. Unlike the project structure, each person has a career path which can be followed.

But there are several disadvantages to matrix structure. First is the dual reporting relationship. Each person has two bosses—the functional boss and the project manager. This can lead to many pressures being placed on a person. Which boss comes first? The one who can hire or fire or the one the person has been temporarily assigned to? There can be conflicts about what work gets done first, functional work or project work. People's priorities continuously change, roles change and sometimes appear undefined. Some may feel they have no control over their lives, they just go wherever they are told to go next.

Project leaders are constantly bartering for resources—people, facilities, and equipment. Conflicts may arise among project managers and between project managers and functional managers. Power struggles may result. Personnel working on the projects don't attain a team spirit. They may work on one project today and another tomorrow or several projects simultaneously, so they never develop the team attitude of all working for a common goal. Each individual may have many goals to accomplish. One project starting early or another starting late may disrupt several other projects. A balancing act is required by all the participating organizations to facilitate all the projects.

Overall, a matrix structure can be costly because it requires more personnel than the functional organization. Most of these personnel are going to be performing administrative duties. Since each project tends to operate independently, efforts may be duplicated if communications are not adequate.

Within a pure matrix structure, the control exercised by the project manager and the functional manager are equal. This does not appear to be how most matrix organizational structures actually work. The balance of power between the two seems to vary and can be quite lopsided. This type of structure can be difficult to manage with its multidimensional information and work flows, number of personnel involved, and opportunities for conflict. A great deal of commitment and work is required to make it work effectively and efficiently.

The matrix organizational structure can be most effective and efficient when personnel and resources must be shared among many projects. However, matrix organizations can be difficult to understand and manage. Blake cites the experiences of the Stanford Research Institute and other such institutions when listing the following characteristics and precepts as the minimum necessary for successful implementation of a matrix organization in an R & D environment (1).

1. A project team performs all R & D and R & D related work.

2. Project teams should contain a minimum of three people. Studies have shown that better results are achieved if teams consist of this minimum staffing level. However, a project team may consist of one individual.

3. Personnel leading a team are given the descriptive titles of "project manager" or "project leader." These are not to be confused with organizational titles. The title is worn only as long as the person is leading the project team. Upon project completion or upon being relieved of duties, the individual may return to his technical or support organizational group and resume whatever professional title he carries there. However, if he is extremely talented as a project leader, he will undoubtedly move on to another project team to manage. Talented and successful project managers will be in demand within the organization.

4. Within a matrix organization, personnel who are senior to the project leader in the hierarchical organization may be reporting to the project leader as part of the project team. This can cause some hurt feelings and bruised egos for those who equate position and title with prestige. It has been found that the more structured the society, the harder it is for those involved to accept the changing assignments and multiple responsibilities associated with the matrix organization. "It has been shown quite conclusively, however, that the matrix organization can be applied successfully in very highly structured societies" (1).

5. It is best to have personnel working on more than one project at a time. Of course, the objection to this is that the worker may be placed in the position of conflicting demands for his time from several project leaders. Results show that in successful matrix organizations, this problem is avoided by allowing personnel to volunteer for membership on a project team. Individuals are then responsible for ensuring that they do not spread themselves too thin and make promises they cannot keep. If they do, it is up to them to somehow meet all of their commitments, working overtime to do so, if necessary. They cannot blame anyone but themselves if they overcommit. They must be able to provide the services to those who are counting on them. A few mistakes made in overcommitment and a person will soon realize his limitations.

6. The project leader in the matrix organization is held fully accountable for the successful completion of the project. To make this happen, he must be given the authority and responsibility to perform the activities necessary. This includes complete control over the expenditure of project funds. This can only work when there is a very clear understanding between the project leader and his team and between the project leader and the line management as to what work is to be performed, the resources that are required, and the time constraints which apply.

7. All persons working on the project must have a clear and complete understanding of the work statement. The work statement must be accepted as a commitment by each team member.

4. Matrix Variations

There can be many forms of matrix organizations, running from very simple to quite complex. Matrices can be layered, one matrix inside another. A company can have an overall matrix structure and, within this overall structure, the various divisions or groups can have their own matrix structure. When a matrix exists within a matrix structure, all the matrices are formal. A mix of formal and informal organizations can also be included in a matrix layering scheme. For instance, a formal matrix may exist for work flow while an informal matrix exists for information flow. There can also be authority and multidimensional matrices where each dimensional slice represents time, distance or geographic area (2).

5. Common Characteristics

Organizations can be purely functional, purely project, purely matrix structures, or some combination of one or more of these. For very small projects, the functional organization may be the choice. For a large project requiring a large number of people and resources, a project organization may be established. For several projects requiring the sharing of personnel and resources, a matrix structure may be set up. Various organizational structures may reside within a company at any one time. There is no right or wrong organizational structure. However, no matter what organizational form is selected, it seems that successful organizations have certain characteristics in common. Karpenko (4) lists these characteristics as follows:

- The organization tends to be purposeful and goal directed.
- Form follows function and power is widely dispersed.
- Decisions are made based on location(s) of information rather than roles in the hierarchy.
- Reward systems(s) are related to the work to be done.
- Communication is open in all directions.
- Some competition is good; however, inappropriate competition is minimized and collaboration is encouraged and rewarded.
- Conflicts are managed and resolved, not suppressed.

B. How to Choose an Organizational Structure

How does one select the organizational structure best suited to their company, division or group? Technology, products, markets, goals, strategies and resources and their rate of change all affect the decision. Should the organizational structure favor one type of communication structure over another? Should project managers be empowered to complete their projects? Are checks and balances required on project managers? Is a high degree of teamwork required?

John Tuman, Jr. (5), stated, "organizational structure should be governed by the strategy and vision established by top management. The structure of the organization should provide for the best arrangement of people and work to carry out the strategy." For research and development that means providing the atmosphere needed for flexibility and creativity.

Communication is paramount to the success of research and development. Two types of communication are needed: There is the work related communication, with one researcher passing on technical information to another researcher for use or study; while the other type of communication keeps researchers abreast of advancements in their technical expertise. This may include symposiums or conferences anywhere in the world. Thomas Allen (3) notes, "to improve R & D productivity and performance, these two types of communication must be managed properly. Organizational structures can achieve either one or both of the communication goals." Further, Allen states,

> the optimal form of organization for research and development is determined by three parameters. The rate of change of the knowledge base determines, in part, the extent to which engineers must be organized in a manner to assist them in keeping current through colleague contact. The interdependency among subsystems and problem areas within a project determines the extent to which intraproject coordination is necessary and the frequency with which it is required. Finally, project duration, or more particularly, the duration of assignment for any specific specialist, will determine the degree to which one must be concerned about the separation of that engineer from his knowledge base. Short term projects have little effect; however, long term project assignments can have a serious effect. (6)

Allen's comments provide a couple of perspectives on how to select an organization. As already has been said, there is no such thing as a good or bad organizational structure; there are only appropriate or inappropriate ones. If an inappropriate one is selected, learn from what doesn't work and improve the structure. Circumstances will change. An organization must be flexible enough to change as required to meet the new challenges.

C. Project Management Organization

It is widely believed and results seem to support that the key to a successful project is proper project management. When is a project management organization used? When is it not necessary? Which organizational structure, of those discussed above, best fit project management? What are the elements of successful project management?

When does it make sense to use project management? There are several criteria to make this determination. According to Kerzner (2), large, one-of-a-kind projects are good candidates for project management. They can be defined and have a specific goal. This type of project should have technologies or methods which are unfamiliar, unique, or infrequently used. They should also be complex with respect to interdependencies among the various tasks required to complete the work. These projects should be critical to the company. There may be times that, because of the importance of the project, management will not feel that the current organizational structure is capable of providing the controls necessary to achieve successful completion of the project. *Critical*, in this instance, may mean that a project's successful completion is tied to the successful growth or possible survival of the organization. It could mean that the project is important to achieving division or corporate goals. *Critical* could mean that the reputation of the company is at risk or possible financial losses could result with the failure of the project.

If a project is repetitive or routine, for example, standard chemistry tests or fracture mechanics tests, a project management organization is not required. Project length also plays a role. Short tests of a few days or weeks may be easily accomplished. But projects lasting several months or years may be better served using project management techniques. The number and type of resources required by the project are also determining factors. Quoting Kerzner again,

> Project management (especially with a matrix) usually works best for control of human resources and thus may be more applicable to labor-intensive projects rather than capital-intensive projects. Labor-intensive organizations may use formal project management, whereas capital-intensive organizations may use informal project management. (2)

One also has to consider top management when making this decision. Just how much authority and control is it willing to give up? Project management gives authority and responsibility for successful project management to the project's manager. Some upper-level managers are not willing to give up this much control. Another consideration is whether personnel are familiar with project management organizations and techniques. Are personnel willing to take on the responsibilities associated with project management?

Not all managers subscribe to the notion that only large or critical projects are worthy of project management organizations. Many believe that the concepts and mechanisms of project management can be applied to smaller assignments even within the functional area. In other words, projects do not have to be large. Even a one-person project may benefit from project management.

According to Blake,

> One of the benefits of the formal functional department/project relationship is that new ideas and approaches are subject, at least indirectly, to scrutiny from all the functional

> areas as well as from the project group. In smaller project teams that happen to be drawn entirely from the same place in the organization, the healthy variety of opinion and experience may be lacking. (1)

Project management naturally leans toward the matrix organization. But, as has already been stated, a functional organization may have matrixing going on. So may a project-type organization. Organizational forms seem to be selected based upon how much authority the top management is willing to delegate. Communication must be considered. How must information flow to allow the effective and efficient completion of the project?

Kerzner (2) has provided several additional factors to be considered when selecting organizational forms for R & D, including

- Clear location of responsibility
- Ease and accuracy of communication
- Effective cost control
- Ability to provide good technical supervision
- Flexibility of staffing
- Importance to the company
- Quick reaction capability to sudden changes in the project
- Complexity of the project
- Size of the project in relation to other work in house
- Form desired by customer
- Ability to provide a clear path for individual promotions

Why would a company want to use project management within its organization? There are several advantages. The project leader is given the total responsibility and authority for the project. Team members work as a team. They have a sense of identity and a common purpose leading to higher morale among their members. Problems are easily identified and, by being identified early, can be prevented from causing more costly problems later on in the project. Project leaders are better able to make trade-off decisions between cost, schedule, and performance because these factors are controlled by the project leader. This is perhaps the biggest advantage. More control over projects leads to lower costs and shorter project duration. Projects have also been shown to have improved quality. Projects are highly results-oriented and usually receive a great deal of management visibility. The end result is a happier customer.

Disadvantages of a matrix organization include the inevitable conflicts with functional managers for resources. Personnel are shifted from project to project. This may be a demotivator for some, leading to feelings of having no control over their work. Matrix project management can be difficult to manage for all the reasons given for these organizations. Team members have two bosses, which can lead to conflicts over work assignments. Sometimes the functional groups will let the projects do their jobs, leading to a lower use of the functional people on the project and to a duplication of functional skill within the project. Because of the complexity of the operation, program costs may be higher, which could lead to lower profit margins.

D. Keys to Successful Project Management

There are several common elements for successful project management. First, project management must be supported by upper-level management and this support must be communicated throughout the organization. Without support project management will not work. According to Blake, the project leader has to have access to the chief executive, even if he never uses it. The authority of the project leader comes from the chief executive, not his immediate line supervisor (1). The line supervisor provides the daily supervision required by the project leader.

The project goals must be understood and agreed to by all project team members. This agreement is obtained at the beginning of the project and should be more than just a passing understanding. Meetings should provide everyone who will be involved with the project with a chance to ask questions and understand the priorities. Functional managers supplying support personnel should understand the commitments being made and the timing necessary to meet them. Formal documentation of these understandings and agreements and responsibility may be necessary.

The project leader must be skilled in project management techniques. He must also be well rounded in research and development. He must be an able administrator capable of handling budgets, procurements, personnel problems, and the host of other administrative duties associated with the job. Being a talented speaker and writer is helpful. The project leader must be aggressive. He must be able to obtain the resources he needs when he needs them. He must have management support, but be able to stave off upper management intervention, when necessary. This requires diplomacy, a necessary tool in the project leader's bag of skills. The project leader must be a team builder and a leader and possess skills for conflict resolution. He must be able to plan and organize. And the project leader must be an entrepreneur. Technical expertise is not enough. Administrative skills are not enough. The project leader must have all these skills if he hopes to have a successful project.

Successful project management uses its functional team members to the maximum. There is a wide pool of talent out there. Not only do the functional areas provide technical support, but they also supply administrative support. When problems arise, personnel can be pulled in as necessary. Project team members have their respective functional groups to assist them when needed. The more opinions and viewpoints which can be accessed, the better the decision that will be made.

Project leader and functional manager conflicts are going to happen. How much control does the project leader really have over the functional manager when the functional manager is providing services to the project? Can functional managers challenge the needs of a project leader? What control does the functional manager retain? Functional managers will undoubtedly be supporting many projects at the same time and they do retain control of where their resources will be used. Decisions will be made which will help some projects while perhaps hurting others. Project leaders do not want to spend all their time explaining themselves and their needs to the functional managers. Functional managers do not want confrontations with angry project leaders. The key is communication. The project leaders must make every effort to keep the functional managers informed of their needs from the very beginning of the project planning. Functional managers do not like surprises. They don't want to be told of needs the day they are needed. Open, honest communication from the start of project planning concerning priorities, goals, schedules, and needs will go a long way to avoiding major conflicts. If a project leader has this type of relationship with the functional manager, then, when those crises do arise where additional resources are needed quickly, he will find the functional manager much more willing to work with him to keep his project on track.

Keeping people informed of problems, changes, project progress, and just general information is important. Regular meetings with those involved on the project should be scheduled. It's not a bad idea to invite management to these meetings because it's a good forum for keeping them informed about the project. Communication is crucial to the success of project management—communication with team members, functional managers, and management.

Problem areas or areas of disagreement must be resolved immediately. Team members must know who to go to with problems or disagreements. Team members have two bosses, their functional manager and the project leader. The lines of communication and reporting for the project must be clear and working. Problems or disagreements with functional managers over resources could drastically hurt the project if not resolved immediately. The project leader must be ready to step in and resolve conflicts.

The common thread through all these elements is communication. Constant communication is the key to success in project management.

E. Summary

In summary, the type of project management organization selected will depend on the project to be managed and company management philosophy. Some upper-level management is willing to give the project leader full responsibility and authority for the project and others will give something less. There are several common elements for successful project management as well as for successful project leaders. Like any management program, project management has its advantages and disadvantages. However, if implemented properly it can improve the quality, cost performance, and customer perception of the project and the organization.

II. ORGANIZING FOR IMPROVEMENT

Organizing for improvement involves applying the existing management structure to carry on the work of continuous improvement. It is not setting up a shadow organization to do TQM while the regular organization does the real work. TQM *is* the real work, with a focus on self-assessment and continuously doing it better. But the TQM process does involve the use of some specialized teams for achieving focus and avoiding dissipation of precious resources through needless overlapping or chasing low-priority problems.

Many research centers have reported using the Crosby model (6) for organizing for improvement. This involves a hierarchical structure similar to that shown in Figure 4.

The effort is led by a steering committee consisting of the organization leadership and may include representatives from lower management or nonmanagement. Reporting to the steering committee are Quality Improvement Teams, who

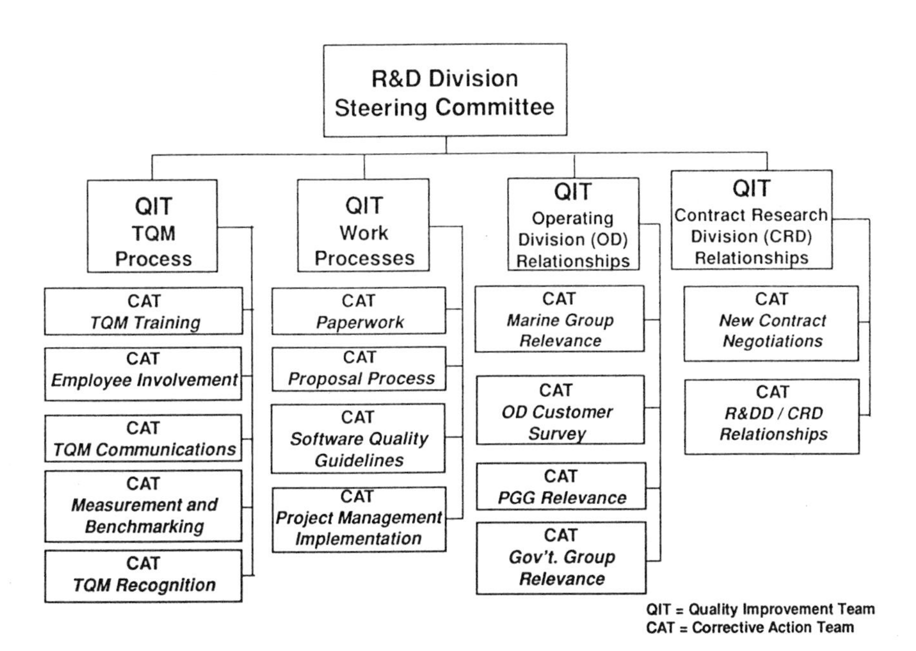

Figure 4 Organizing for improvement.

focus on key improvement areas and who charter Corrective Action Teams to carry out the work of analyzing detailed problems and developing solutions.

A. Steering Committee

Juran (7) calls this group of top-level managers the Quality Council and their job is to spearhead any significant changes in the way the research organization is run. The reason is simple—they are the ones responsible for research operations anyway. No change would be possible without their support, and significant change can only be possible with their full participation. It has to be "their process." Some of the things this council is responsible for are:

1. *Determining readiness of the organization for change*. If the employees and managers of an organization generally believe the organization is doing fine, there may not be enough incentive to change, at least not substantially. It is up to the council to determine whether the time is right. While companies often use a customer survey to determine if change is required, an employee survey in addition to the customer view would provide a more complete picture. The internal survey questions should center around potential key areas for change such as work processes, creative environment, and leadership, but not be worded to exclude the possibility of finding other areas that need help. The customer survey asks about research center responsiveness, technical capability, and innovation. Above all, the center needs to gauge its relevance to its customer's business. Some tools to help in this effort are described in Chapters 4, 5, and 11.

2. *Defining the mission of the laboratory*. This is no easy task. It involves some high-level soul searching, because all strategic planning will be conceived and executed to support the mission. Some mission statements tend to be followed with some detail to explain what the words mean, but many are designed to be memorable. The Air Force's "Fly and Fight," comes to mind. And, although it is certainly concise enough to be remembered easily, one might wish to lengthen it slightly to include, ". . . and Win," to be absolutely clear. The mission can be thought of as representing the frame of a picture and establishes the boundaries within which activities take place (8). It says "what is" to be done and, while it provides definition about what is an appropriate area of investigation (i.e., high-energy physics), it also implies by exclusion what isn't (i.e., astronomy). The mission is relatively stable and may properly be dictated by top management and their sponsors.

3. *Creating the initial overall vision for the laboratory*. The vision provides the picture within the frame and tells "how well" the "what" defined by the mission statement will be done. While the vision may have [possibly should have (8)] originated from the bottom up, it needs to be stated by the leader and should be shared by all employees at all levels. Continuous improvement is hard work and everyone has to feel that it is worthwhile. Focus groups of employees and

lower-level managers are useful to obtain feedback on "trial balloons" until everyone is comfortable with the vision.

4. *Identifying key management change areas in the plan.* These major change areas will be managed by Quality Improvement Teams. Which areas need to be worked on will be indicated by the results of customer and employee surveys and by a breakdown of the mission statement. Each key change area may have its own vision statement which needs to be decomposed into measurable parts (see Chapter 12).

5. *Establishing a mechanism for improvement.* Assign responsibilities, create Quality Improvement Teams, provide resources. Review results, measure, take action, review again. Delegate much of the details to the Quality Improvement Teams, but retain responsibility for overall resource allocation and setting policy.

6. *Providing training.* Provide concepts of quality improvement to everyone, but detailed tools selectively where needed and as they are needed so that they can be immediately applied. Just in time, but not just too late. Recognize that training does not end with an overview of the "seven-step process," but that the need is pervasive. It goes well beyond even the more rigorous problem-solving tools such as statistical design of experiments and Quality Function Deployment. Excellence requires updating of skills directly applicable to the organization's core technologies and will require ongoing investment in seminars and other forms of continuing education.

7. *Constantly working on building a supporting culture.* Of all the tools and concepts, consistent personal example is the heart of quality improvement. Management's persistent determination to keep open two-way communication, to emphasize the need for continuous improvement, to treat employees with respect, and to provide a stimulating creative atmosphere will bring the desired culture into being almost regardless of other tools employed. Lord Nelson is quoted as saying, ". . . no captain can do very wrong, if he places his ship directly alongside that of the enemy" (9). A similar idea can be applied to managers who need only to "walk the talk" to cause the transformation in culture. It is the personal example of management, the leaders of the organization, that will make or break the team. It is a long-term process and the managers have to keep the pressure on—themselves first, then the rest of the organization. They have to insist on continuous improvement and adjust the rewards and recognition accordingly. The adage of "what gets measured, gets done" could be amended to "what gets rewarded, gets done."

B. Quality Improvement Teams

The task of the Quality Improvement Team (QIT) is to focus on one or more of the key areas for change. This team operates on a management level to analyze and measure progress toward the vision for its assigned area(s). The team needs to identify the present state of the organization with respect to the area and then determine what the desired future state is. It may be necessary to brainstorm what the inhibitors are to achieving the desired state and to prioritize them for assignment to the problem solvers, the cross-functional Corrective Action Teams. One approach might be to set up two teams for the same problem. One would be assigned to work on the present state to improve the system as it currently exists. The other would be set free to visualize what the future state should be and to lay out a plan for reaching that level.

The QIT must define suitable metrics to determine progress, either toward improving the current state or achieving the future state. More information on metrics is given in Chapter 12. Corning's QITs measure themselves against the Malcolm Baldrige National Quality Award criteria and develop key results indicators for each of their areas of interest (10).

One of the reasons for having a formal organizational structure for the improvement process is communications. Each QIT needs to be aware of problems scheduled for solution by other Quality Improvement Teams to prevent overlaps.

Usually, one member of the QIT serves as a sponsor to the Corrective Action Team (CAT). The sponsor acts as an ongoing communication link between the two teams. While the sponsor is not continually active with the CAT, or is so close as to inhibit work, he or she can operate in a mentor role, interpreting the expectations of the Quality Improvement Team and helping to "run interference," when needed. The sponsor can also give guidance on what is likely to be the response of the QIT to a direction the CAT wishes to pursue.

The Quality Improvement Team gives authorization, resources and identifies specific training needed for its CATs. They monitor the CATs progress and provide feedback through the sponsor.

C. Corrective Action Teams

1. Cross-Functional Teams

It is well known that the more complex problems are multifunctional. Juran comments on the inability of Quality Circles to make a significant difference at the macro level. They weren't designed to operate there—circles are designed to work locally, usually at a nonmanagement level. But the big problems cut across boundaries. The fundamental reason why management is said to be responsible for 80 percent of the problems is that they have to fix the "system." The Japanese instituted cross-functional problem solving for about ten years before they introduced it to the line workers in the form of quality control circles.

Before the Corrective Action Team can even begin, it has to establish a clear understanding of the problem. A series of interactions with the sponsoring QIT may be required until the problem is properly defined and scoped. The resultant agreement should be in some form of a charter to, say, examine the process for proposing contracted research to reduce the overall time required to create and approve a proposal.

Once the assignment is clear, the team follows some derivative of the "seven-step process" described in Chapter 11. Or it may use tools such as Quality Function Deployment or Area Activity Analysis described in Chapter 5. As the work progresses, the team needs to be sensitive to the politics involved. Since the problem usually cuts across organizational lines, the CAT has to keep the affected area managers apprised as to what the team is doing. As soon as the team notices a different manager's "turf" is affected, that manager's interests need to be represented to the team. The team has to recognize that resistance to change occurs in many ways and for different reasons. The reason given for opposition is not always the real reason. The story is told of a machine shop where the operator's parts used to be collected at one end of the lathe or mill. The finished parts would be picked up periodically by a person and transported to the storage area. An idea was proposed to install a conveyor system to move the parts automatically from each machine directly to their destination. One of the operators objected on the grounds that the parts might be damaged and the idea would not save as much as it would cost. As it turned out, that operator's lathe was located so that during shop tours, the tour would stop in the vicinity of his lathe. Often, the visitors would pick up one of this operator's finished parts and admire his workmanship. Obviously, the operator did not want the system to change and deprive him of the occasional recognition. Juran (7) remarks on this difference between the "stated" reason and the "real" reason for resistance to a change.

As with all teams, the cross-functional Corrective Action Teams want to feel what they are doing means something—that they will make a difference. But a key lesson is to not give the team too large a task; don't ask them to "solve world hunger." One team can not do that anyway and they will only become discouraged. The problem has to be scaled down by the Quality Improvement Team to a reasonable size and have a better than even chance of being successfully resolved before handing it off to a CAT. One research center has instituted CATs to improve the relevance of its work to the needs of each of three main business groups. It would appear that this challenge will be of an ongoing nature and may require several smaller projects to make any headway. These initial CATs may eventually be upgraded to Quality Improvement Teams.

2. *Family Improvement Teams*

The Family Improvement Team uses the same tools as the cross-functional team, but its scope of investigation is within a functional area such as one department or section. The supporting Quality Improvement Team is made up of managers of that department. A charter might involve improving the methods for sample control within the Chemistry Section. The Area Activity Analysis or Quality

Function Deployment are particularly suited for Family Improvement Teams, but the Seven-Step Process for problem solving is also used.

D. Trainers

Consultants have several advantages over in-house trainers, particularly for "jump starting" the organization. They have developed their material and have had experience using it with other audiences, so they can often predict the questions. They are more readily viewed as the experts, since they fit the standard definition of being more than 50 miles from home with a briefcase. But consultants are expensive and not as available as in-house people. In-house trainers will develop their material using their own situations and examples. Looking at SPC charts from a piston ring production line can be a turn-off to a fluid dynamics research group. Exxon reported their single most important decision was to

> . . . custom design the training programs, develop in-house trainers and facilitators, translate the quality jargon into research terms, and teach quality tools and principles by way of examples and problems that are meaningful to researchers. (11)

There is, however, a disease called "guru-itis." That is, deciding one knows it all. There is still much to be learned from others. Consultants have seen a lot of different situations and can help others avoid making the same mistakes.

But what should be taught and how much? Westinghouse conducted awareness training the first year, then followed with soft tools, i.e., brainstorming, Pareto analysis, and cause and effect diagrams. Then, they went to hard tools like design of experiments and Taguchi techniques (12). At first, simple team skills, group dynamics and real basic problem solving will suffice. Additional skills will eventually need to be added—Just in Time.

It all takes time to do. Exxon reported it took them about two years to build their infrastructure, develop a conceptual understanding of the process, and learn the tools and techniques (11).

E. Facilitators

Good facilitators are an essential tool for getting a team off to a good start. There are four phases of a team's life often referred to as "forming, storming, norming, and performing." The facilitator's role is particularly important during the first two. During forming, when the team begins to set its meeting protocols and to understand its charter, the facilitator helps to build good meeting habits and to provide a refresher course in the use of tools. In phase two, the facilitator's sensitivity to group dynamics and an ability to keep things under control is invaluable as the individual members go through a sometimes emotional power struggle and a lot of prejudices, personal and professional (and some even related to the problem), work their way to the surface. High-performance people tend to be very passionate about their work, regardless of a sometimes gentle nature presented to the public. If the team really cares about its results, there is likely to be some intense feelings. The facilitator has a very important role in helping these to be worked through in a constructive manner. Things will settle down and the "norming" phase will see less need for the facilitator's help except during the occasional slow periods that seem to occur particularly when gathering and assessing data or wrestling with proposed solutions. There will come a point when the facilitator needs to back out entirely and this needs to be recognized by everybody, especially the facilitators themselves. They are people too and want to be appreciated, so the team or the sponsoring QIT should make provisions for facilitator recognition. Organizing some of the facilitators into their own "support group" is useful for an exchange of ideas and to identify improvements to the Quality Council.

F. Individual Involvement

One would think, with all that is talked about and written about teams, that the only way to improve is with a team. Not true. Not everyone gets to be on a team, nor should they. Cross-functional team members are picked for their unique skills in helping the team. Problems given to these teams are those that appear to be significant to top management. Family team members work on problems within a work area, but often those are selected by area management. Not everyone will get to work on a team, but it is very likely that everyone has at least one good idea. A formal mechanism needs to be in place to give those ideas a chance to be considered. Managers will protest that they listen to their people, but not all do and fewer do well. There are times when even the most conscientious manager just doesn't have the time to give each of the employees an opportunity to speak up. Employees feel that if the idea won't save a million dollars or win a Nobel prize, no one will be interested. Research managers might think their mechanisms for patents or new-idea disclosures are sufficient, but a lot of worthwhile ideas have to do with research process and culture and not with a new widget. Of course, many of the suggestions will be about changing the paint schemes in the rest rooms or moving the water cooler, but that's OK. The idea is quantity not quality. The intent is to improve a thousand things by one percent rather than one thing by a thousand percent. The system for processing ideas should be simple—avoid adding bureacracy—and provide for a rapid response to the suggestor. A decentralized review by the affected organizations helps to keep from adding staff, but an element of accountability, such as a centralized log of ideas generated and responded to, is needed to keep things moving.

G. Recognition

When it comes to deciding how to reward people for coming up with good ideas or participating on teams, management tends to "choke up." They are very aware of the inequities that creep into any program involving financial remuneration and are leery of starting a new one. They should not be so concerned. If they were familiar with the works of Maslow (13) and Herzberg (14), they would know that money is not an effective motivator. Lack of sufficient money is an effective demotivator; but the benefits of giving someone a raise for doing a good job lasts only a short time, then the person feels they are earning every penny. What is important is that employees feel that they have a positive connection with their upper managers. If people are inspired by their leaders and feel their leaders know, respect, and appreciate what they are doing, the people will perform way above the call of duty—to the point of ultimate sacrifice in some cases. With this powerful potential, managers have only to invest their time to motivate their workers.

This idea was tested in an informal telephone survey of several companies. The more successful ones used few or no financial rewards. The reasons given by one leader in the TQM movement was that financial rewards tend to focus people on big changes and tend to work against teams. It was that company's belief that ideas and continuous improvement should be a part of every job every day and should not involve a separate financial reward system.

An employee survey at one research center asked what were the important items of reward or recognition. The results should not have been surprising. Pins and parking places were OK, but what really counted was personal recognition from management. A newsletter article with pictures would be very nice. The most expensive item? Offsite team dinners. Not a lot of investment to keep the backbone of the organization happy. Or is it? Maybe not monetarily, but it takes time and effort to keep abreast of all that is happening, particularly in "flat" organizations with several direct reports. A neglectful parent may try (unsuccessfully) to make up for their lack of time with the family by using money and expensive toys to assuage their guilt. So too might the manager want to make up for a lack of personal involvement with the employees by throwing bonuses or raises to those lucky survivors who have learned how to survive in a "sink or swim" work environment. If the organization succeeds, it's through no fault or cause of that manager and the company might be better off saving that particular executive's wages to provide more bonuses. To quote Tom Peters, "when a person becomes a manager, they give up all honest forms of work. All they have left is how they spend their time." That time is best spent being a leader, teacher, and coach to the employees.

REFERENCES

1. S. P. Blake, *Managing for Responsive Research & Development*, W. H. Freeman, San Francisco, 1978
2. H. Kerzner, *Project Management: A Systems Approach to Planning, Scheduling, and Control*, Van Nostrand Reinhold, New York, 1989

3. T. J. Allen, ''Organization Structure, Information Technology, and R & D Productivity,'' *IEEE Transactions on Engineering Management*, Vol. EM-33, No. 4, p. 212, IEEE, New York, 1986
4. V. N. Karpenkop, ''An Approach to Management of High Technology Projects,'' Project Management Institute Seminar/Symposium, San Francisco, September 17–21, 1988, p. 547
5. J. Tuman, Jr., P. E., ''Project Management for Turbulent Times: Creating the Rapid Response Proactive Organization,'' Project Management Institute Seminar/Symposium, San Francisco, September 17–21, 1988, p. 46
6. P. B. Crosby, *Quality Is Free: The Art of Making Quality Certain*, McGraw-Hill, New York, 1979
7. J. M. Juran, *Management of Quality*, Juran Institute, Wilton, CT, 1981
8. E. E. Jones, *Quest for Quality in the Church: A New Paradigm*, Discipleship Resources, Nashville, 1993
9. R. Southey, *The Life of Nelson*, Naval Institute Press, Annapolis, MD, Pub. 1830, reissued 1990
10. E. Seward, ''It All Began with a Customer's Request,'' *Research Technology Management*, September–October, 1992, p. 28
11. C. M. Eidt, Jr., ''Applying Quality to R & D Means 'Learn-As-You-Go,''' *Research Technology Management*, July–August, 1992, p. 24
12. J. B. Moreland and S. S. Hines, ''Total Quality Management in R & D,'' 1992 Symposium on Management for Quality in Research and Development, Juran Institute, Wilton, CT, 1992
13. A. H. Maslow, *Motivation and Personality*, Harper & Row, New York, 1970
14. F. Herzberg, *Work and the Nature of Man*, World Publishing, New York, 1966

APPENDIX: OBSERVATIONS ON THE USE OF PROBLEM-SOLVING TEAMS

Robert J. Cartwright
Babcock & Wilcox
Alliance, Ohio

Background

Although quality teams had been in existence at this research center since 1982, experiences related in this appendix are based primarily upon direct observation of team activity over the past ten years. Most of the teams were in the technical service and support sections with a few in the research technical sections. This provided a foundation for study and comparison of a broader spectrum of methods and motivations.

None of the teams were involved with production-oriented responsibilities or goals. The teams were departmental and cross-functional. Members in the departmental teams sometimes known as Family Improvement Teams, Participative Task Teams, or Quality Circle Teams consisted of service, technical, and a few professional people. Their problem-solving license was limited to the team member's immediate work area and scope of responsibility. The problems accepted for study and improvement were suggested by team or departmental members and approved by a governing board (Quality Team Advisory Board), consisting of middle- and upper-level managers.

Membership was strictly voluntary and team size varied from four to twelve members per team. Immediate supervisors were encouraged to attend meetings and to be members and in some cases team leaders. For the most part, team leaders were selected by the area manager and the facilitator and served for many projects. Sometimes, the team could vote a replacement for a leader who wanted to step down.

Problem solutions were delivered to the board and selected managers in a semiformal presentation. The presentations could cover one project if it were a major concern or special subject, or an accumulation of smaller accomplishments. In most cases reviews to management were scheduled for each team at least once each year.

The cross-functional teams were different from the quality circle teams in a number of ways, although problem-solving techniques were similar. The cross-functional team license for problem solving was much broader and covered interdepartmental subjects. Membership was conscripted and teams were formed by the section manager, who also usually selected the project. The team leader may have been the section manager or was occasionally selected by the section manager and facilitator from members of the existing team. Team membership consisted of lower-level management and professional staff of between six and ten members. Membership could change from project to project, but for the most part, a new member was added only to acquire broader knowledge levels in areas where special expertise was required for the newly-selected project. Team progress was reported to upper-level management in various ways—review luncheons once a year, letters to parties affected by the problem solution, and monthly activity reports.

The facilitator for both types of teams had similar responsibilities. The primary one was to assure an orderly and timely attack of the problem and to assist, insist, persuade, and train everyone involved to ensure that they adhered to a proper methodology for problem solutions. He also had to provide the communications required between the team and management, consultants, vendors, etc. Another responsibility was to provide assistance for establishing and maintaining cooperation and coordination to satisfy management protocol and the proper chain of command. Close attention was directed to the type and scope of problems selected and their conformity to the "license" of the team. Certain domains of problem categories such as personnel, wages, and corporate policy were outside team activity guidelines.

The problem-solving methodology was, in general, selected by upper-level management. Much latitude was given to the facilitator to improvise the implementation of the process and customize it to the needs of the groups using them. Some tools for problem solving and improvement were specific with respect to method and organization but had been demonstrated in a production line application. This created a problem when attempting to translate methods and meaning to the world of R & D. Yet, the principles behind these steps were valid. The tools for both team types were identical with regard to function but their implementation required a much different finesse when applied to research ideas and opinions as opposed to support group work.

Several gains from using teams were immediately obvious: Teams provide participants the latitude for self-determination for what is accomplished in self- and job-related improvement. They also learn how to set and complete realistic objectives. A new awareness is born with regard to what barriers the other person faces in the common field of economic survival. All of these transitions are necessary beginning steps toward merging the all too often mutually destructive "*them*/*US*" into a compatibly productive "*we*."

The factor that seems to most determine the effectiveness of the problem-solving process and the tools associated with it was the "human factor." This is present to some degree for any process that requires cooperation among individuals and where efforts must be interleaved, focused, and directed toward common goals. The proportionate impact on the outcome is weighted much more heavily toward balancing personalities and motivational roots when the solution of problems requires a higher proportion of combined human intellects and a lesser portion of physical resources.

Human factors may manifest themselves in any one or a combination of ways which can contribute to the "team personality." The team absorbs attributes from each of the individuals involved. As the team learns and grows, these attributes form traits of the team personality. This indicates that, to varying degrees, it may be impossible to efficiently solve problems with knowledge or logic alone. Perceptions of any subject are limited to past experiences which are tainted by personal likes and dislikes, learned traits, and social or economic backgrounds. All of these perceptive associations are continually being modified by new levels of information and education, either formal or informal, and are relinked to form new libraries of experience which are referred to as the "learned processes." This is a simplistic view of human interaction with problems. Many books have been written to give insights into human behavior under a wide range of circumstance and environment. The question is, "How does one cope with change in a technological environment and merge it with the complex personality of a team where few of the human factors can be measured meaningfully or accurately by conventional means, no matter their importance?"

Problem/Opportunity

Teams have exhibited a growth or incubation period of different time intervals to adjust with each succeeding problem. This initial pause between merely accepting the assigned problem and embracing it as a real opportunity became shorter and shorter as the teams matured with each accomplishment in problem solving. Change alone can seem to be a problem, especially if people feel they do not want it or cannot prevent it from happening. A major element that affected the ability of members to adjust to this new environment of "I *can* have an impact on the outcome," was the attitude of their immediate supervisors or peers regarding acceptance or nonacceptance of team activity. Without exception this factor accelerated or delayed the "team spirit." This, in turn, affected each member's ability to change from having a "problem with the problem" to helping with problem solutions. When this supervisor and peer support was present, little time was needed to evaluate the problem as an opportunity.

When management support was not demonstrated or was questionable, team attitudes were predictable. Members reacted to the mixed signals and dropped out, lost momentum, or at the very least, required special motivation to continue. When circumstances have deteriorated to this point, the experience will be denigrated by everyone involved. This type of failure becomes a skeleton in the "incentive closet" that will come back to haunt or impede the motivation of anyone who in the future wishes to reestablish improvement programs.

One team member who had been told by his immediate supervisor (also a team member), "time will be better spent on the job he is paid to do," left the team even though upper management approved of the team activity. From that point on, he could not be convinced that participation in anything that "changed" current activity would not come back to adversely affect him. In his mind, change could only harbor problems, not opportunity. Ironically, his supervisor became more active as the program grew in later years.

The uncertainties which erupt just by way of rearranging priorities or disrupting "status quo" tend to suppress a feeling of adventure or opportunity for most people when timing, magnitude of change, or control of events, appear to be out of their hands. Teams will be more likely to see problems as a chance to be innovative if they feel they are shareholders in the process with solid support and approval of management as a partner.

Acting Upon/Being Acted Upon

Most circumstances that involve changing the current way of doing things leave people feeling as either movers or being moved. It can make a very subtle difference in their interpretation of events and their part in them and will determine which role they assume when confronted with a new move. Are they self-motivated or are they being directed by forces over which they have no control? This is an attitude that will very likely become contagious within the team. The positive drive of "We are in control," or the negative "They will not let us change anything," will take hold in every project. Either attitude will be very difficult to reverse once it sets in.

Several teams started with four or five members and eventually grew to ten or twelve. A common trait began to manifest itself as the team grew in size: "Let George do it." Getting all members to act and feel like meaningful contributors became difficult and the team's productivity began to slide. There was an atmosphere that there was nothing the team could do that would impact the problem. This condition continued until, at each meeting, all members were assigned a part to be completed for the next meeting. Each member would be polled for input regarding the particular phase of the problem that person was assigned. By creating a new atmosphere where each team member was recognized and expected to contribute without regard to numbers, the individual was not lost in a crowd. Up to a point, more members could then translate into more momentum and better prospects of taking control of the problem.

Peer Pressure

Peer pressure by departmental members who preferred to have no active role in team activity or in the process for change, but offered negative input from outside the circle, at times became a factor. Intentional or not, it had its effects on the comfort level the team carried in its approach to the problem. "Ghost" members bred a presence whose unofficial influence negated energies expended by legitimate members trying to solve the problem. Common among these external inputs was the comment that "nothing is being accomplished" (in a suitable time frame established by those who were not working on the problem). This produced frustration in the teams because the team members were not mature enough in the process of team problem solving to properly cope with it. Instead of maintaining a time perspective, they felt pushed by the impatience and comments from outsiders. The team wanted to take short cuts in the process and jump to a solution. This view that problem solving must fit a certain schedule lest management support may be reduced or the team's efforts will not be recognized, can undermine a willingness to stay with the problem.

Management counteracted these influences by publicly supporting the team's activity in a quality team newsletter. When approval was given for the team's recommended solution, a new respect was felt immediately. Having to study the issue brought a new awareness and respect for the problems a manager must struggle with. The problems no longer looked quite so simple. A major benefit of proper training in problem analysis was the change from impatience and a tendency to jump to solutions to a willingness to take the proper time backtracking for more data, developing more accurate theories, and devising solid methods to validate them. This also fostered a new sense of self-respect because the team was officially entrusted with a share of the load. The team had the tools in their hands to formulate and institute corrective action.

Informed/Uninformed

Almost everyone at times has been thrust into a situation where "something must be done now." Usually, this is with little or no knowledge as to what brought the situation to its current status and not a clue as to what is expected as an outcome. The sensation of being off-balance when balance is sorely needed does not improve the odds of a proper reaction. No one likes surprises like this; even less when the results can have an abrupt impact on their future well-being. This is particularly true of team activity during stressful changes within the company, such as downsizing, economic uncertainties, or major policy changes. During uncertain or threatening times factors that make up the character of each individual and that would only be background instincts in good times, become overly sensitized and likely to highly affect reactions. Which factors become sensitized will determine whether the individual or team will rise to the occasion and excel, or will cause the effort to fold. In short, how well the problem-solving process goes depends upon the work climate perceived by the individual and is directly related to "people tools" rather than technical ones. These tools should be considered as much an asset of the team or company as any other tool on the books.

In good times or bad, certain fundamental tools of human interaction must consciously and continually be addressed and never be assumed to exist within the company or team. Whether one feels motivated to take a constructive part in any venture in large part depends upon whether they feel they have the best information available or whether they feel left out. When asked what are the three most important elements that determine the value of a property, the response is "location, location, location." This can be restated when evaluating the prospects of a successful team. The three most important attributes that determine potential for team success are communications, communications, communications. The need is so simple, yet fundamental, and is ofter overlooked or assumed to be happening when it is not.

Members' View of Themselves

A team was working on a problem involving the quality and timeliness of information between groups within the section. The emphasis was to help the whole section look like and operate as a single entity in the eyes of customers outside the section. Prior to attacking the problem, the team members were asked to state reasons they felt the section was not looked upon as a cohesive unit by their customers. Some responses were "we are already performing that way," "we cannot operate any differently," "functions have to be the way they are," and "as a service group we are doing what is expected and have limited control." After much discussion, the facilitator suggested getting data about how members within their own groups see themselves prior to tackling links outside the section over which they had less control. A survey was constructed to examine where individuals and groups viewed themselves within the section with respect to the customer services they were expected to provide. In general, the survey questions were geared to ask how much each individual knew about where they fit in the organization, how their job contributed to group, departmental, and sectional goals. Also, did they know what these goals were?

The survey results pointed to a very limited knowledge of how or where group members fit into the local picture or that there *were* any larger goals they should be aware of. Most thought that autonomy in their own job was not only sufficient, but expected. Very few had previously been asked for input and most had little idea how their work related to other groups within the section or even whether they should ask.

The incorrect assumption that the problem was caused by inconsistent information coming from customers outside the section was easy to live with. Everyone had accepted it as fact and a normal state of being without question. Dogma was easily substituted for reason because communication on the local level was assumed.

There were complaints that it was all a waste of time or that the team wouldn't influence the results of anything because members were being pushed in a wrong direction. These complaints could be traced back to misinformation or a lack of information.

Management Support or Threat?

When a team is in its initial development stages, leaders may be challenged as team members try to establish their individual territory or a ''pecking'' order within the team. If an immediate supervisor or manager is a team member, his perceived attitude toward the projects has the potential to be a strength or weakness in the team structure. The outcome is usually predictable.

In the case of a supervisor member who had not accepted the new culture of giving subordinates an equal share in the ownership of a problem, positive results were difficult to attain. The supervisor's dominance destroyed a good percentage of incentive by other members of the team. If left unchecked, the team could have become a personal tool of the supervisor. This would have led to minimal progress and would have undermined an already shaky relationship between management and the team. In the team's eyes, this new concept of participation was just another way for management to keep tabs on them while giving the illusion of cooperation.

To compound matters, the facilitator was walking a tightrope where protocol would not and could not permit contradiction of the manager especially in front of the team even though the process was being compromised. This type of redirection is best attempted privately, at a more proper time and place. This is a very difficult situation until, in some cases, the supervisor accepts or somehow grows into the new partnership and is willing to learn how to switch hats gracefully. Adjustment of this magnitude is, however, a difficult but very productive accomplishment.

In other instances, managers had no difficulty grasping the hands of new partners and usually had worked on this premise in their departments prior to their introduction to quality teams. Where this spirit of cooperation existed, teams were very strong from the start and in most cases there was no change in momentum, whether the manager was team leader or not. Teams of this character did a great deal to strengthen the partnership and enhance credibility between managers and their subordinates.

Where switching of leadership roles could not take place, either the supervisor would quit the team or the team's progress was much less than what it might have accomplished. Part of the energy that could have been used to attack the problems was spent resolving conflicts.

Scope of the Project

Another threat to a new team is the size of the proposed project. It is not uncommon for the team to be initially overloaded by a project that is too large or too early for a team lacking experience in the problem-solving process. More experienced teams know how to break the task into smaller bites. A number of small successes will eventually give the team the confidence to be comfortable with larger and more complex problems.

Experienced/Inexperienced

Past experience forms a library from which people can draw when encountering similar circumstances. When anything new crops up, they make a quick review of their library to face the new with *something*, rather than *nothing*. They then resort to building walls or steps depending upon how they feel about the change and their chances of successfully coping with it. A person's ability to cope or function in these circumstances is dependent upon how well they feel they are prepared to resolve or accept differences when they are initially exposed to them. Those previous experiences help them adjust to the new situation. Progress or growth can be measured by the willingness to readily grasp experiences that are new, whether they are wanted or not, approved or unapproved. A comfort zone is established more quickly when people feel they have the tools to deal with change regardless of the turn of events or subject involved. From this basic level of comfort evolves a willingness to examine change. People recognize the hazards, but they also appreciate the opportunities.

Human Factor

People paired with problems or more correctly the right people paired with the right problems—sounds like a fitting motto for productive change. If there is such a thing as a "sure fire" ingredient for solving problems, this is it. But qualified people to implement change do not just show up at the right time any more than business success just happens.

This is probably one of the most difficult obstacles in the list of requirements necessary to successfully institute change on a scale that will transform an entire organization. Combining talents that are focused to a single purpose requires much more than money. Changing a perception or a way of life can at times induce trauma. It is difficult enough on an individual basis; on a scale involving a complete team it may be a dramatic shock. It may also require a dramatic shock. For instance, when an individual (or organization) is informed that they must change their lifestyle or they are dead (or out of business), a need has been established to which everyone can easily relate. Such a dramatic need demands a review of various types of resources required to cope with the need (survival) and an acceptance of the resultant change.

Willingness to Participate

A number of surveys have been sent out to various team members and non–team members to test attitudes toward participation for improving the day-to-day work and how past efforts in this area were regarded. Response was very much in favor of improvement but attitudes about participation and past efforts toward team activity varied drastically. There were very few fence sitters. This did not exactly come as a surprise, since two distinct types of suggestions had been encountered: the areas of participation and improvement. "They should" and "we can" were divided along the lines of non–team members and current team members.

There seems to be an inherent internal mechanism in each individual that affects their acceptance or rejection of anything new. All will participate in the work environment, some more than willing, some because they have to. Everyone who is a member of an organization involved with change shares in its problems one way or another. There are those who will grapple with the problem in any manner they can to resolve it, and those who insist on becoming a part of the problem. If people do nothing to contribute, they may be doing something by omission that deprives or degrades the team's efforts. Projects that to some degree involve human sensitivities and personal interactions tend to amplify or exaggerate this effect most.

Need to Be Productive

No one wants to be a part of any activity that is considered wasteful. Everyone has experienced and easily detected economic waste, but to the professional or career-oriented individual, intellectual waste is even less tolerable. A sure way to kill motivation is to reach an impasse on a project and not have a plan or the leadership capable of followthrough to get back on track or to revise the current plan. This usually results in "make work" efforts that lead to dissatisfaction among team members who are normally highly productive.

A team which consisted of both supervisors and technical support personnel had been quite successful, yet came to such an impasse. Ostensibly, because of normal job-related pressures, a supervisor left the team. He stated off the record, "further participation is a waste of my time." The impact this had on the rest of the team was manifold. Some otherwise reliable members quit the team because they felt if it was not worth the supervisor's time, it certainly should not be worth theirs. Others took it as a show of lack of management support. Still others remembered that the team had past successes and recognized that the loss of one member should not cancel the future effectiveness of the team. Others resolved to continue rather than fail at something they had started. It took a careful reassessment of team goals and a reinforced determination of the remaining members to continue with the project. This was a hard lesson that all members have an impact not only on the outcome of the project but on the attitudes of each other.

Supervisor Participation

Participation by supervisory personnel at all levels was encouraged to lend weight and credibility to the team. Communications between managers and personnel was strengthened inside and outside the meetings. Most team members responded with peak innovative skills in the case of impasse if they were encouraged to draw upon their talents and the environment was conducive to everyone being recognized as the carrier of a valuable area of expertise. All team members, regardless of rank outside the team, will recognize and value the intellectual assets of other members if this valuation of worth is recognized in them as well. This mutual assessment takes time to establish itself across the team.

The team learned that the quantity of members was not nearly as important as all the members having mutual respect and a drive to participate in a productive activity which met a challenge. When this feeling of self-worth and recognition is an accustomed pattern of activity, acceptance of "highs" and "lows" in process or progress is less likely to swing the individual or team off course. These slow times come in most problem-solving cycles. To the mature team, it is an axiom that progress will rarely be swift.

Getting to the Bottom of It

Nearly everyone in the work-a-day world has been asked about the goals they have set for some aspect of their lives. This is also a question quality teams must ask if they want to focus their enthusiasm and joint talents to full advantage. How many times do individuals attempt to envision goals and the steps to attain them without first determining where they presently are? One of the first things team members should be asked is to examine in very specific detail how the process is currently being handled. An easy tool to do this is a flow chart. When there has been insufficient effort to determine the magnitude of a problem or root causes have not been clearly understood, attempts to arrive at a problem solution usually met with marginal results. Most attempts appeared to have resolved only symptoms. Further exploration would have revealed the problem was actually caused by a less visible or recognizable part of the process.

One particular team was well into the solution stages of a project. They had superficial solutions to half a dozen different services which appeared to have a bearing on doing business in their machine shop. But, after a year, they just could not get all the parts together in a meaningful and logical manner. As it turned out, a very vague problem description or goal was given at the origin of the project. The initial exploratory data was not given to some members working on different aspects of the same problem. Through the encouragement of a new facilitator, they were able to start over with a first step of analyzing what was currently being done in each area. It became apparent that the big problem was in fact many smaller problems. One smaller problem was selected that seemed likely to have the greatest potential for improvement. A study by the whole team was done to focus all effort and direct all team resources to the one problem by analyzing what was currently being done, along with an evaluation of what they were being asked to do by their customers. From this preliminary study, a more definitive route was set for a more realistic approach to an attainable goal.

Another mistake to avoid is impatience when scoping out the problem. Most people find it very easy to nearsightedly speed up problem solutions. Rarely does this produce correct answers because the time has not been taken to consider approaches to the problems as simple as asking customers the right questions. Part of this stems from a tendency to try to save time on data gathering and to let preconceived notions be accepted as facts.

A case in point deals with a project to determine why many lab renovation projects did not meet their target for estimates. Last-minute changes were not only costly, but critical time schedules suffered. It was assumed the project managers were not sure what they wanted or were not aware of the magnitude of what they were asking.

The process that past projects had followed was analyzed. After much data was gathered, a flow diagram was made indicating that certain steps in the initial phase should have been considered and were presumed to have been. Actually, they were either omitted altogether or were not done effectively. Even during construction of the flow diagram, it became apparent that many of the missing steps involved lack of understanding between the service group and the project manager.

A major stumbling block was that the questions asked during the onset of the project were not the right ones. From the flow diagrams and a questionnaire circulated to project managers, an interview sheet was created to help the project manager ask the right questions of the customer. It also provided a cross-check of sorts for the manager in making cost estimates.

Get the Data

One of the largest hurdles most teams have encountered is getting used to acquiring factual data to replace dogma. On a project that involved handling heavy metal sheets, a large amount of data was gathered which indicated questionable conditions had existed for a number of years. A "jib" crane had been used that had its capacity exceeded. The manual method of handling the steel plates as a truck was driven out from under them posed serious risk to the worker whose job was to stabilize them as they fell from the truck. Factual data were gathered for a presentation to management.

Team members were reluctant at first to make the presentation since it would be an admission that this practice had been the norm for a long time and "fault" would generate repercussions. At this point, someone postulated a circumstance where the expected accident would happen within the next day. They were sitting on all this data and could have prevented it. Whose court is the ball in? Where is the "fault" now? How would one balance the feared repercussions for past practices? Needless to say, the presentation was made promptly and the problem was resolved to the team's recommendations immediately. Instead of repercussions, the team was commended. For this team it was a turning point in their trust of the intentions of management. It changed a dogma about certain work areas being ''sacred cows'' that no one dared speak about, let alone try to do something about. It also changed the attitudes of workers and of some lower-level supervisors. It encouraged them all to be better extended team members and to be aware of concerns to upgrade the work space to the advantage of everyone. This resulted in more problems being resolved by direct communications without having to institute a quality team project.

Opportunity to Grow

Team participation offers a lot of potential for technical and personal growth. One of the first teams of this study was in a work area where very poor communications existed between laboratory personnel and management and, for that matter, even between co-workers. The work ethic between employees was one of ''job security'' by not sharing information, only doing what they were told, and not associating with others because no one could be trusted. In this crippled environment, anyone expressing an interest in the work area of another person, even if the intent was to help, was looked upon with suspicion.

When the team was first formed, a great deal of difficulty was encountered when attempting to get team members to express opinions, let alone initiate conversations about work practices that involved their areas of expertise. After a number of small, benign projects coupled with impromptu dissertations about how teams really are subsets of the larger team and how it is not a case of ''them and us'' with management, attitudes changed slightly. After a few small successes, the facilitator was no longer looked upon as a spy, their co-workers for the most part were accepted as possible helpers, and management was accepted as a distant partner.

From this point on, the team grew stronger and healthier with each project demanding more support from each other and management. The members have reached a growth level where they are convinced that ''if the boat sinks, they all get wet.'' Within their meetings, a marked difference has taken place in the level of technical productivity, willingness to communicate, offers of help, and even volunteer participation in management reviews of projects. It was a most satisfying moment to see people who, a short time before, would hesitate to say ''good morning'' to a manager, now enthusiastically explaining the inner workings of a project.

The same attitudes changed in every context. Many of the team members expressed an improved approach to organizing and handling problems at home using methods learned and applied on the quality team.

Often, the intangible gains far outweigh the benefits that can be measured by conventional means. There are continuing improvements even though they are rarely tracked and recorded as such. This has been a pattern with most teams. A new awareness of the team approach to solving work problems has stimulated new growth in the individuals who took the challenge of change seriously. This example of extreme personal change is not present to the same degree in all teams but is a striking example of win/win in the growth and development of both individuals and teams.

Keeping Teams Motivated

Everyone appreciates improved economic circumstances, random as this may seem, but the one reward that is always there at the culmination of a project is a bonus in self-esteem. This may sound trivial if one is not accustomed to accomplishing anything. Self-recognition is an important item since how people see themselves is generally how they come across to others. The confidence level they have when confronted with new challenges is dependent at least in part on past successes and the role they have had in them.

For example, one research center went through a very low economic time and financial recognition was minimal. During this period, there was a marked distinction in performance levels. Those teams who had developed a pride in their work and were mature in their willingness to do their best regardless of the rewards prospered. Their work was their signature and trademark. Teams that had good track records in the past not only survived these slow times, but made good contributions to improvement in their work environments. Those teams who had not developed the trait of inner credibility had little or no reserve to survive because they felt economic times would limit anything being in the pot for them. Interest in team survival was a low priority.

Inner motivation is not always easily described or detected and is usually not recognized fully unless it is called upon at a crucial time. Then the response is real and significant. An important concept that their work had real value was conveyed to teams as a part of their training. Regardless of the economic times, credibility is the glue that bonds teams and their partners, along with other elements of the company and its customers. It is much easier to rework or rebuild a physical item that is broken than to restore lost credibility between humans.

4

INDUSTRIAL R&D STRATEGIES

Philip A. Roussel and Tamara J. Erickson
Arthur D. Little, Inc.
Cambridge, Massachusetts

Kamal N. Saad
Arthur D. Little, Inc.
Brussels, Belgium

I. THE EVOLUTION OF THIRD-GENERATION R&D

The purposeful management of research and development is a complex and delicate balancing act. General management wants R&D to serve multiple purposes: supporting existing businesses, helping launch new businesses, and deepening or broadening the company's technical capabilities.

To respond to these various management wants, R&D engages in types of work characterized by different technological uncertainties and differing time frames. The uncertainties and time frames depend on the nature of the technological weapons chosen to fight the particular battle—the maturity of the technologies involved and the degree to which they are mastered by the company.

As we look at today's industrial scene, we see three generations of R&D—and technology—management in practice. Recognizing which generations of R&D management are practiced in your company provides a foundation for change—when change is appropriate. How do you distinguish among them? You recognize them by their distinguishing traits: by the R&D management philosophy in place, by the way in which R&D is organized, by the way R&D/technology strategy is formulated, by how R&D is funded and how resources are allocated to R&D, by how R&D targets are selected and R&D priorities are set, and by the way R&D results and progress are measured and evaluated.

A. First-Generation Management of R & D

First-generation management of R & D is a holdover from the good old days of the 1950s and early 1960s. It is characterized by the lack of a strategic framework for the management of technology and R & D. The annual budget provides the total framework for R & D. General management possesses scant insight and provides little guidance. The company's future technology is decided largely by R & D alone.

The operational context can be described as fatalistic. R & D is an overhead cost, a line item in the general manager's budget. General management participates little in defining programs or projects; funds are allocated to cost centers; cost control is at aggregate levels. There is minimum evaluation of the R & D results other than by those involved in R & D. There is little communication from R & D other than to say, "Everything is going fine." There is only a modest sense of urgency: "Things are ready when they are ready." The communication and cooperation breakdown between R & D and the businesses is clear, and especially severe in radical and fundamental R & D.

Some classic examples of "managementspeak" versus "researchspeak" characterize the we/they dialogue about R & D that still exists in many companies:

- Business people believe that "R & D does not understand business," and researchers believe that "targeting stifles motivation."
- Business people think "researchers are uncontrollable," and researchers think that "administration smothers creativity."
- To the business managers' complaints that "results are always 'just around the corner,' " the researchers retort that "breakthroughs cannot be forecasted."

1. The Management and Strategic Context

In first-generation management, the management philosophy is characterized by failures of confidence in the relationship between general business management and R & D management.

The intuitions of R & D managers dominate. They decide what, when, by whom, and why in isolation from the broad business context. General management often remains aloof and sometimes does not even know that fundamental research is being carried out at all, let alone what it costs. Adventurous work is done "under the table" for fear that if general management finds out, the research will be discontinued.

In first-generation management, support for radical R & D is analogous to a courtroom where R & D is the advocate and general management is the prosecutor. However, there is no judge, so the two parties enter into plea bargaining. R & D management proposes the what and the why and gives an indication of the when and how much but makes only loose commitments regarding timing and total cost. General management can only decide whether it is willing to approve next year's budget.

In the world of incremental R & D, the R & D manager often does the bidding of general and functional managers from areas such as marketing and production. Functional managers decide the what, when, why, and how much to spend, leaving R & D management to figure out the how and by whom. Companies practicing first-generation management of their incremental R & D have much to gain by carefully reviewing all incremental R & D work being done, for example, to improve product quality or to reduce production costs and then evaluating whether the benefits are worth the costs. The conventional wisdom is that such work is "always good for you." This truism may obscure the real reason the work is being done—because momentum became inertia. The work was started because it was needed; it has been continued because it does some good; it is maintained because "we have always done it."

In first-generation management, R & D is typically organized into cost centers by scientific or engineering discipline. Much R & D is centralized at the corporate or divisional level, and incremental R & D is distributed to the business units. Incremental R & D is often grouped first into activity centers such as clinical development in pharmaceuticals, field development in pesticides, design and prototype development in machinery companies, "bread-boarding" in electronics, and piloting in chemicals. Then it is further broken down by technological discipline or expertise center—clinical medicine, biology, rotary equipment, software development, corrosion, and so on.

Project management—meaning a discrete set of activities with a specific objective, a resource plan, a time frame, and a budget—is not explicitly recognized in first-generation management. Responsibility for activities is assigned to one line manager or another in the hierarchical R & D organization. In such companies, the matrix-type organization, with a project manager as a full partner, is avoided as an unnecessary complication and an incursion into the realm of line management. As a result, responsibility for achieving the R & D objective is obfuscated as it is handed over from one line manager to another—for example, from an electronic design laboratory to the engineering department. As responsibility is passed from one department to another, advancing the R & D objective may get unequal attention.

In either case, continuity and accountability are elusive. The R & D strategy problem in first-generation management is rooted in the difficulty in defining technology in ways that both technologists and business managers understand and feel comfortable with. General managers working in a first-generation mode tend to see technology in terms of scientific and engineering disciplines—at the incremental end, mostly in terms of what a technology is and much less in terms of what it does for the business. In such companies, the importance of technology is judged from the perspective of the technologist. Novelty is one major consideration; the expertise of veteran technologists is another.

Technological uncertainty is taken as an uncontrollable given that will decline over time. But technological uncertainty is difficult to assess. Business and market uncertainties are subjects for the business and marketing managers to worry about, and only when the work is sufficiently advanced do technological and business uncertainties merge. No wonder then that the conventional wisdom is to "develop the technology first and link it to the business later," or to "let the general and functional managers worry about the business aspects and leave technology to the technologists," much as one might suggest leaving war to the generals. The danger in the approach is clear. Polaroid, for example, came out with a new instant-movie-camera film system just as videocassette cameras were coming on the market.

2. *Operating Principles*

In first-generation management, the funding of R&D—the aggregate levels of expenditure and where in the corporation they are accounted for—is at the discretion of general management even though the CEO and general managers at the division and business levels typically have limited insight into the company's future technological needs on which to base funding decisions. R&D expenditure is merely a line item in the budget. Furthermore, where the funds are accounted for is often a matter of affordability, convenience, or both, giving rise to accusations of "robbing Peter to pay Paul."

R&D resource allocation, in contrast, is at the discretion of R&D; only the R&D managers at various levels know how the funds are expended. R&D budgeting is done by means of a top-down cascade, and each level defines how it will spend the part of the budget that falls within its direct control. There is little upward visibility. More often than not, availability influences how resources are used; business objectives and needs have little influence in the short and medium term on the resource configuration.

Scientists, engineers, and R&D managers working in radical fundamental R&D in the first-generation mode view targeting, milestones, and dates as the imposition of rigid linear logic on a process of idea generation and exploration that thrives on creativity, intuition, and spatial reasoning.

There is no denying the power of creativity, intuition, and spatial reasoning, or their untold benefits. The problem for companies using the first-generation approach, however, is that later, when R&D yields results, the linkage between those results and the business's needs may be haphazard. General management is uncertain how to influence the situation. Furthermore, if targeting is not acknowledged as useful, priority setting is not even considered.

Targeting and priority setting are more acceptable in incremental R & D because technological uncertainty is not a significant factor and the time frame is typically more immediate. The business targets for incremental R & D are selected by the general and functional managers, and R & D objectives and resources are subsequently defined by R & D. A first attempt is made to match the two during the annual budget cycle, but the real trade-offs between needs and resources are made during the course of the year; priority setting is operational, not strategic.

Measuring results, and thus evaluating R & D progress, tends to be ritualistic and perfunctory in the first generation. To begin with, result expectations are not defined rigorously from the outset. "We seek a novel chemical entity," "we are aiming at an innovative new widget," or "we want a lower-cost manufacturing process" are not precise enough definitions of results against which to evaluate success or progress.

In these circumstances, progress reviews tend to focus on scientific and technological problems and how they might be resolved. In fundamental and radical R & D, activities are peer-reviewed periodically, typically every six months or once a year; and the emphasis is on technological achievements since the last review, in light of the effort expended and obstacles still to be overcome. Progress and results of incremental R & D are reviewed in terms of what has been done during the past three or six months, whether it was done on time, and whether the cost centers involved were on budget.

B. Second-Generation Management of R & D

Second-generation R & D management is a transition state between the intuitive and the purposeful styles of management. It is practiced by companies that have recognized the reinforcing interrelationship among organizational functions and thus seek to introduce greater order into their management.

Second-generation management provides the beginnings of a strategic framework for R & D at the project level and seeks to enhance communications between individual businesses and R & D management. It makes the business or the corporation the "external customer" for R & D practitioners alongside—and as important as—the "internal customer," R & D management. Long-range plans and annual budgets in these companies recognize projects as distinct and discrete activities. Management also recognizes explicitly the differences among the strategically distinct types of R & D and tries to set a course to differentiate them in strategic and operating policies.

Second-generation management is most distinctly differentiated from the first generation by business and R & D management's cooperation in the joint consideration of individual projects—projects' cost over their lifetimes, their impact on the businesses, their uncertainties, their management, and their execution.

For individual projects, the results can be splendid. But the consideration of and decisions about projects on a project-by-project basis, however beneficial for each individual project, still omits the strategic dimension dealt with in third-generation management—the interrelationships among projects within a business, across businesses, and for the corporation as a whole.

The purposeful management of R & D at the business and corporate (multibusiness) level is still missing. The portfolio concept remains absent. The spirit of partnership between general and R & D management continues to be project focused. Even though the managerial and strategic principles of project evaluations in a strategic context may be clear for senior managers, they do not readily penetrate the company's operational levels.

1. The Management and Strategic Context

Management philosophy in the second generation is characterized by a relationship in which general management seeks to balance the advocacy and championship of R & D against strategic goals without destroying motivation.

By establishing a supplier/customer relationship between R & D and business managers, business managers hope to become "bottom-line responsible" for what they choose to spend on R & D and thus more cost/benefit conscious and willing to involve R & D management to a greater extent in cost/benefit assessment. They also hope to make R & D managers work hard to demonstrate their relevance and to be more responsive to the needs of their business and corporate clients.

Although the management philosophy seeks to institute a commercial environment, it does not go so far as to install an open market. The "customer" cannot buy outside services if equivalent services are available within the corporation.

Like management in the first-generation mode, managing in the second generation tends to centralize fundamental and radical R & D and to distribute incremental R & D to the businesses. But second-generation R & D management has a major advantage: It clarifies and acts on the discrete, project nature of R & D, makes active use of matrix management, and puts professionally trained or experienced project managers in charge of significant programs and projects. These project managers are assigned the tasks of planning, mobilizing resources, and ensuring that projects are carried out on target as well as on time and within budget.

Management in the first-generation mode focuses on the difficulties of matrix management, the diffusion of authority, and complex communication. Management in the second generation accentuates the positive and adopts a pro-active attitude. It recognizes the multidisciplinary nature of R & D, the need for continuity, and the need for dedicated professionalism in managing the complex relationships called for by most R & D projects of any size. The project manager is responsible for what is to be done, when, and at what cost. The line (or resource) managers are responsible for deciding who to assign to the team and for the quality of the output.

Second-generation R & D management attempts to link R & D and technology to the needs of the business on a project-by-project basis. It allows R & D to challenge the appropriateness of business objectives and pro-actively to suggest how R & D and technology can constructively interact to produce a business plan the quality of which neither alone could construct.

Still, it formulates R & D plans on a project-by-project basis, separately and independently for each business and for the corporation. The process fails to deal adequately with activities not directly related to existing businesses but important on a corporate level. The process is unable to optimize R & D resources for the businesses or for the corporation as a whole. The process offers no mechanism for deciding between, say, allocating a sum of resources to a cost reduction project in business A or allocating the same sum to the development of a new product in business B, even though the benefits to the corporation may be significantly greater in one case than in the other.

2. *Operating Principles*

General funding parameters for fundamental research are established at a level the company feels it can afford, typically a largely arbitrary percentage of the R & D budget. These funds are provided centrally, by the corporation or the division.

Funding for radical R & D is often shared by the business and the division or the corporation in order to share risks. Funding levels are determined by the needs of existing businesses and those of new businesses and technologies important to the corporation, after assessment by the divisional or corporate staff responsible for new business and technology.

Funding of incremental R & D is typically through the business. The level is negotiated between business managers and their R & D counterparts, usually in the context of the annual budget.

Resource allocation and priority setting also vary across the R & D spectrum. Corporate central R & D management, where it exists, allocates resources and sets priorities for fundamental R & D. Resources for radical and incremental R & D projects are allocated and priorities are set through joint decisions by customers and suppliers. However, the make-or-buy decision is usually left to R & D managers, who base their choice on how flexible or inflexible their internal resources are and what work can be accommodated internally.

Management in the second generation usually tries to measure the results of R & D by using quantitative approaches, such as net present value, return on investment (ROI), and payout measures, for each project of significant size. General managers often find it hard to quantify the benefits early in the life of projects, before the technological uncertainties have been dealt with. Their quantitative characterizations are usually barren, less for lack of competence than for lack of discipline.

Often this management runs into a market-intelligence gap. The marketing people say, "If you can tell us what you expect to achieve, we can tell you what the market might be." The R & D people say, "If you can tell us what the market will value in five years, we will be in a better position to give the market what it wants." The gap frustrates everyone. Marketing people generally have little or no idea what the market needs will be five years hence; they are paid to worry about the market this quarter, this year, and perhaps next year. The R & D people, on the other hand, usually have little direct access to market information; they are paid to do R & D.

As R & D work advances and the time to commercialization shortens, the market–intelligence gap gradually closes, and management finds it can begin measuring results in terms that everyone feels comfortable with: cost, benefits, time, and the like. But early in the life of radical projects, management is reduced to measuring progress mainly against technological milestones.

Second-generation management finds quantitative approaches to measurement even more daunting when it comes to fundamental research. The time frame to results is even longer, the uncertainties greater; and early on the benefits can be described only in the broadest of technological and business terms. Even the project's costs are difficult to estimate beyond the immediate future, since the project could be discontinued at any time and what may need to be done tomorrow is a function of today's results.

Companies working in the first generation are fatalistic in their attitude toward evaluating progress. Their counterparts working in the second generation worry about how to reach decisions to accelerate or decelerate an effort and to abandon an effort when appropriate. They establish formalized peer review systems that involve the best talent within the company and the best external talent they can find.

R & D managers communicate the results of these progress reviews regularly to their customers. However, as is often the case in commercial relationships, the vested interests of each side differ. The result is difficulties in deciding what to do when projects do not progress satisfactorily. A spirit of partnership is clearly still missing.

C. Third-Generation Management of R & D

Third-generation management seeks to create across business units, across divisions, and across the corporation a strategically balanced portfolio of R & D formulated jointly in a spirit of partnership between general managers and R & D managers (see Figure 1).

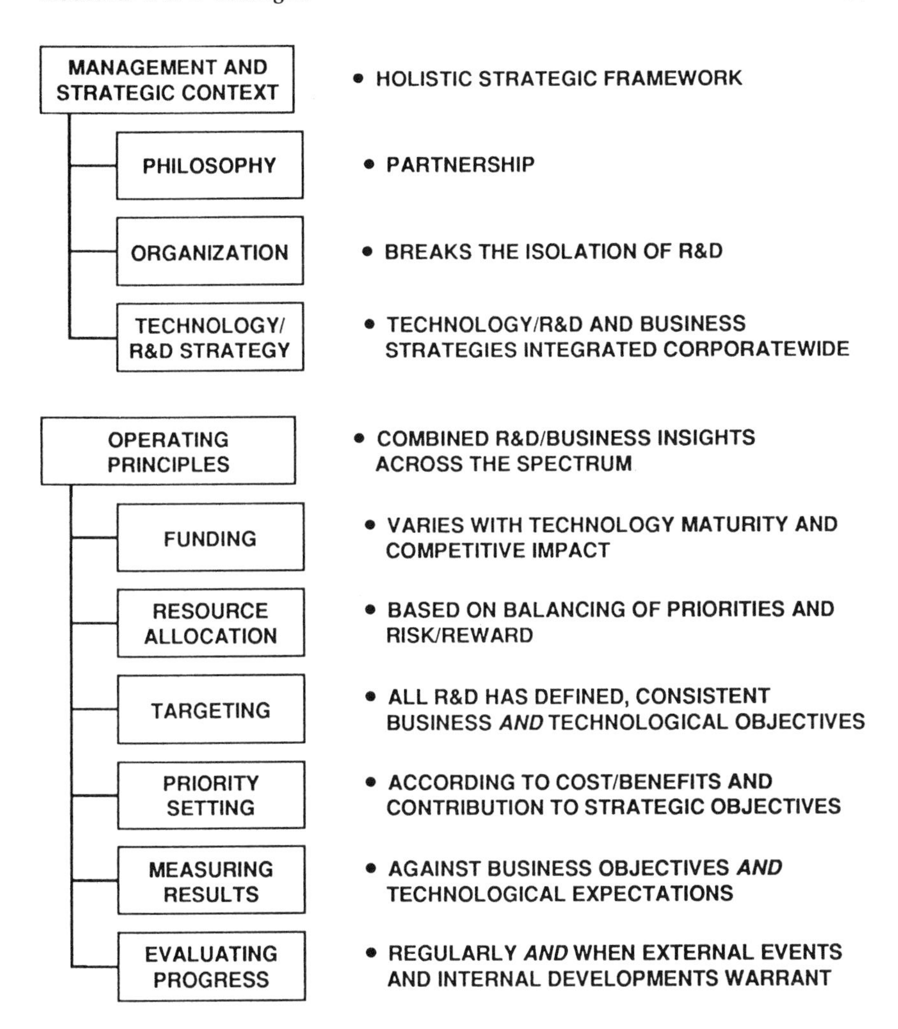

Figure 1 The characteristics of third-generation R & D management (strategic and purposeful).

R & D seeks to respond to the needs of existing businesses and to the additional needs of the corporation while at the same time contributing to the identification and exploitation of technological opportunities in existing and new businesses. General management in the third-generation mode institutes a strategic and operational partnership between R & D and the other vital functions in which R & D challenges and helps define the company's real technological needs, both today and tomorrow, in additional to helping meet those needs. What appear to be examples of third-generation management practices and successes can be found in IBM's development of the Proprinter and Merck's introduction of several major new pharmaceuticals in a very short span of time.

1. Management and Strategic Context

How does third-generation management differ from first- and second-generation management? In terms of management philosophy, third-generation management creates a spirit of partnership and mutual trust between general and R & D managers. They jointly explore, assess, and decide the what, when, why, and how much of R & D.

Although general and functional business managers may not contribute insight into the "how" and "by whom" questions (particularly when they have no technological education or R & D experience), R & D managers in companies employing third-generation management find it useful to inform their colleagues on these topics as a means of motivating their moral support. The partners recognize that, although each has a unique contribution to make to the managing of R & D, bringing the different perspectives together when preparing and making important decisions enhances the quality of these decisions.

Furthermore, companies working in the third generation take a holistic view of the full range of their R & D activities. On the one hand, they recognize the different strategic dynamics and the different sources and levels of uncertainty along the spectrum. On the other hand, they find it immensely valuable to understand and take into account the interrelationships among the activities concerned.

Companies working in the third generation seek to organize their R & D in a way that breaks the isolation of R & D from the rest of the company, in order to promote the spirit of partnership between R & D managers and their general or functional management counterparts. By concentrating scarce resources and rare skills, these companies organize to promote sharing where it matters. They exploit technological synergies by integrating the R & D and technology plans across businesses and across the corporation by coordinating plan execution and by sharing experiences and information between distributed centers.

They design their communication networks to ensure a continuum across the R & D spectrum and forward to the market. They believe in the matrix as a powerful way of managing R & D, and they seek to make their project managers full partners with their R & D line-manager counterparts. These companies work to formulate integrated corporate/business/R & D/technology strategies that take account of synergies and trade-offs between projects across businesses and corporate programs, particularly when technologies are shared by different parts of the corporation (see Figure 2).

Such companies select targets by setting their fundamental research in a business context, confident that providing researchers with a sense of business

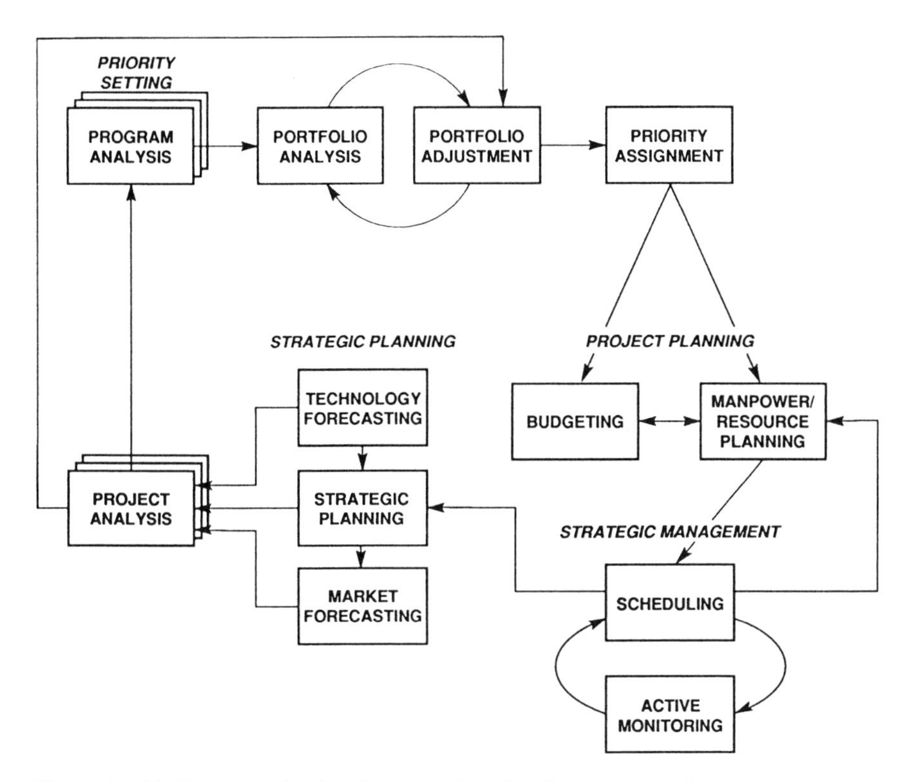

Figure 2 Linking strategic planning to project planning and execution.

purpose is a motivating factor and need not be inimical to creativity. They ask themselves questions such as:

- How relevant and important to the company would the successful completion of program A be within the X years it might take?
- Do we have the critical mass of resources required? Are they in-house or available externally?
- Given their relative importance, likely time frames, and resource availability, should we proceed with all the programs proposed or only with some?
- Which programs should get priority?

2. *Operating Principles*

Third-generation management establishes funding levels for the short-, medium-, and long-term needs of the businesses and the corporation. It seeks to afford what they need, rather than simply funding what they think they can afford. Funding policies are flexible.

If there is strategic space for R & D on emerging and young technologies, third-generation management provides it at the corporate and divisional levels. Speculative technologies are typically funded at the corporate level when the risk is high enough to cause a business unit to shy away from strategically desirable investment.

With incremental R & D, management encourages customers within the company and their internal suppliers to test regularly the charges they respectively receive and make against market rates for similar work. When management is smart, it also periodically "zero-bases" its incremental R & D and technical service budgets.

Companies operating in the era of third-generation management of R & D have a resource allocation principle for radical R & D that requires a strategic balancing between priority projects and technologies across business and corporate needs and opportunities. They allocate resources to fundamental research on the basis of a combination of technological merit, business relevance, and critical-mass considerations. They do not shy away from conducting trade-offs across businesses and among types of R & D.

For example, an incremental R & D project to improve process efficiency in the short term in business A may be more or less worthy than a radical R & D project to bring out a new product in the mid-term in business B. Third-generation management takes a corporate perspective. It assesses not only the direct rewards of each project but the "springboard potential" to its businesses and to the corporation in the form of "stepout" potential, technological synergy, and knowledge buildup. It assesses not only the strategic importance of each project to its business but also the strategic importance of the business to the corporation. It also assesses the nature of the skills and resources needed and their relative availability or scarcity. Only then does third-generation management decide whether to choose between projects or to accommodate both projects by increasing resources.

Finally, third-generation management works hard to maintain flexibility in internal resources. It does this by encouraging the use of multidisciplinary approaches, by making maximum use of external resources, and by always considering the "buy" alternative before investing to "make" internally.

It sets priorities regularly, for both radical and incremental R & D, between projects and technologies, according to their costs/benefits and contributions to business and corporate objectives, by their time frames, and by associated risks—all on a corporate portfolio basis. It also reassesses priorities whenever external events or internal developments warrant.

In the third generation, guidelines for measuring results and progress are rooted in the principle of management by objectives. Companies operating in this generational mode always examine the business implications of their own as well as external technological developments. The desired technological results are specified at the outset in light of business objectives. Progress is reviewed and results to date are reevaluated against expectations whenever significant external technological or business events warrant such review—not only in light of internal project developments and certainly not simply on an arbitrary time schedule.

This synopsis of the management processes involved in third-generation R & D may strike some as "Utopian." It isn't. It requires management will, intelligence, commitment, and no more. Its characteristics are elementary.

II. PLANNING THE TECHNOLOGY PORTFOLIO

Portfolio analysis and management is a subject generally associated with investment accounts, mutual funds, business positions and strategies, risk/reward relationships, and the balance of cyclical/anticyclical markets. Two common threads link these applications: Ultimately they are all expressed in financial terms, and their purposes typically are to search for the optimum point of equilibrium between risk and reward, stability and growth.

The optimum point of equilibrium may also be thought of in its inverse expression, the optimum point of compromise. The definition of "optimum," of course, varies as widely as do the ambitions, competence, vision, and culture of individual companies. One open-sea sailor will compromise speed for comfort; another will yield comfort and even safety for speed; others will seek the "optimum" balance of all.

These concerns for portfolio content are clearly prudent, but are they sufficient? This chapter presents evidence that they are not.

A. Beyond Risk and Reward

Although companies typically express their business portfolios in terms of financial risk and reward, many consider other portfolio issues, at least intuitively. In people planning, for example, everyone would recognize that a major marketing department populated entirely by 50- to 60-year-olds promises danger; its "people portfolio" is out of balance. Similarly, "opportunity portfolios," "geographic portfolios," "time-frame portfolios," and others tend to be addressed implicitly in good business planning.

But one portfolio of burgeoning importance is rarely addressed: the technology portfolio. The widespread failure to recognize, plan, and manage technology portfolios is at the root of many of the competitive ills of the last decade, particularly among American companies. In the U.S. passenger car industry, for example, the uncorrected dependence upon obsolescent technological renewal resulted in loss of competitive position. While this may be an extreme example, it is unfortunately not an isolated one, and it offers powerful lessons for all of industry.

B. The Technology Portfolio

A technology portfolio is that assemblage of technologies and plans (including R & D plans) for new or modified technologies that expresses—with reference to many portfolio variables—the current and future technological strength of a company. Thus, a company's technology portfolio plays a critical role in its business portfolio.

For example, Figure 3 depicts the business portfolio of a U.S. oil company that operates four major businesses: exploration and production, polypropylene, polystyrene, and refining. (In every figure in this chapter in which circles are used to denote investments, the area of the circle signifies the relative size of the investment.) The company competes in industries whose maturity ranges from growth to aging. At the time at which this business portfolio was analyzed, the company regarded its return on investment—a measure of competitive strength—as reasonably satisfactory. Its five-year plans promised continued satisfactory returns. Management's confidence in the company's future was palpable.

However, a technology analysis performed for the company transformed confidence into dismay. The analysis projected the results for the company's busi-

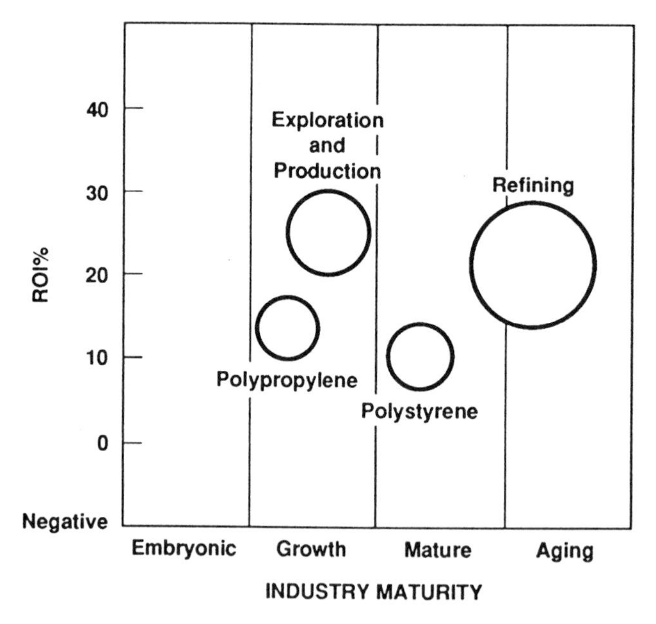

Figure 3 A U.S. oil company: return on investment versus industry maturity, 1987.

nesses over the five-year planning span if all their technology plans were fulfilled (Figure 4). The industries in which the businesses competed would, of course, grow more mature. But all the businesses would lose competitive position, and thus earn lower returns. This loss would occur because the company's technology plans would fail to maintain technological competitiveness in the face of strong competitors, whose continuing investments in improving technologies were likely to erode this company's competitive position. Figure 4 demonstrates the need for superior technology analysis and planning in the competitive context.

C. Balancing Portfolio Elements

Effective technology portfolio analysis and planning is far from simple. Table 1 lists some of the many variables that should be accommodated. The meaning of the first two elements, size of technology investments and industry maturity, is self-evident. The remaining elements are discussed below.

1. Technological Competitive Position

Authur D. Little's definition of technological competitive position is set forth in Table 2, and the concept is illustrated in Figure 5, which presents an analysis of the current technology portfolio of a U.S. chemical company. This company maintains investments in nine businesses whose industry maturities range from

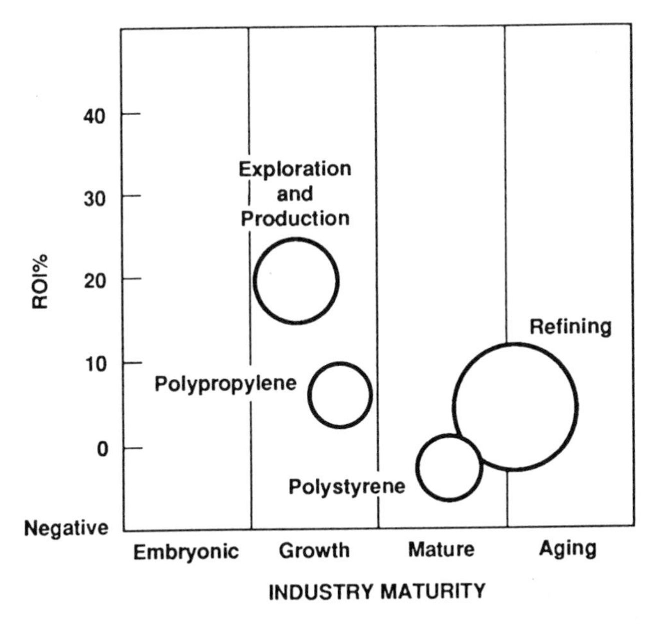

Figure 4 A U.S. oil company: return on investment versus industry maturity, 1992.

Table 1 The Technology Portfolio: Planning Variables

- Size of technology investments
- Industry maturity
- Technological competitive position
- Technological maturity
- Competitive impact of technologies
- Durability of competitive advantage
- Risk
- Reward
- Market "newness"

embryonic to aging. Most are attractively placed in technologically favorable to strong positions. But four businesses—smaller but still representing investments of tens of millions of dollars—are not as satisfactorily positioned.

This company did a superior job of technology analysis and planning. Figure 6 projects the likely change in portfolio from 1987 to 1992. The company planned to exit from three businesses for two reasons: reliable means to improve those businesses' technological competitive positions were not available, and the company's resources could be applied more profitably to liquid crystal polymers, nylon resins, caprolactam, and a new investment in a sophisticated process to convert caprolactam directly to molded or extruded products. In this instance, 1992 looks rather more promising than 1987.

Table 2 Definition of Technological Competitive Position

Dominant	Powerful technological leader. High commitment, funds, manpower; creativity well recognized in industry. Sets pace and direction for technological development in industry.
Strong	Able to express independent technical action, set new directions. Technological commitment and effectiveness consistently high, and plans executed creativity, effectively, on time.
Favorable	Able to execute plans at an average pace and quality. Has strengths that can be exploited to improve technological position. Typically not able to provide sustained technological leadership, except in developing niches, but can keep XYZ corporation competitive.
Tenable	In catch-up mode. Unable to set independent course. Time and/or quality of execution frequently slip.
Weak	Low quantity and/or quality of technical output. Often a short-term, firefighting focus. Products, processes, time frames for which R & D is responsible typically slip badly.

TECHNOLOGICAL INVESTMENTS
1. Conductive polymers
2. Liquid crystal polymers
3. High-performance adhesives
4. Nylon film
5. Nylon resins
6. PCB composites
7. Caprolactam
8. Nylon apparel fiber
9. Polyester resins

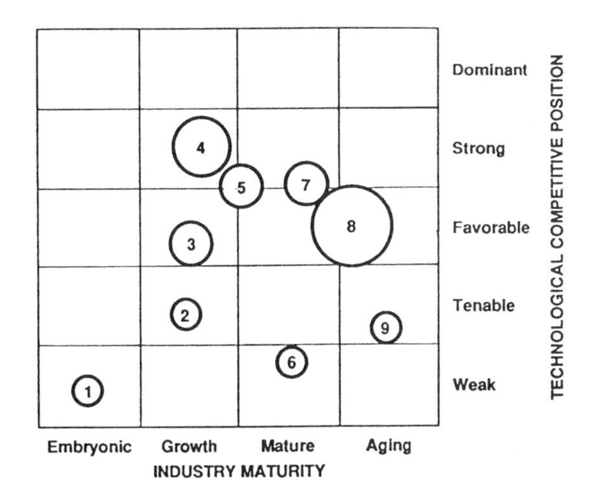

Figure 5 A U.S. chemical company: technological competitive position versus industry maturity, 1987.

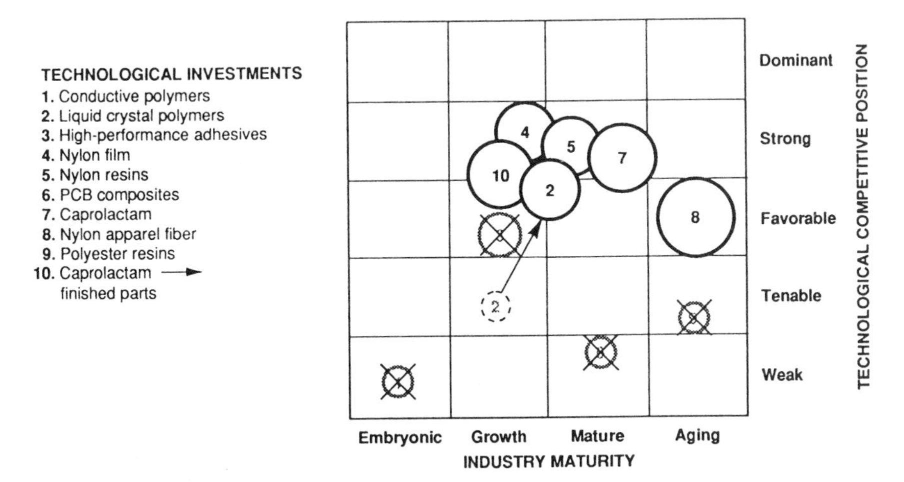

Figure 6 A U.S. chemical company: technological competitive position versus industry maturity, 1992.

2. *Technological Maturity*

The same chemical company's technology portfolio can also be analyzed with reference to size of investment, industry maturity, and technological maturity. Our definition of technological maturity is illustrated graphically in Figure 7, with examples from the technical universes of medicine (above the line) and catalysis (below it). The technological maturity cycle is expressed on the horizontal axis by the age of the technology and on the vertical by the diminishing scope for additional technological advance with increasing maturity. An embryonic advance with increasing maturity. An embryonic technology offers great, often unimaginable scope for additional scientific and technological advance, while an aging one is virtually exhausted. In Table 3, the characteristics of R & D are arrayed as functions of technological maturity. As technology moves from the embryonic stage through the growth and mature stages, it typically gains predictability in terms of technical performance, financial reward, and R & D costs—but the commercial advantage it offers becomes commensurately briefer.

Figure 8 applies the concept of technological maturity to the same U.S. chemical company, projected to 1992. As the company had projected the elimination of businesses with aging technologies, it found this portfolio distribution satisfactory.

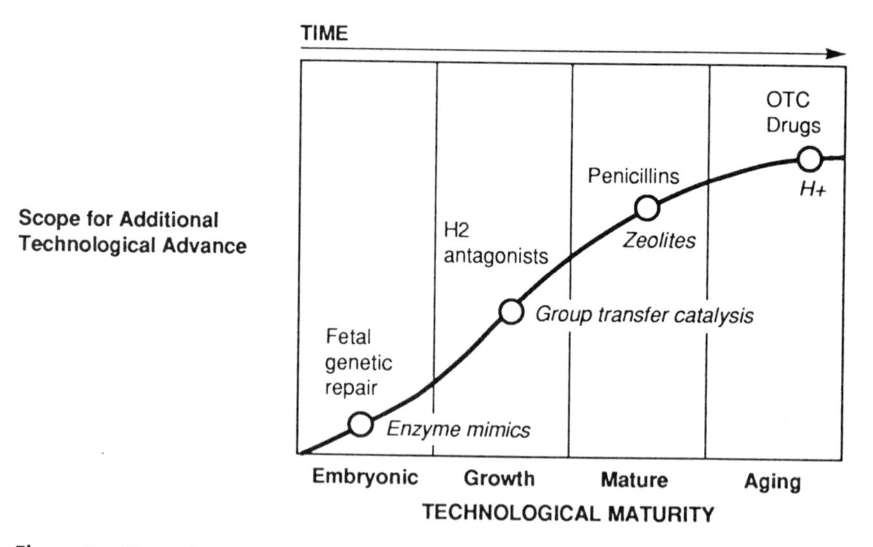

Figure 7 The technology maturity cycle.

Table 3 Characteristics of R & D as a Function of Technological Maturity

Technological maturity	Time to commercialization	Knowledge of competitive R & D	Predictability: Technical	Predictability: Reward	Predictability: R & D Costs	Durability of commercial advantage
Embryonic	7–15 years	Poor	Poor	Fair	Poor	High
Growth	2–7 years	Fair/moderate	Fair	High	Moderate	Moderate
Mature	1–4 years	High	High	High	High	Fair
Aging	1–4 years	High	Very high	Very high	Very high	Short

3. Competitive Impact

The competitive impact of the technologies practiced by an industry or a company can be classified as base, key, or pacing, as defined below.

Base technologies are those that a company must possess and practice well, but that are so well developed and available to all competitors that they cannot provide competitive advantage. Examples are the thermal pasteurization of milk, most distillation processes, and most process instrumentation.

Key technologies are those in an embryonic or growth stage of development that offer substantial scope for innovation and provide opportunity for technological differentiation from competitors. An example is group transfer catalysis in polymerization, practiced recently by Dupont.

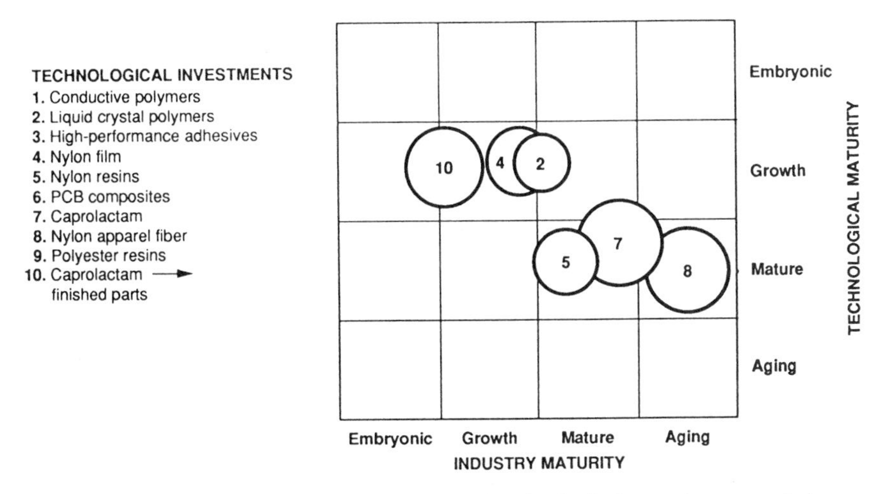

Figure 8 A U.S. chemical company: projected technological maturity versus industry maturity, 1992.

Pacing technologies are those emerging technologies not yet fully expressed in practical terms and often still in scientific exploration, but that offer the potential of radical change, often altering the basis of competition of an industry. There are many examples in biotechnology and superconductivity.

4. *Durability of Advantage, Reward, and Risk*

Figure 9 offers a symbolic representation of the relationships among four variables:

- The competitive impact of technologies from base to key on the horizontal axis.
- The durability of technological advantage gained (if any) on the left vertical axis.
- The reward from investment in the technologies on the right vertical axis.
- Technological risk, also on the horizontal axis.

While these relationships are only symbolic, they are in general accord with business judgment and with experience across industries. Figure 9 suggests, for example, that investment in technology 7, representing a key—i.e., differentiat-

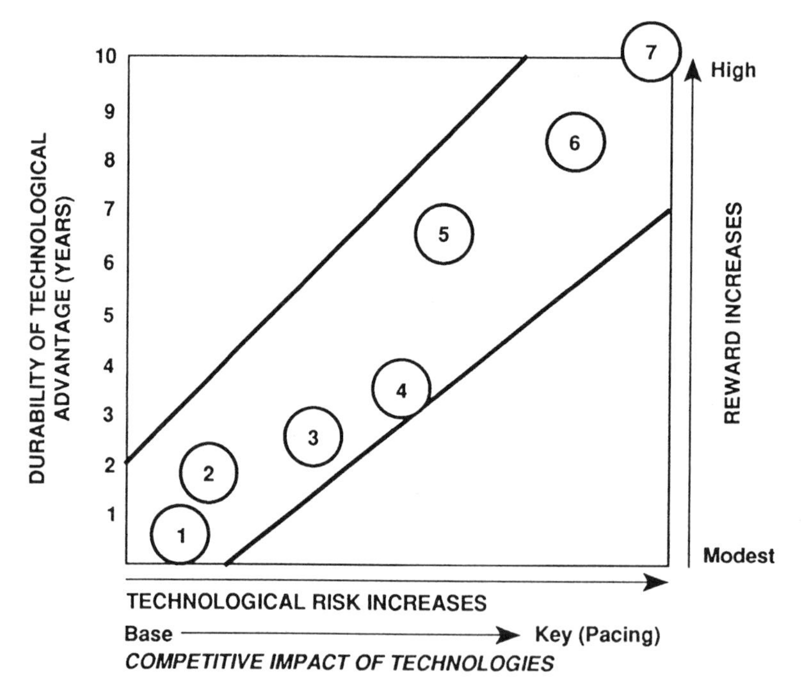

Figure 9 Durability of technological advantage and reward as functions of competitive impact and risk.

ing and probably proprietary—technology, differs greatly in results to a company from investment in base technology 1.

5. *Market Newness*

The "newness" of the markets a company is addressing refers to the degree of familiarity of the company with those markets—i.e., whether the company understands the bases of competition and competitive dynamics (in which case a market is old to the company) or whether it is inexperienced in them (in which case the market is new to the company).

Figure 10 shows the complex interaction of market newness with other key variables in the technology portfolio of an international chemical company. The competitive impact of the company's investment in technologies is depicted across the top horizontal axis, from base to key (and/or pacing). Technical risk, similarly, increases from left to right. The newness of the markets to the company is shown on the vertical axis. Business risk, more than technical risk, increases from top to bottom. Technological competitive advantage, the durability of technological advantage, and the potential reward all increase as depicted on the lower horizontal axis from left to right.

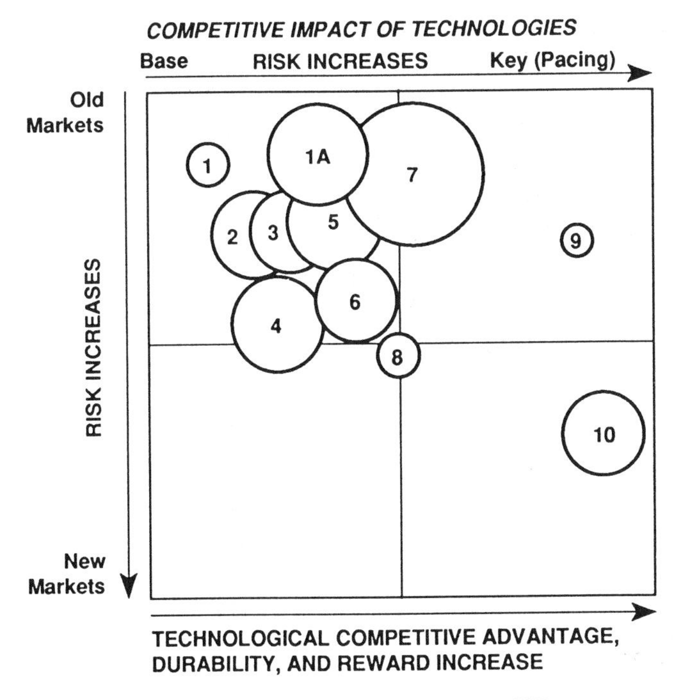

Figure 10 An international chemical company, 1987.

6. *The Optimum Balance*

This multivariable representation makes clear the opposing forces that technology planning must take into account. Low risk is in conflict with high reward and with a thrust to diversify markets. The inevitable decay of technologies from key to base increases competitive exposure and tends to reduce earnings. But balancing that decay may demand new investments in higher-risk technologies. What is the "optimum" balance, the "optimum" compromise among these conflicting, sometimes hostile forces? The answer is specific to each industry and company.

For the international chemical company portrayed in Figure 10, analysis of its 1987 technology portfolio led to some drastic surgery (Figure 11). It determined to divest two businesses (1 and 2) that were in poor competitive positions and defied renewal, and one business (10) that enjoyed advanced technology but was demonstrably weak in the competitive arena.

The company's plans called for major investments in technology (including R & D) in four businesses (11, 12, 13, and 14), all in markets in which the company felt competent and confident to compete, and all embodying technologies that promised significant competitive superiority.

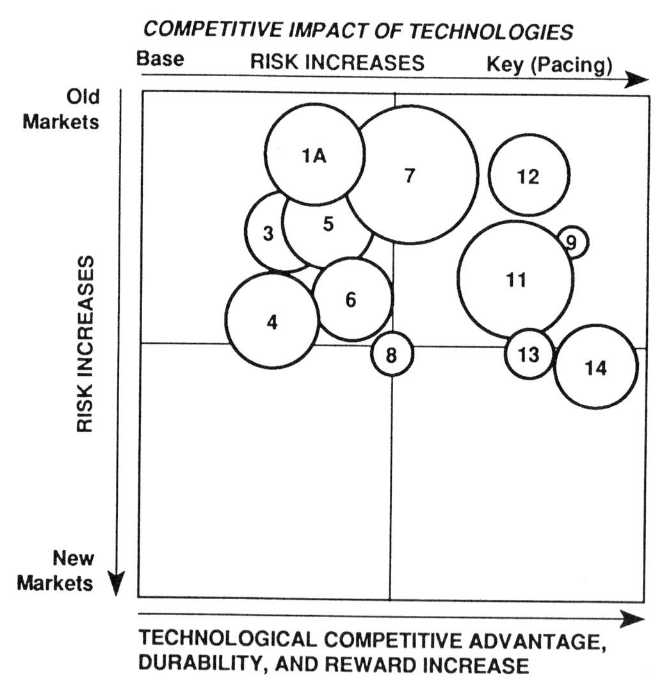

Figure 11 An international chemical company, 1992.

D. Key Questions for Management

We do not intend in this chapter to offer an answer to the question: "What is an optimum technology portfolio?" We can, however, suggest several key questions a company should address as a precondition to developing a superior portfolio:

- What is your company's current technology portfolio?
- What will it become with time?
- Is it (and will it be) satisfactory, "optimum" for your company in your industry (or industries)?
- If not, what can and should be changed to make it so? At what costs? In what time frames? At what risk/reward relationships? In what priorities?

Superior, even outstanding, technology alone will not ensure superior company performances. But it is equally clear that in many industries, inferior technology will, in time, ensure inferior, possibly even fatal, company performance.

ACKNOWLEDGMENT

The information in this chapter is excerpted from an Arthur D. Little book, *Third Generation R & D: Managing the Link to Corporate Strategy*, which was published in 1991 by Harvard Business School Press; and "Prism," Arthur D. Little's quarterly management journal.

Arthur D. Little, the international management consulting firm headquartered in Cambridge, Massachusetts, helps corporations, institutions, and governments meet the challenges of today's complex and rapidly evolving global marketplace. The company was founded in 1886 and today has more than 2300 staff members working in 26 offices worldwide.

5

PLANNING FOR QUALITY

George W. Roberts
Babcock & Wilcox
Alliance, Ohio

David T. Farrell
Ernst & Young
San Jose, California

I. ORGANIZATIONAL PLANNING

A. Identifying Customers

Chapter 4 discusses project identification for industrial research laboratories where customers can readily be identified—they are the business units. For some institutions however, the question of "Who is the customer?" may cause difficulty. In the case of government-funded basic research laboratories, there may be no immediate user of the knowledge, causing the search for a customer to take unusual turns. Bodnarczuk (1) discusses the issue of customers for the high-energy physics (HEP) community. His opinion is that

> for the basic research performed at Fermi National Accelerator Laboratory, the internal customer is a single scientist, the laboratory director, who has ultimate approval authority in matters experimental and otherwise. The external customers are university-based collaborations of physicists for whom Fermilab provides the necessary particle beams for doing HEP and a wide variety of other support services for experimental activities in all phases. Consequently, Fermilab's research program is an extension of the training that universities provide for graduate students in HEP, with one of its major products being PhD physicists.

Priority for project support in a National Laboratory may be determined by the overall direction that the scientific community is going with respect to the work proposed. It is to the principal investigator's advantage if the project seeks to answer questions that have not previously been proposed and if the questions are of some interest to the scientific community. The laboratories have a need to demonstrate their continued leadership in their assigned mission areas. This leadership will constantly be challenged by jealous competitors throughout the world, and the results of any work will be critically reviewed as to its value to science as well as its technical integrity. If a proposal can be shown to be in line with the overall interests of the funding agency, will be headed up by an extremely competent principal investigator with a proven record for success, is deemed to have a reasonable measure of success and the results will completely destroy a current theory of science, the chances of being accepted and funded are great. Assuming, of course, that the funds are available and there is no serious political opposition. If industrial researchers long to get away from the administrative morass that seems to dictate what and how research is supported, they must realize that, in today's society, a number of bureaucratic impediments must always be overcome in order to do research. To the researcher in the National Laboratory who does not think he or she has to please a customer, there is always someone else who holds the purse strings. Whether that person evaluates the value of the proposed research directly or through agents such as peer review teams from competing laboratories, there will be a value judgment applied to every proposal and every report of research work. Those that judge are the customers.

B. Critical Needs

Part of an overall strategy of serving the customer is to determine what that customer's critical skill needs are, for the short and long term. The research institution can then position itself for technical work areas that are appropriate to the general needs of the customer division or sponsoring agency. Funding levels may need to be set to support a given level of effort in critical skill areas needed by the potential users of the technology. For an industrial laboratory, this strategy permits research to focus on a portfolio mix of potential "game-changing" innovations, while keeping staffing levels appropriate to respond to quickly developing needs by the customer divisions. As discussed in Chapter 4, however, it is imperative for the modern industrial research institution to maintain an acute awareness of its customer division's markets and needs and to take a proactive approach to identifying the proper mix of research projects. For a government-sponsored basic research laboratory, positioning could mean having the staff and facilities to be able to continually push back the limits of mankind's technical ignorance in areas that generally have meaning to the technical peers of that institution and to the sponsoring agency.

One industrial laboratory performed a survey of its customer operating divisions to determine what the core technologies and critical processes were, as perceived by the operating division management. The results of this critical needs survey were published throughout the company. The performance of this survey was aided by the fact that the corporation had established a number of "councils" that were concerned with various broad functional areas such as engineering, manufacturing and quality. Representation came from middle and upper managers and senior technical people and each council had an executive assigned to it by the corporation president. Each council sponsored a number of technical committees that represented technologies vital to the corporation's well being. The technical committees included technical specialists from each of the operating divisions who were able to share experience and concerns and to recommend research in specific areas. The technical committees were a vehicle to provide the technology "push" that complemented the market "pull" for new technology and products.

The operating divisions asked their representatives to assist in responding to the survey. A listing of basic and critical technologies was identified. "Basic" was defined as families or groupings of technologies along scientific or organizational lines. "Critical" was defined as absolutely required to design, manufacture, sell, or support the product now or in the future. Table 1 shows those

Table 1 Critical Technologies Identified by Nine or More Operating Divisions

Basic Technology	*Critical Technology*	*# of ODs*
Chemistry and Corrosion	Analytical Chemistry	9
	Water Chemistry	9
	Cleaning	9
Codes, Regulations, and Standards	ASME	9
Materials	Metallurgy	10
	Failure Analysis	11
	Heat Treatment	11
	Nonferrous	12
Nondestructive Testing	Penetrants	11
	Ultrasonics	12
Plant, Control, and Process Systems	Controls	12
Structural Mechanics and Stress Analysis	Finite Element Analysis	9
	Fracture Mechanics	9
	Elevated Design Temperatures	12
	Shock and Vibration	11

technologies identified by nine or more operating divisions. Although nine was an arbitrary number, it represented half of the divisions responding to the survey. The resulting list represented a means to determine if there was an adequate research base to support the major technical needs of the corporation. The data was further refined to identify which of these critical technologies were also "core" technologies—that is, those that set the corporation apart and ahead of its competition. This information could then be used to identify the strategic "intent" of the corporation. The research division management could now review their customer's strategic technology needs and determine an appropriate skill mix within the various technical sections.

The critical skills survey provided an indication of where the business units could cooperate among themselves to sponsor research, seminars, users groups and other mutually beneficial activities. Other representations of the survey data showed where certain basic technologies were not represented by a technical committee, thus stimulating revised charters or formation of new committees. The study also revealed where a division was unique in its need for a critical technology and stimulated that division to take appropriate action to protect itself. Finally, it enabled the management of all the operating divisions to reconcile their R & D programs against business needs in the context of their critical technologies.

C. Quality Function Deployment

1. Introduction

Quality Function Deployment (QFD) is a structured process for listening to the voice of the customer, translating customer requirements into measurable counterpart characteristics, and "deploying" those requirements into every phase of design, procurement, manufacturing, delivery and service, involving every needed company function.

In QFD, all operations of the company are linked to, and driven by, the voice of the customer. This linkage brings quality assurance (QA) and quality control to all relevant functions in the organization and shifts the focus of quality efforts from the manufacturing process to the entire development, production, marketing and delivery process. QFD provides the methodology for the entire organization to focus on what the customer wants, rather than fixing what the customer doesn't like, and puts the emphasis on designing in quality at the product development stage, rather than problem solving at later stages.

The words "quality function" do not refer to the quality department, but rather to any activity needed to assure that quality is achieved, no matter what department performs the activity.

QFD identifies any areas where design or manufacturing requirements may be missing or inadequate to meet customer requirements, where technical evaluations agree or disagree with customer ratings, and where new product designs or manufacturing processes may be needed to satisfy customer requirements. It also provides a valuable historical reference source that can be used to enhance future design and engineering technology, and to prevent design errors.

The methodology of QFD was first formalized at the Kobe Shipyard of Mitsubishi Heavy Industries in 1972, and has been utilized extensively by Toyota since 1977, following four years of training and preparation. During the first ten years of its use, Toyota credits QFD with the following results: a 20 percent reduction in startup costs of a new van in 1979, a 38 percent reduction by 1982, and a cumulative 61 percent reduction by 1984. During this period, the product development cycle time was reduced by one-third, with a substantial reduction in engineering changes.

Present users of QFD in the United States include: AT&T Bell Labs, Black & Decker, Chrysler, DEC, DuPont, Eaton, Ford, General Electric, General Motors, Hewlett-Packard, Johnson Controls, Kraft Foods, Proctor & Gamble, Rockwell, Scott Paper, Sheller-Globe, Texas Instruments and Xerox. Organizations such as these consistently report benefits including:

- Measurable improvement in customer satisfaction/market share
- Reduction in time-to-market (product development cycles)
- Significant reduction in the number of engineering changes
- Start-up costs reduced
- Identification of competitive advantage and marketing strategy
- Reduction in warranty claims and customer complaints
- increase in cross-functional teamwork

The "House of Quality," a matrix format used to organize various data elements, so named for its shape, is the principal tool of QFD. (Figure 1) Although "house" designs may vary, all contain the same basic elements:

Whats: The qualities or attributes the product or service must contain, as required by the customer. Whats are compared to competitors' qualities and ranked by importance.

Hows: The technical means for satisfying the whats, including the exact specifications (how much) that must be met to achieve them.

Correlation matrix: An evaluation of the positive and negative relationships between the hows, sometimes called the "roof."

Relationship matrix: An evaluation of the relationships between the whats and the hows. It identifies the best ways to satisfy the customer and generates a numerical ranking used as a guide throughout the development process.

These "Houses of Quality" are used in each of the four primary phases of QFD:

1. *The voice of the customer*, in which customer requirements are identified and translated into design specifications or product control characteristics in the form of a planning matrix.

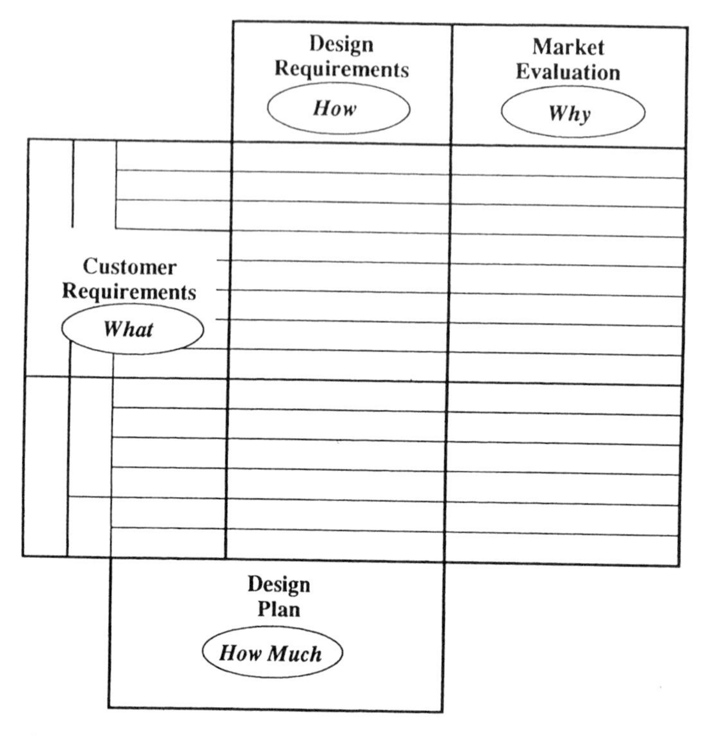

Figure 1 Basic "House of Quality" design.

2. *Deployment matrix*, which translates the output of the planning matrix, design specifications, into individual part details which define critical component characteristics.
3. *Process/Quality Matrix*, which defines the process for making each component and establishes critical component parameters.
4. *Operating instructions*, which define the production requirements for each component/operation.

Rather than present a conceptual treatise on QFD, it is the objective in the limited space available here to provide a measure of practical guidance so that the reader may be encouraged to try it out on a project of modest scope, thereby gaining the skill and confidence to apply the methodology on a broader scale.

2. *Getting Started*

Effective application of QFD is likely to require significant change in the way an organization operates. Substantial training and planning are involved. There are new tools to be learned. Cross-functional teamwork capability will be put to the test. Purchasing, sales and engineering organizations will be involved in product or service design and development to a degree never previously experienced. All of these factors suggest that success, even of a limited project, is dependent on a high level of management commitment. As with any significant change effort, effective sponsorship by key management personnel of each affected organization is essential. Attention should be given at the outset to the principles and methodologies of Organizational Change Management, which are beyond the scope of this chapter.

To maintain sponsor commitment, senior management personnel should be involved in the selection of initial projects. The selection team should identify a wide variety of potential projects and formally apply an agreed-upon set of evaluation criteria (for example, market share, product life cycle, cost reduction potential, revenue enhancement potential, market share implications, need for time-to-market reduction, competitive position, organizational commitment, potential for success, alignment with business plan/strategy, etc.).

With the project selected, it is now time to select the QFD team leader and members. Care should be taken to include participants from all key organizations involved from the product (or service) concept/design through to delivery and after-sale service. It is likely that the core team will be augmented by additional people at different stages of the QFD process, as additional detail is required.

The core team's initial activity should include the development of a team charter, including mission, objectives, guidelines for team conduct, work plan and a timetable. The team charter should be shared with the executive group to assure alignment on mission, objectives and expectations. Consensus around a team charter also clarifies the resources that will be required, and avoids mid-project surprises that can derail the effort. The team should also plan for periodic status reporting (at a minimum, upon the completion of each QFD phase) to the executive group to maintain sponsor awareness and commitment.

Time spent early in the process in team building and member training is an investment well worth the effort. It is important that team members understand the QFD methodology at the outset, have an agreed-upon road map of the process they are going to follow, and possess basic team skills. In addition, competence in the related methodologies and tools described immediately below will substantially enhance the effectiveness of the team, and the success of the QFD project.

3. *Links to Other Methodologies and Tools*

It should be noted that QFD is not a stand-alone methodology. Many of the other methodologies and tools with which quality and improvement practitioners are familiar come into play in the execution of a QFD project. Some of those include:

The "Seven basic tools": Brainstorming, Cause & Effect Diagrams, Checksheets, Pareto Diagrams, Graphs, Histograms, Scatter Plots, and the like, will be useful throughout the project to collect, analyze and prioritize both ideas and data

Statistical methods: To analyze variation in a process, product or component and assess process stability, control and capability to meet customer requirements

Design of experiments (Taguchi methods): To optimize multiple process variables through efficient and effective experimentation

Benchmarking: To perform competitive analysis and to identify industry ''best practices''

Process Improvement/Innovation: To enable the team to produce actual and significant improvement to the processes, subprocesses and activities necessary to meet or exceed customer and business requirements and expectations

4. *Voice of the Customer: The Planning Matrix*

The planning matrix is not only where QFD begins, it is the foundation on which every step which follows is based. In this phase we abandon previous practices of assuming that we know what the customer wants and needs, or, worse yet, a marketing philosophy predicated on the notion that we can ''sell'' the customer on what we already have to deliver. We begin by really listening to the customers' expression of their requirements, in their own terms. And we probe for a level of detail heretofore unidentified. A variety of sources and ''market research'' tools should be considered, including: follow-up letters to customers, observations of the customer using the product, interviews, focus groups, ''buff'' magazines, independent product reviews, trade shows, sales contacts, printed and/or telephone surveys, etc. Also, the past history of unsolicited customer feedback should be reviewed, such as customer returns, complaint letters and accolades. Product or service dimensions around which customer input is sought can include any or all of the following: performance, features, reliability, conformance, durability, serviceability, safety, environmental aspects, aesthetics, perceived quality and cost. Examples of the application of these dimensions in the case of a toaster are shown in Table 2.

Also, customer requirements can be expressed at multiple levels of detail, commonly called primary, secondary and tertiary. The customer requirements, so expressed, become the vertical axis of the top portion of the planning matrix, as shown in Figure 2.

While collecting customer requirements information, the relative importance of each characteristic should be identified (using a numerical rating scale) to enable prioritization of those characteristics and to be used in conjunction with the competitive evaluation to assure favorable performance measures on those characteristics most important to the customer. This data will be entered on the right side of the top portion of the matrix.

Next, the final product control characteristics (design requirements) that must be assured to meet the customer's stated requirements are identified and arranged

Table 2 Dimensions of Quality for a Toaster

Dimension	*Possible Customer Requirement Statement*
Performance	- *Speed of making toast* - *Always produces requested darkness*
Features	- *Handles a wide variety of sizes* - *Up to 4 slices at a time* - *Handles frozen foods, eg waffles*
Reliability	- *No breakdowns* - *Works well with fluctuating currents*
Durability	- *Long service life* - *Hard to damage*
Serviceability	- *Easy to clean* - *Easy to repair - can do it yourself* - *Availability of spare parts*
Aesthetics	- *Fit with a variety of kitchen decors* - *Variety of color choices* - *Clean lines, smooth design*
Safety	- *Child can operate* - *No shock hazard*

across the top horizontal row of the matrix (see Figure 3). These are measurable characteristics that must be deployed throughout the design, procurement, manufacturing, assembly, delivery and service processes to manifest themselves in the final product performance and customer acceptance.

Since not all of the relationships between customer requirements and the corresponding final product control characteristics will be equally strong, the next step involves building a ''relationship matrix'' to display all of those relationships (using either symbols or numerical values) to further enable focus on those characteristics that are most highly related to high-priority requirements. It is a way to validate that design features cover all characteristics needed to meet customer requirements. An example of the relationship matrix is shown in Figure 4.

The relationship matrix also serves the important function of identifying conflicting requirements, for example: high strength vs. low weight. When such conflicts are identified, the application of Taguchi methods as mentioned above is indicated to optimize design characteristics. While analyzing the relationship matrix and its implications, it is important to keep open the possibility of modifying or adding to the list of product control characteristics.

Primary	Secondary	Tertiary
Performance	Speed of making toast	Fast
		Not slowed down by multiple slices
		Not slowed down by thick slices
		First slice as fast as subsequent slices
	Always produces requested darkness	Should have a "warm" setting
		Darkness exactly like dial setting
		Darkness not affected by thickness
		Darkness not affected by temperature
		Can darken one side only
Features	Handles a wide variety of sizes	Will warm rolls and muffins
		Toast regular bread
		Will handle oversize breads
	Handles frozen foods eg, waffles	Handle any temperature of material
		Frozen items done as well as bread

Figure 2 Customer requirements.

The next step is completion of the Market Evaluation, including both the Customer Importance Rating and Competitive Evaluations, on the right side of the top portion of the matrix (see Figure 5). This section displays the comparative strengths and weaknesses of the product in the marketplace around those requirements most important to the customer. The data for the Competitive Evaluation can come from the same sources as the customer requirements data, from independent or internal product testing, and from independent Benchmarking of product characteristics as described briefly above. As such, the comparisons can be based on both objective and subjective data.

The company's and competitors' present performance data for the final product control characteristics is entered in the lower section of the matrix in the columns for each relevant characteristic as shown in Figure 6.

The present performance data is next used to establish performance targets for relevant characteristics that, when achieved, will position the product as highly competitive or best in class. Target data is displayed immediately below the corresponding competitive data as shown in Figure 7.

The final element of the Planning Matrix is the selection of those product quality characteristics that are to be deployed through the remainder of the QFD process. Those that are a high priority to the customer, have poor competitive performance, or require significant improvement to achieve established target

Customer Requirements	Final Product Control Characteristics										
	Thermostat Accuracy	Heating Wire Thickness	Heating Wire Coverage	Variable Power Delivery	Airtight Case	Interior Dimensions	Horizontal Layout	Variable Heating Area	Shut-off	Humidity Sensor	Exterior Surface
Fast											
Not slowed down by multiple slices											
Not slowed down by thick slices											
First slice as fast as subsequent slices											
Should have a "warm" setting											
Darkness exactly like dial setting											
Darkn... ckness											
Darkne... mperature											
Can darken one side only											
Will warm rolls and muffins											
Toast regular bread											
Will handle oversize breads											
Handle any temperature of material											
Frozen items done as well as bread											

From Figure 2

Figure 3 Final product control characteristics.

levels should be taken to the next level of QFD analysis. Those characteristics are indicated at the bottom of the Planning Matrix, also shown in Figure 7, and become the input to subsequent matrices described below.

5. *Quality Characteristics: The Deployment Matrix*

The deployment matrix begins with the outputs of the Planning Matrix, in particular the overall product quality characteristics, and defines their deployment down to the subsystem and component level. In the process, the component part characteristics that must be met in order to achieve the final product characteristics are identified and the matrix indicates the extent of the relationship between the two, as shown in Figure 8. It is the critical component characteristics that will be deployed further and monitored in the later stages of QFD.

It should be noted that a focus on high-priority characteristics is crucial from this point forward. Experience has shown that a number of QFD teams have lost focus or energy at this point due to the apparent complexity of the process or the sheer number of charts or possible correlations. To start, work to identify the

From Figure 3

From Figure 2

Customer Requirements	Thermostat Accuracy	Heating Wire Thickness	Heating Wire Coverage	Variable Power ...very	Airtight Case	Interior Dimens...	Horizontal Lay...	Variable Heatin...a	Shut-off	Humidity Sensor	Exterior Surface
	Final Product Control Characteristics										
Fast	○	●	●	●	○	○	△	○			△
Not slowed down by multiple slices		●	●	●	○			●			
Not slowed down by thick slices		●	●	●	○			●			
First slice as fast as subsequent slices		●	●	●	△						
Should have a "warm" setting	●			●							
Darkness exactly like dial setting										5	
Darkne... ...ckness								5			
Darkne... ...emperatu...		4	4	5	2			3			
Can darken one side o...	2		1					2	5		
... warm rolls and muffins	3	3	3	4		4	5	5			
Toast regular bread	4	5	5	5	2			1			
Will handle oversize breads	4	5	5	5	2	4	5	5		5	
Handle any temperature of material	5	3	3	5	2					5	
Frozen items done as well as bread	5			5	2					5	

● – Strong Relationship ○ – Medium Relationship △ – Weak Relationship
or
5 – Strong Relationship 3 – Medium Relationship 1 – Weak Relationship

Figure 4 Relationship matrix.

three to five most critical finished component characteristics and then no more than the same number of component part characteristics, and concentrate on these. When the results have been deployed through the remaining steps with success, you can go back and address the next higher priority part characteristics.

6. *The Process Plan and Quality Plan Matrices*

In the previous step, critical component part characteristics were identified. We are now ready to identify the process used to produce those parts, the steps in the process that are critical to those characteristics, the appropriate control points in the process to assure conformance to requirements, and the process monitoring plan. This data is displayed in the Process and Quality Control Plan Matrix (Figure 9).

As with previous matrices, symbols or numerical values can be used in the Process Plan Matrix to show the strengths of the relationships. Control points are established at the steps in the process that are critical to meeting component

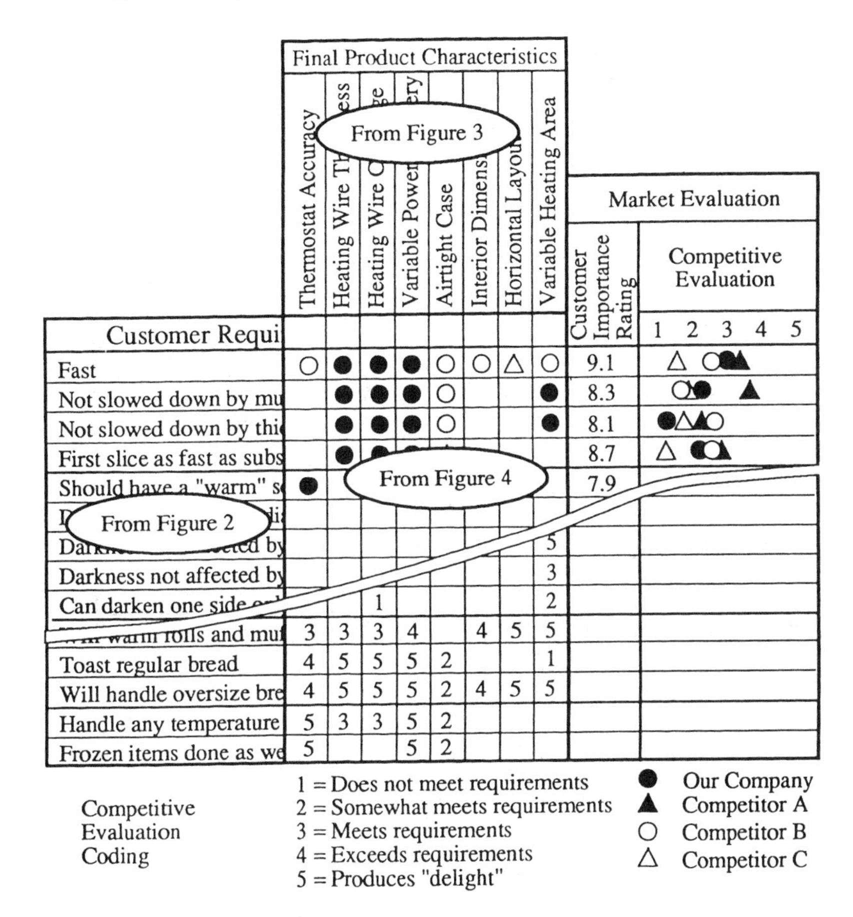

Figure 5 Market evaluation.

characteristics. They establish the data and the strategy for achieving product characteristics that meet high-priority customer requirements.

In the Quality Plan portion of the Matrix, the process steps can be displayed in flow chart format, and the control points and checking methods for each control point can be taken to a more specific level of detail. It is the latter information which forms the basis for developing the final QFD document, the Operating Instructions.

7. *Operating Instructions*

Unlike previously described matrices, the operating instructions in QFD do not have a single prescribed format. They may be designed to meet the specific characteristics of the process and needs of the process operators.

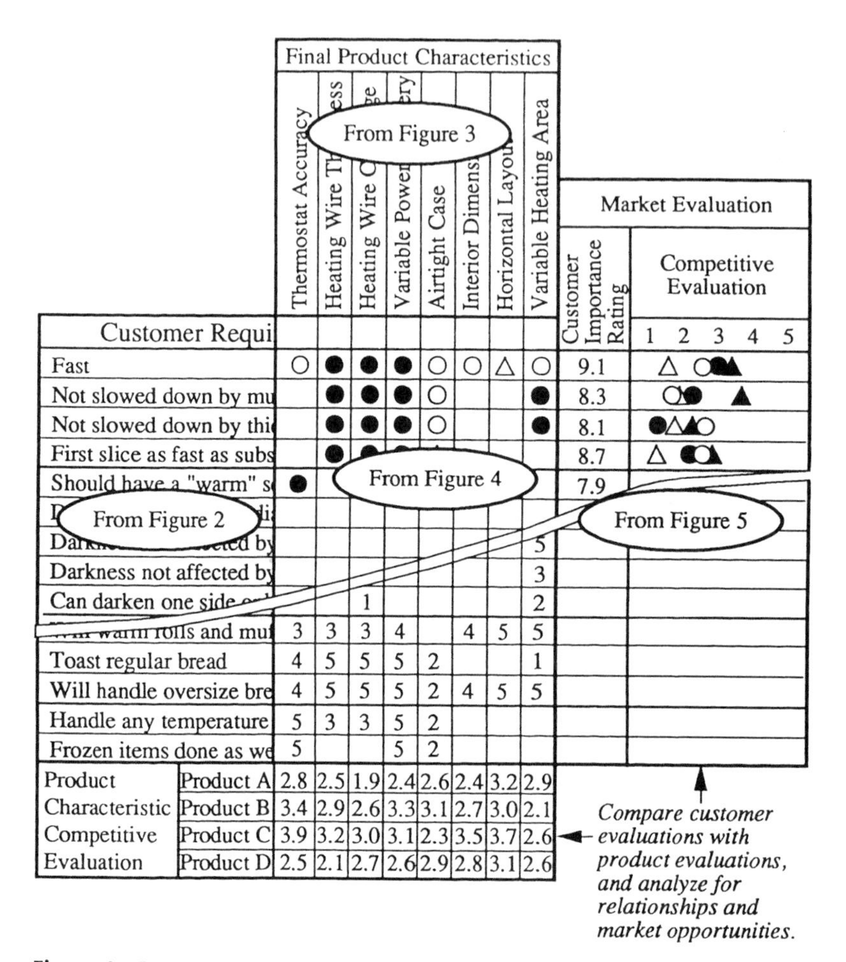

Figure 6 Product characteristic competitive evaluation.

The essence of this final step, other than implementation, is to deploy the results of the quality plan to the people who will be executing it. The instructions should be in sufficient detail to provide needed information, in a useful format, on what to check, when to check it, how to check it, what to check it with and what parameters are acceptable. This step makes the final connection between the work of the operator and his or her ultimate objective, satisfaction of the customer's requirements.

8. *Summary*

QFD is one of the most rapidly growing methodologies in the quality arsenal. Its successful application requires management's willingness to invest the "3

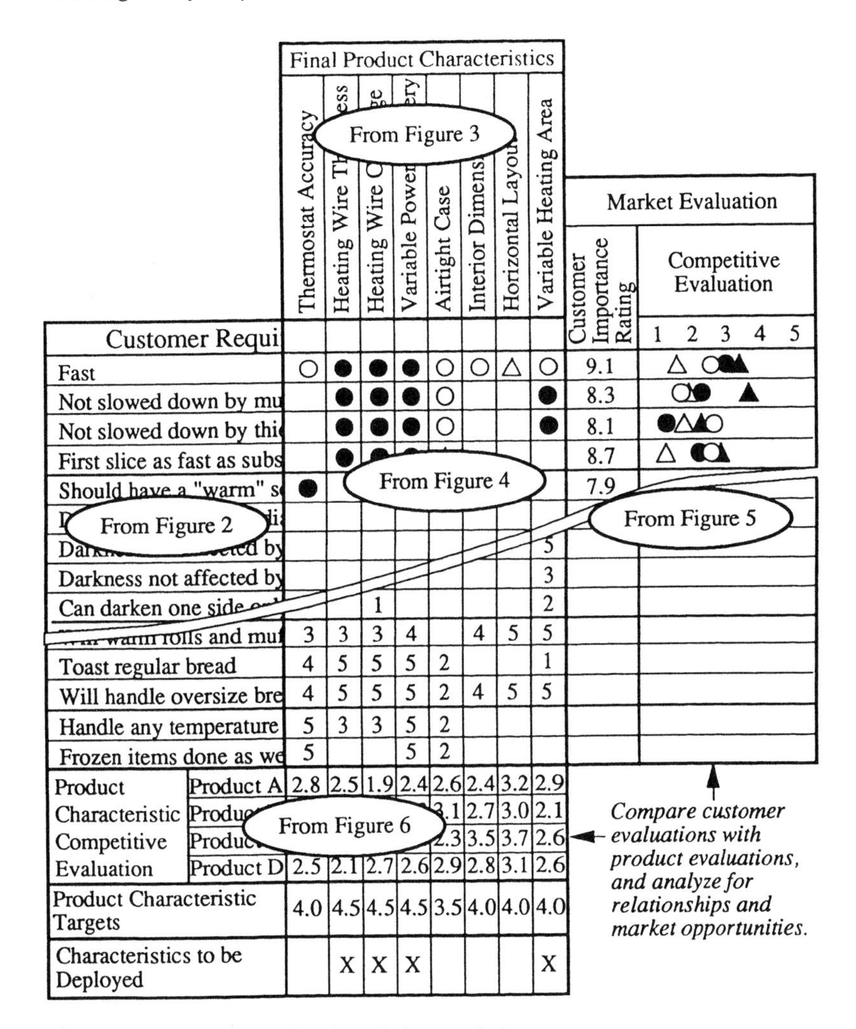

Figure 7 Target data and selected characteristics.

T's'': time, tools and training. And there are a number of issues to be considered before adopting the process. Training and implementation time and costs can be significant. Is there sufficient sponsor commitment to sustain the effort over time? How successful has the organization been at other cross-functional team efforts? How are you going to incorporate on-going process improvements into the QFD documentation system? QFD can be complex to administer, who is going to be responsible for that task? Product revisions, although fewer in number, require greater effort to integrate into the entire system. The increased effort at planning

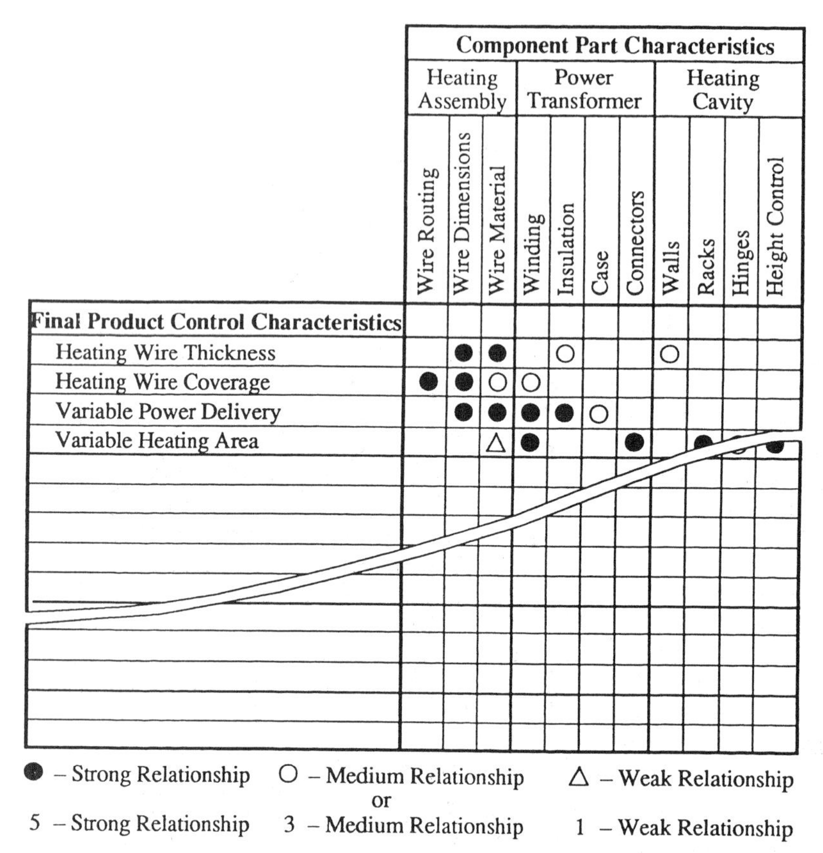

Figure 8 Deployment matrix.

stages, rather than implementation, requires a "culture shift" in many organizations.

As Ted Kinni (2) put it, "Simply put, QFD is work. On the other hand, its growing popularity indicates that faster and less expensive development cycles, improved quality and reliability, and greater customer satisfaction are well worth the effort" (2).

D. Area Activity Analysis

Area activity analysis is a methodology designed to analyze activities in the business process to determine their contribution to meeting customer requirements. It is a tool that identifies opportunities and problems within an area. It provides a disciplined approach to identifying and monitoring key measurements. Area Activity Analysis was originally called Department Activity Analysis

		Heating Assembly							
		Product Control Characteristics					Process Monitoring Plan		
No.	Process Steps	Wire Length	Wire Thickness	Ductility	Radius	Control Points	Check Points	Monitor Method	Freq.
1	Mount Wire Stock					Alignmt	Spacing	Visual	100%
2	Set Counter	●				"0" Meter	Meter	Visual	100%
3	Set Die		●		○	I.D.	Temp.	Histogrm	1/hr
4	Draw Wire		●	○	○	O.D.	Temp.	$\overline{X}$ & R	5/hr
5	Cut Wire	●				Length			
6	Shape to Specifications				●	Radius			
7	Mount to Heating Frame								

NOTE: *The above data drives the Operating Instructions.*

● – Strong Relationship ○ – Medium Relationship △ – Weak Relationship
or
5 – Strong Relationship 3 – Medium Relationship 1 – Weak Relationship

Figure 9 Process and quality control plan matrix.

(DAA) in Harrington's *The Improvement Process* (3). The change to "Area" instead of "Department" was because of the different ways in which organizations use the "department" label.

1. Team Formation

In a typical TQM process, there are two types of teams: Cross-functional teams have members from more than one organization and usually address major change or focus areas such as the general issue of innovation or relations with other corporate entities. The intra-organizational or "family" team deals with issues within one laboratory or department of the organization. Such issues might be storing, tracking, or disposal of test samples. It is this type of team for which the Area Activity Analysis (AAA) is particularly suited. AAA is needed because most areas do not really understand just why they do things that they do. They don't know who their suppliers are, nor who their customers are. Even if they do know, it is unlikely that they have ever been party to setting and measuring specific requirements. Improvements can not be made without understanding these items.

The analysis team should be formed from within a natural work group such as a research technical section or a group within that section. In most cases, the groups using AAA are those that report to the first level of management or supervision. It can be used at any level, though, and accomplish the same results of understanding, analysis, and improvement. The team needs to develop a mission statement for the area. "Who are we" and "what do we do" are questions not lightly answered and can be expected to take some thought. Most, if not all, of the areas planning to use AAA have been in operation for some time. Only a few will already have their mission defined by their management. Even if it has been defined, it is important that the people develop their version of the area's mission. The mission will set the stage, will put in perspective, just what the area is all about. This is what needs to be captured in a sentence or two. It should be noted that a mission is not a vision for what the organization wants to be in the future, nor is it how they are going to do something. Neither is it a list of operating principles, which are part of the policies and procedures.

2. *Major Activities*

Once a mission statement has been established, the team needs to identify the major activities of the group. Activities are the actions that are taken by the area members to produce an output—what is being done to fulfill the mission. Activities are normally made up of a series of tasks. For example, an activity might be "report preparation." The tasks involved are things like compiling notes, creating data plots or tables, typing the document, proofing it, getting approvals, making copies and distributing them. What is called an "activity" should not be too broad or too narrow. If there are more than six identified, they might need to be reviewed for the level of detail of each one. No activity should be listed that takes up less than 5 percent of the group's total effort. A sample mission and activity listing is shown in Figure 10. Note that the group has all signed the form indicating their concurrence. It is useful to have the next higher management level also concur.

3. *Requirements*

The next step is for the team to identify customer requirements. But who is the customer? Often, the customer will be an internal customer, someone who directly receives the group's output. For a support section that performs mechanical designs, the customer may be a research technical section that needs a new test facility. Further down the line, the customer for the technical section may be a design engineer in an operating division of the corporation. The operating division

AREA ACTIVITY ANALYSIS	*MISSION AND MAJOR ACTIVITIES*

Function Name: Technology and Quality

Area Name: Quality Assurance | Area No: 179

Area Mission:

The R&D Division QA Section's Mission is to provide independent assurance of compliance of internal and external customer quality requirements and to be a technical resource in quality and quality assurance to R&DD Management, project leaders, McDermott Operating Divisions, and external customers.

List Major Activities of Area and the percent of total time for each:

Audits - 30%
Code Compliance - 30%
Project Administration - 10%
QA Section Administration - 10%
QA System Development - 2%
Training - 6%
Supplier Control - 5%
Identify and Resolve Problems - 2%
Administrative Support for Division TQM Process - 5%

Manager's/Preparer's Name: GWRoberts/CTJones	Date:9/9/	Extension: 7402/7709

Area Approval:

Figure 10 AAA mission statement and activity listing.

designer needs pilot plant data to substantiate calculations prior to committing to her final design for a power plant. *Her* customer may be the plant manager for a large utility about to invest millions to service an area needing more electricity—an area in which the support section mechanical engineer lives. And so the chain goes on from customer to customer. But for the mechanical designers in the support section, the immediate customer is right down the hall. Should this customer be treated differently from the operating division's utility customer? Realistically, the answer to that question will be "yes." But internal supplier–customer transactions have a significant bearing on the efficiency of the organization and there has been much less effort to find out what that internal customer needs. This may be done with a meeting with the customer or it may require a survey of several customers so that requirements may be combined. However, requirements should not be vague or too general. They must be able to be measured. To be clear, the requirements should include:

What the customer expects you to provide, i.e., a report to meet a certain format or software with a certain level of performance.

When the customer needs the output. There should be a window, i.e., not before a certain date nor after a certain date. This could be measured in days or minutes.

4. *Measurements*

What are the measurements by which the output will be judged? There are three types:

Effectiveness: The extent to which the outputs meet the needs and expectations of the customer. They may include measurements of accuracy, cost, performance, timing, and usability.

Efficiency: The extent to which resources are minimized and waste is eliminated in producing the output for the customer. They may include resources expended per unit of output, percent value added time of total time, and processing and cycle time.

Adaptability: The flexibility of the process to handle the changing business environment along with the customer requirements. It is managing the process to meet today's special needs and future requirements. Adaptability is a measurement largely ignored, but it is critical for gaining a competitive edge. These measurements could include number of special requests per month and percent of such requests granted by direct contact employees.

5. *Default Action*

What will happen if what is provided is not what the customer expected or what was agreed to. This is the default position, the fall-back plan. Will the report be rewritten? Will the delay cost the customer money or time or have a downstream impact? One has to be prepared for this, for it will happen. Of course, this is part of the preparation when supplier and customer agree on the requirements of what, when, and how good. Specifying the conditions and solutions for default is one of the most overlooked items in customer/supplier relations.

There must be a means for getting feedback from the customer as to the quality of what is being provided. The supplier should not expect the customer to measure output (though they will if the supplier has a history of not meeting their requirements), but the supplier should ask for customer feedback with regard to meeting their requirements and to provide this on a regular basis. It may well be that a simple change in what the supplier does could make the customer's job much easier.

If possible, the group should meet with the customer, confer, and agree on all of these assumptions (see Figure 11). Such a conference may not be possible with multiple customers, but may be accomplished by survey. The assumptions should be reviewed with the customer at least once a year.

6. *Value Added*

The next step is to analyze the activity to determine which parts are value added and which are not. There are three basic types of ''value-added'' activities. Their definitions are:

Real value added: This is the work the group does to the input to make it an output. It is the work that the customer is willing to pay for. This could be things like constructing the test apparatus, logging test data, and writing the report.

Business value added: This is the work that is necessary for the way the group does business, but not what the customer would pay for. This may be things like updating administrative procedures, saving company records, or obtaining special permits or licenses.

No value added: This is the work that is caused by the process being poorly designed and inefficient. The customer doesn't want to pay for it, nor is it necessary for the business. This includes things like storage, repairs, all reviews, and approvals.

Figure 12 shows how these three relate to each other through a flow diagram.

For the activity under analysis, brainstorm all the tasks required to do the activity. Although the activity may seem simple, there is a lot of seemingly ''little'' things that are necessary to get the job done. Each of these tasks should be identified by the value-added category to which they belong. Then estimate the average length of time required to do each task. Figure 13 shows how this data may then be compiled.

AREA ACTIVITY ANALYSIS *OUTPUT REQUIREMENTS AND MEASUREMENTS*

Note: Use a separate page for each activity listed on the Mission and Major Activities page

Activity: Activity Report	Area No: 179	Date: 11/4/	Prepared by: GWR/LMM

Output

What have you and your customer agreed to:

Customer: Director, Technology and Quality

Requirements: Complete, correct, and understandable report

Measurement: Number of times report needed clarification, and number of grammatical errors

Feedback: Notes or phone calls from Director. Errors detected by report recipient or manager.

Default: Correct report

Customer: Director, Technology and Quality

Requirements: Efficient preparation of report

Measurement: Total number of hours needed to prepare report. Target less than 20 hours per month

Feedback: Monthly accounting reports

Default: Improve efficiency

Note: Use another sheet if there are more than two customers and/or requirements for this activity.

Customer Approval:

____________	____________	____________
____________	____________	____________
____________	____________	____________
____________	____________	____________
____________	____________	____________

Figure 11 Output requirements definition.

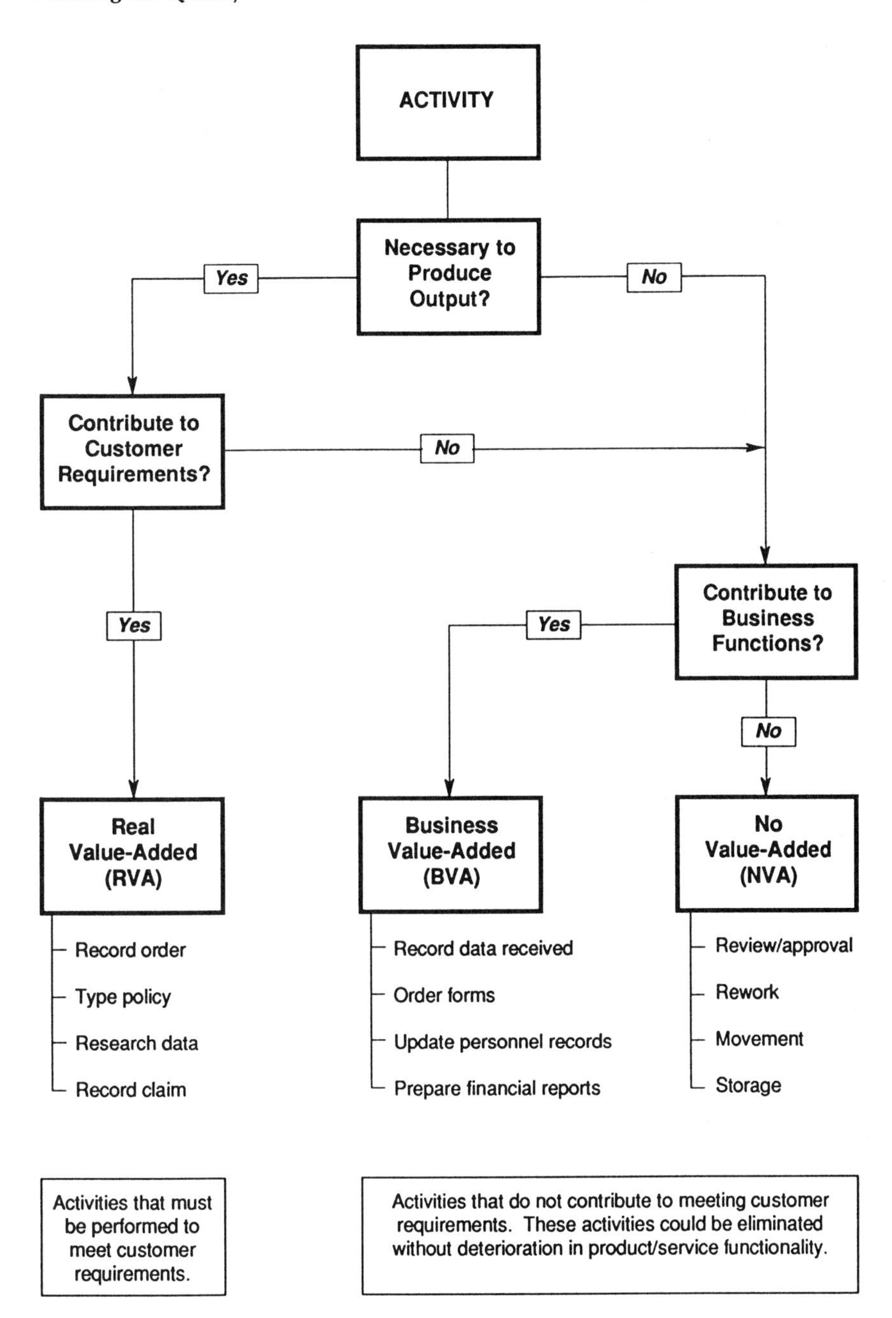

Figure 12 Flow diagram for value-added assessment.

AREA ACTIVITY ANALYSIS *QUALITY CONTROL GROUP — RECEIVING INSPECTION*

Procurement Process for Acceptable Items	*Avg. Cycle Time Hours*	*RVA*	*BVA*	*NVA*	*Avg. Process Time Hours*	*Task % Process Time*
Review purchase requisition for correctness	0.3		X		0.3	5.7%
Apply appropriate Quality Notes and sign off	0.1	X			0.1	1.9%
Forward to Purchasing	0.1		X		0.1	1.9%
Purchasing process time	52					
Review draft P.O. requirements, sign, and date, if correct	0.3			X	0.3	5.7%
Indicate necessary changes, if incorrect	0.1			X	0.1	1.9%
Purchasing process time	1					
Return to responsible purchasing agent	0.1		X		0.1	1.9%
Purchasing process time	28					
Compare P.O. requirements to documents received	0.25	X			0.25	4.8%
Check any mill test reports, C of Cs, etc. against requirements of the Code and the P.O.	0.25	X			0.25	4.8%
Document review on Q.A. copy of MTR	0.25		X		0.25	4.8%
Take documentation to Inspection area to complete inspection	0.5		X		0.5	9.4%
Compare P.O. requirements to items received	0.5	X			0.5	9.4%
Document inspection on Q.A. copy of P.R.	0.5		X		0.5	9.4%
Apply identifying marks to items and/or fill out green tags	0.5	X			0.5	9.4%
Release completed items to material handler	0.25		X		0.25	4.8%
File documents in project file	0.25		X		0.25	4.8%
Total Time in Hours	266.3				5.3	100%
Process Time as a % of Cycle Time					2.0%	

Figure 13 Calculating time to perform process and assigning value-added categories.

7. *Costs of Poor Quality*

The next step is to estimate the costs of poor quality associated with each task. Essentially, these costs consist of three elements: prevention, appraisal, and error.

Prevention costs are the things that are done to avoid errors. This may include training, improved forms and tools, improved process control, and so on.

Appraisal costs are those associated with detecting errors—proofreading reports, inspections, audits, peer review, supervisor's review, and the like.

Error costs (sometimes referred to as failure costs) come from having to correct an error. All rework, repair, retest due to mistakes, and time to investigate cause of problems fit in this category. These are very expensive costs. Often these "failure" costs are subdivided into "external failure" costs from problems detected outside an organization. For corporations, these would include liability and warranty costs. "Internal failures" are those detected before the product or service is delivered to a customer outside the organization. While expensive to correct, they are preferable to external failures.

An area should always have some prevention costs because the tasks can always be improved, and that improvement is what makes a better job for the customer and ultimately saves money for the work area. But if there are substantial areas for improvement, there will be appraisal costs to check to see if there are errors. The first order of business is to drive external failures inside the organization where they can be dealt with before the customer is directly affected. For this reason, an increase in appraisal costs may be necessary to reduce external failure costs. Then, an increase in prevention costs is needed to reduce both appraisal and failure costs. It has been shown in industry that error costs can be ten to thirty times more expensive than prevention costs. Juran shows a more optimum distribution of costs as prevention = 10 percent, appraisal = 40 percent, and combined failure costs = 50 percent of the total (4).

By reviewing the tasks required to perform an activity, they can be placed into each of the cost categories. However, there is one group, the Real Value Added, that are not costs of poor quality. They are the costs to get the job done. If the time associated with each task has been estimated, it can be used to calculate the costs of poor quality using the matrix shown in Figure 14.

As can be seen, Real Value Added (RVA) is purely the job the group does that the customer pays for. It is not a quality cost. (It may be a target for overall cycle time or process time reduction, however, as will be discussed later.)

Business Value Added (BVA) is the work that must be done to get the job done, but not work for which the customer would (willingly) pay. Parts of BVA, such as keeping company records, are job only. BVA also includes specialized training, which is a prevention cost. BVA may also be inspection, reviews, and

AREA ACTIVITY ANALYSIS			*VALUE-ADDED AND QUALITY COSTS*
Note: Use a separate page for each activity listed on the Mission and Major Activities page			
Activity: Large Job Estimates	Area No: 179	Date: 10/27/.	Prepared by: GWR/LMM

Work Accomplished In Area

How many hours per week spent on this activity? 30.5 hours/task

Allocate this time into three value-added categories

RVA 20.5

BVA 0

NVA 10.0

Total 30.5 hours/task

Value-Added Versus Quality Costs Matrix

Using the matrix below, further allocate the value-added times to the appropriate "Job Only" and/or "Quality Costs" category.

		Quality Costs		
	Job Only	*Prevention*	*Appraisal*	*Errors*
RVA	20.5	No	No	No
BVA	0	0	0	No
NVA	No	No	4	6
Totals	20.5	0	4 (13.1%)	6 (19.7%)
Total	30.5	This should be equal to the hours per task for the activity (as above).		

Figure 14 Quality costs calculated.

audits (appraisal) that the group needs to do because that's the way it has always been done or because requirements have been imposed by regulators.

No Value Added (NVA) is certainly not the job the customer wants to pay for. It is not prevention, but it could be appraisal. In this case, appraisal is necessary because the process is letting errors get to the customer and the group must inspect or review the work to keep the impact of the errors internal. Whether appraisal is BVA or NVA is not so important as is the fact that appraisal should be reduced, if not eliminated. The error part of NVA clearly has to be eliminated.

Why go into all this detail? Most work areas are where they are because not enough attention has been paid to what is being done. As a result, the group does a lot of things that may have been necessary at one time but no longer are. The people get used to a level of performance and soon it seems like normal. This intense level of analysis provides a means to dissect the operations of the group so that the waste and errors can be seen and eliminated. Most areas will be measuring these costs in terms of time, but they can be converted to dollars through the local accounting department. The use of dollars is a common denominator and can be quite revealing since some areas carry large amounts of overhead.

8. *Summarize the Measurements*

Not all activities can have one of every kind of measurement, but the area as a whole will more than likely have at least one of every kind if not two or three. Figure 15 is a means to help summarize this.

9. *Evaluate Suppliers*

The area should now consider itself as a customer and look the other way toward their suppliers. What would they expect? The same that their customer expected of them. In that sense then, they could just repeat what was described in the Customer section. Most of it will be repeated, but with a supplier orientation.

Each of the activities listed on the page with the Mission have one or more inputs and suppliers. These suppliers (a company, an organization, a person) should be clearly identified and the input from this supplier specified. Then the requirements must be accurately stated to be sure the input is what is desired. The supplier should always be included in this determination to make sure there is agreement on the requirements.

Each section previously discussed is now addressed in the context of the group's needs from its supplier. Requirements should specify what, when, and the measurement to be used to determine acceptability. Then review what the group's default position will be if the supplier fails to meet the requirements and how to feed back performance information to the supplier. Figure 16 shows an example of requirements defined for a project leader being audited (also a customer for the final report) who must supply information to the auditor.

Activity Formal Internal Surveillance or area Project Leader

	Customer	Supplier
Is this a requirement from your customer or for your supplier?	Yes ☒ No ☐	Yes ☐ No ☐

Effectiveness:

Requirement: Identify data affecting problems in time to prevent loss of data.

Measurement: The number and severity of findings (or lack thereof) identified during the audit.

Efficiency:

Requirement: Perform all surveillances within the project surveillance budget and as timely as possible.

Measurement: Hours spent versus hours estimated.

Adaptability:

Requirement: Provide flexible response to changing needs.

Measurement: Number of "first opportunity" surveillances completed versus number of surveillances required. Number of surveillance delays versus total number of surveillances.

If this was for the whole area, then you should have at least two measurements each for effectiveness, efficiency, and adaptability.

Now look at the time it takes to get the job done.

Cycle time: 40 hours

Process time: 22 hours

Since you have already done the following for each activity, use this part to summarize the area.

Value-Added Versus Quality Costs Matrix

Using the matrix below, further allocate the value-added times to the appropriate "Job Only" and/or "Quality Costs" category.

		Quality Costs		
	Job Only	*Prevention*	*Appraisal*	*Errors*
RVA	7.5	No	No	No
BVA	3.25	6.5	0	No
NVA	No	No	.75	4

Figure 15 Compilation of output measurements.

10. Set Targets

As a result of the Area Activity Analysis, a list of performance measurements can be prepared. The team should select three to five of the most important measurements for posting in the area. Keep these Team Improvement Charts

PROJECT AUDIT — SUPPLIER: PROJECT LEADER

Input Needed	*When Needed*	*Measurement*	*Default*	*Feedback*
1. Define scope schedule and cost limitations of the project	• At time of estimate	• Changes exclusive of external factors	• Re-estimate audit • Adjust audit	Default action and explanation
2. Access to well-organized project records and personnel	• During audit	• Non-productive audit time due to records problems or personnel not available. Number of findings due to record problems.	• Extend audit time • Notification of added cost and schedule • Issue findings where appropriate	Default action and explanation
3. Participation in audit and meetings	• During audit	• Availability of PL during audit	• Extend audit time • Notification of added cost and schedule • Issue findings where appropriate	Default action and explanation
4. Response and action for CARs and open items	• After audit, before sign-off of report	• Number of cycles required for closure	• Return responses with comments	Default action and explanation

Figure 16 Input requirements for suppliers.

simple and large enough so that they can be read from a distance. Each chart should show at least six months of data and should have its targets reset.

As shown in Figure 17, two types of targets are used: first, the performance levels that the customer expects and then tighter targets called "challenge targets." The challenge targets provide the team with interim goals between the customer-expected performance level and the ultimate standard of error-free performance. This scheme eliminates the tendency of most companies to stop all efforts to improve an activity just because the target has been met. It also means that management must look at targets in a new way. Management should expect all customer targets to be met 100 percent of the time, but they should not expect the challenge targets to be met initially.

II. PROJECT PLANNING

A. Project Work Plans

As work is authorized by the R&D organization, a project leader or principal investigator is normally identified to carry out the requirements of the research function. This key individual has the primary responsibility for establishing the technical and quality standards for the project. He or she has the overall

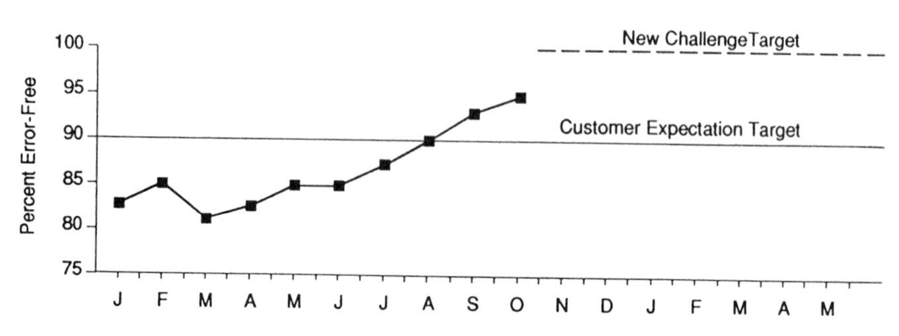

Figure 17 Performance indicators.

responsibility for the technical performance of the project and is in the best position to schedule the required checking, inspection, and documentation functions (5). The project leader (assisted by a representative from the QA organization if one exists) should evaluate the proposed research activity and determine, in light of customer or sponsor requirements and the nature of the work, which of the research center's good engineering and scientific disciplines are required for that specific project.

The experimental concept for a project can be described in a project technical plan. If the research is performed for an outside customer, the project technical plan may be replaced by a statement of work as a part of the contract proposal. In either the project technical plan or a contract statement of work, the elements to be included in that planning function should be similar.

A suggested list of topics to be covered by the project technical plan is:

1. Introduction with purpose of the project and background
2. Description of the activity; that is:
   a. Type of experiment or analysis
   b. Philosophy, basic principles, and limitations involved in the choice of the test or experiment design concept
   c. Parameters to be investigated
3. Test Apparatus
   a. Description of the test section or experiment to be used
   b. Identification of materials, parts, or components having an effect on the results
   c. Identification of measurements having an effect on the results
   d. General identification of special processes, e.g., welding, heat treating, brazing, nondestructive examination to be used which may affect the test results
4. Description of the overall test program and identification of specific test approaches to be used or developed during the course of the investigation

5. Description of data acquisition and reduction methods
6. Description of data evaluation methods

The project technical plan may be used as a formal means of agreement between the customer and the research center, but at the least, it is a way of informing technical sponsors of the proposed direction for the project. While the amount of detail seems extensive, not all will be applicable. Obviously, a greater level of detail is required to plan the qualification of a reactor design than is needed for an examination of the concept of Fourier transforms or an early experimental approach to determine neutrino behavior. The level of detail will vary but should be sufficient that someone technically competent in this type of work will be able to read the plan and understand where the project is headed *as is known at the beginning of the work*. It should be recognized from the start that such a plan is destined to be changed, probably often, as progress (or lack of it) dictates.

B. Quality Standards—ISO 9000

Over the years, there has been a significant movement toward applying formal quality assurance methodology to the research process. Certainly by the early 1960s, many research centers had quality control programs for their fabrication shops, and some government contracts such as the Minuteman and Apollo programs employed inspectors for their qualification testing as well. But the first appearance of specifications defining quality assurance measures to be taken for the research process itself seems to have been the Department of Energy Reactor Development Technology Standard RDT F2-2 (6). The application of these controls was then extended to the commercial nuclear laboratories by the Code of Federal Regulations, 10CFR50 Appendix B (7) and its primary interpretation documents ANSI N45.2 (8) and later, ASME NQA-1 (9). Later, these concepts were incorporated into the Good Laboratory Practices issued by the federal government for the medical industry. Some professional societies had begun to issue standards for laboratory work (10), and the nuclear industry seems to have been the driving force for these as well. During the late 1980s, pressure was brought to bear on the Department of Energy funded laboratories to add the type of controls then common among commercial nuclear laboratories. The impression was that, although the commercial nuclear industry had maintained tight controls over all of its work, including research (largely due to public pressure and litigation, particularly after Three Mile Island), many of the projects at the government labs had been shielded from this oversight and were therefore not as well controlled. In response to this apparent deficiency, the DOE began to impose formal quality assurance requirements on their contractors, the operators of the U.S. National Laboratories and the other laboratories that were part of the nuclear weapons program. The first lesson DOE and their labs learned was the same one learned by the commercial nuclear laboratories—most of the standards did not make sense to research. The standards used a manufacturing paradigm with examples from the fabrication shops, but had little to say to researchers. Frustrated with waiting for the consensus standards process to produce a workable document, the DOE issued a department order (5700.6C) (11) that generally addresses quality assurance for all of its subcontractors. Later, a supporting document, DOE-ER-STD-6001-92 (12) was issued which defines the application of the DOE Order for basic research laboratories.

1. ISO 9000

The International Organization for Standardization (ISO) established Technical Committee TC-176, in 1979, to take on the issue of quality in pursuit of ISO's goals of developing a common set of manufacturing, trade, and communication standards (13). TC-176 succeeded in publishing the ISO 9000 series of standards in 1987 in the belief that these documents provided a comprehensive set of management concepts and models for external quality assurance requirements. The design of the ISO 9000 series is such that the ISO 9004 (14) and its developing daughter documents are pointed inward to the organization and lay out guidelines for a total quality management process. The ISO 9001 (15), 9002 (16), and 9003 (17) documents are designed for external use as contractual quality assurance documents—their application is guided by ISO 9000 (18) and its daughter documents. The guidance of ISO 9004 should provide little problem for research laboratories for two reasons: first, it is a guidance document and not intended for contractual use, and second, the concepts of TQM are generally quite flexible. Many laboratories have already published their successes in TQM and the necessary adaptations are becoming familiar to researchers. The ISO 9001, 9002, and 9003 documents are likely to be more troublesome because they are designed for contractual use and are the basis for third-party certification, with its attendant problems of interpretation. As stated before, the research community has not been without experience in attempting to force-fit manufacturing quality standards to research. Some laboratories had programs in place to meet requirements of 10CFR50 Appendix B since 1974 (5). But the first organized effort to address and perhaps redress the problems inflicted by misapplication of these standards began in 1979, when the Energy Division of the American Society for Quality Control established its R & D Committee. Its charter was and is the "promulgation of both the theory and practice of Quality Assurance as it is applied to the field of energy research and development." A more intensive effort was begun by the American Society of Mechanical Engineers in 1990. Then, in Baltimore, approximately thirty people from the Department of Energy (DOE) and from DOE's research and development community attended a "rambunctious meeting where the diverse, undigested opinions were passionately expressed, flung as it were upon a discordant heap to be sorted out later" (19). The group's composite knowledge, background, and experience (research and

QA professional) assured that it was not just a matter of "we're different," but that there was a real problem borne out by practice. By 1993, after several intense meetings, the working group had substantially agreed on the definition of the problem. Although directed at the NQA standards, the commonality of the problem with any standard using a manufacturing paradigm is clear. The problem was partially expressed thusly:

> Applying NQA standards to research and development *support* activities is applying those standards to activities that are comparable to activities for which the standard was intended. But applying the NQA standards to research and development *science* activities uncovers a mismatch. The standards' logic and internal structure do not match the logic and internal workings of research and development science activities. The thought processes prevalent in research and development science activities are not the same as the thought processes associated with construction projects." (20)

The evolution of the ISO 9000 series from its predecessors, the U.S. Department of Defense quality systems standard MIL-Q-9858, the North Atlantic Treaty Organization's (NATO's) AQAP series, and the British Standards Institute's BS 5750, has carried with it the baggage of the manufacturing paradigm. The antecedent to the nuclear standards is also MIL-Q-9858 (which says something about the soundness of the design of that venerable document for manufacturing and construction), so it is not unexpected that the research community might find a common problem with all of them. Table 3 provides a matrix that compares requirements of ISO 9001 with other regulatory requirements common to R & D.

The apparent success of the ISO 9000 concept has been demonstrated in the increasing attempts at application of them to all industries and all types of activities. But the standards' weaknesses (similar to the problems with NQA) is evidenced by a trend toward local/industry specific adaptations (21). Technical Committee TC-176 is working toward making the ISO series as adaptable as possible to all activities. They have established Generic Product Categories (hardware, software, processed material, services) and will generate guidance standards to deal with special needs. Their intent seems pointed toward a one-size-fits-all certification of activities not only involving quality but including industrial safety, health, finance, and personnel.

The American National Standards Institute, Accredited Standards Committee Z-1 on Quality Management and Quality Assurance (authors of the ANSI/ASQC Q90 series version of ISO 9000) has chartered a writing group on Quality Systems for Research. This group views the issue of research quality as having two component parts, science and support, and that both parts fall within the general approach described by J. M. Juran, that of planning, controlling, and improving quality (22). It would seem then, from the experience of the ANSI and the ASME groups, that the traditional constructs of the ISO 9000 series can be readily adapted to the "support" activities of the research institution, but that the "sci-

Table 3 Comparison of ISO 9001 with Other Regulatory Requirements

ISO 9001 Requirement * Addressed By Regulatory Standard NA - Not Addressed	Corresponding Regulatory Requirements		
	RDT F2-2	DOE-ER-6001-92	10-CFR-50
4.1 Management Responsibility			
Quality Policy	*	*	*
Organization	*	*	*
Responsibility & Authority	*	*	*
Verification Resources & Personnel	*	*	*
Management Representive	NA	NA	NA
Management Review	*	*	*
4.2 Quality System	*	*	*
4.3 Contract Review	NA	NA	NA
4.4 Design Control			
General	*	*	*
Design & Development	*	*	*
Planning	*	*	*
Activity Assignment	*	*	*
Organizational & Technical Interfaces	*	*	*
Design Input	*	*	*
Design Output	*	*	*
Design Verification	*	*	*
Design Changes	*	*	*
4.5 Document Control			
Document Approval & Issue	*	*	*
Document Changes/Modifications	*	*	*
4.6 Purchasing			
General	*	*	*
Assessment of Subcontractors	*	*	*
Purchasing Data	*	NA	NA
Verification of Purchased Product	NA	NA	NA
4.7 Purchaser Supplied Product	NA	NA	NA
4.8 Product Identification and Traceability	*	NA	*
4.9 Process Control			
General	*	*	*
Special Processes	*	NA	*
4.10 Inspection and Testing			
Receiving Inspection and Testing	*	*	NA
Positive Recall or Urgent Production Needs	NA	NA	NA
In-Process Inspection and Testing	*	*	*
Final Inspection and Testing	*	*	*
Inspection and Test Records	*	*	*

Table 3 Continued

ISO 9001 Requirement * Addressed By Regulatory Standard NA - Not Addressed	Corresponding Regulatory Requirements		
	RDT F2-2	DOE-ER-6001-92	10-CFR-50
4.11 Inspection, Measuring, and Test Equipment	*	*	*
4.12 Inspection and Test Status	*	*	*
4.13 Control of Nonconforming Product			
Control	*	NA	*
Review and Disposition	*	NA	*
4.14 Corrective Action	*	*	*
4.15 Handling, Storage, Packaging, and Delivery			
General	*	*	*
Handling	*	*	*
Storage	*	*	*
Packaging	*	*	*
Delivery	*	*	*
4.16 Quality Records	*	*	*
4.17 Internal Quality Audits	*	*	*
4.18 Training	NA	*	*
4.19 Servicing	*	NA	NA
4.20 Statistical Techniques	*	*	NA

ence'' activities must be dealt with somewhat differently, using the vernacular and concepts familiar to the practitioners of the art. The focus is on establishing responsibility for directing the work and developing a comprehensive research plan. As envisioned by the ANSI group, the plan needs to include what is to be done (see Project Work Plans above); describe how the work is to be performed and documented; give assessments to be performed on the work; and indicate how the results are to be transferred. Other sections of the ANSI document address the ''Institutional Quality Management Program,'' a work scope more amenable to a standard like ISO 9001. Here, the institution is expected to identify its mission, customers (one group of whom are the researchers), and stakeholders. Among other things, the institution should plan to meet its customers and stakeholders needs and adopt a strategic planning process that sets goals for the continued improvement of its *support for quality*. It should provide support for research in the form of human and material resources, establish an environment that fosters creativity, and support worker empowerment. It should establish a program of assessments and use the results as a basis for improvement. Finally, the institution is expected to create an environment of continuous quality improvement through leadership, goal setting, and training.

2. *Technology Development Planes*

As the reader reviews the options for dealing with quality assurance issues, the question will naturally arise, "what is appropriate for my work?" Research varies widely and there has been much consternation over attempts to overly control the quality of research (23). Several attempts have been made by various professional organizations to define standards for application of quality assurance measures and all have struggled with the question of degree. The first of these was a guidelines document written by the American Society for Quality Control Energy Division that classified research as basic, applied, and benchtop/analytical. The guidelines then provided a matrix of suggested applications for each type based on the perceived risk involved in use of the data emanating from the research (24).

The ASME Nuclear Quality Assurance Working Group on Research and Development has defined research as falling into three categories: basic, applied, and development, and approaches application of assurance techniques based upon the category of work (20).

The difficulty with categories is that they don't easily accommodate the transient state of knowledge that is being used to acquire more knowledge. There is a cycle of knowledge that is continually repeated and that takes people from an intellectual state of ignorance to where they have acquired knowledge that they can routinely use. Figure 18 illustrates this cycle. People begin in a state of ignorance, but without knowing that they are ignorant. Then, some need causes them to recognize their ignorance and causes them to investigate that unknown area. The need could be mandated by many reasons including an overall personal curiosity about such things, the opportunity to obtain funding (crass though that may seem), a need described by some other entity—organizational or personal such as a peer, or as the result of some previous activity in which they were involved that has posed some new and interesting challenges. They then review existing theory and conceive of possible approaches to obtaining more information. They establish a new or modified theory that should be tested and begin to generate calculations that seem to support or deny the hypothesis. More calculations may be required to develop a design for an apparatus that can reproducibly measure associated phenomena that will allow them to further prove or disprove their theory. Having accomplished that and if the opportunity exists for direct application, they may take a great leap to the design and qualification of an artifact, process, or methodology that makes use of the new-found knowledge. The attempted application of that knowledge creates new questions that make them again aware that they are ignorant and so they begin anew. This process could be described as a knowledge improvement cycle and could be superimposed on the Spiral of Quality defined by Juran (4) since the cycle leads to increasing levels of knowledge. The state of knowledge acquired by this improvement cycle is peculiar to each investigator or research team, however, and does not necessarily represent the overall state of knowledge for the technology. Competitors have a way of surging ahead at least temporarily despite one's best efforts.

Technology development has been described as starting with basic research, progressing through applied research, then through development and finally into application or translation into a design of a particular piece of hardware, or process, or technique. The combined effects of many researchers improving their knowledge and enlightening the scientific world as well propels the technical community along that plane as shown in Figure 19. Depending on what problem is ultimately being solved, one enters a technology plane at various points and, depending on what new problems are encountered, one may have to double back to an earlier point on the plane.

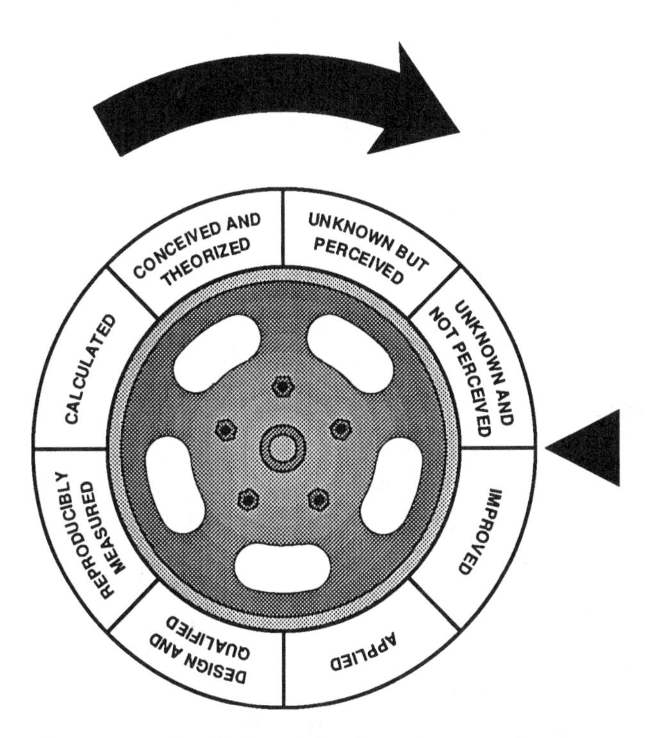

Figure 18 Scientific knowledge improvement cycle.

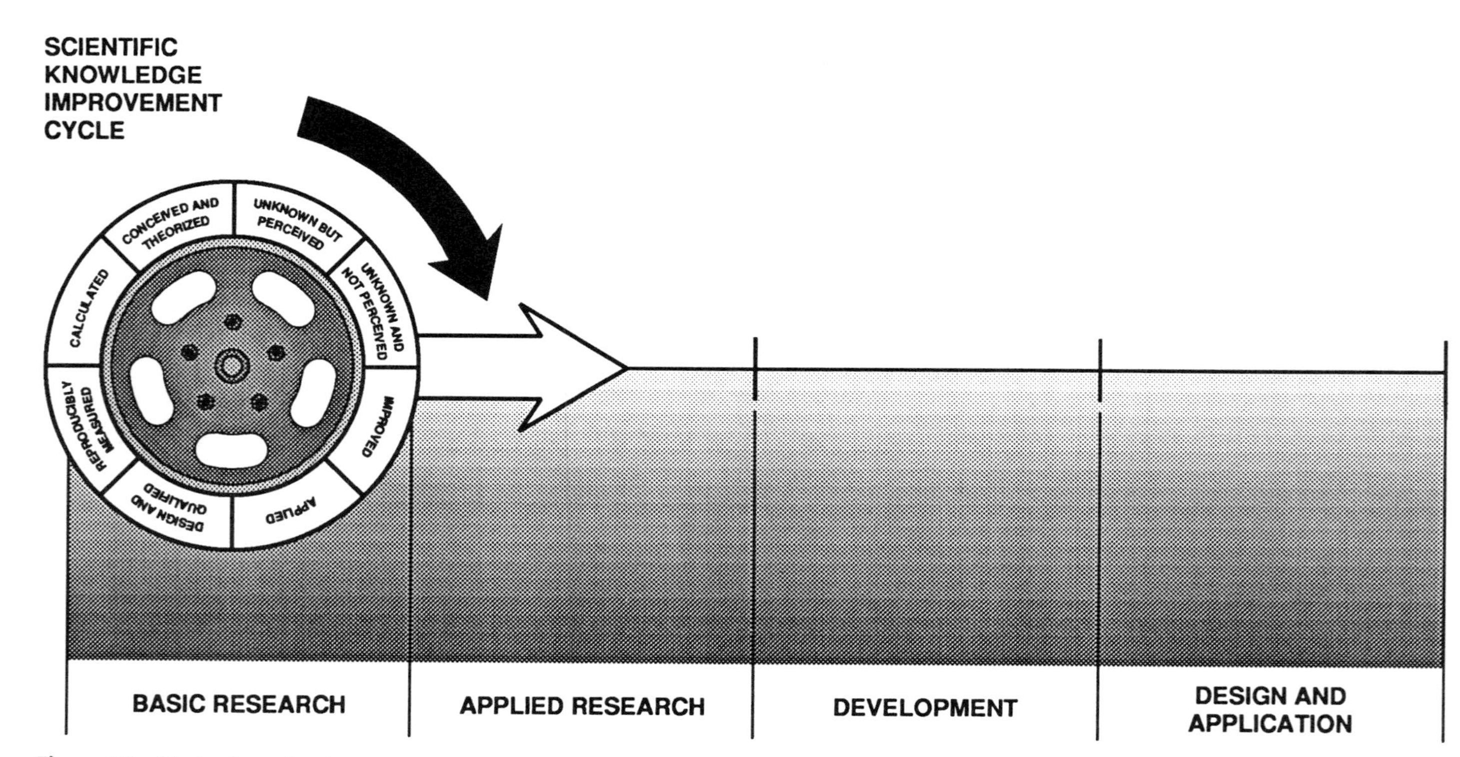

Figure 19 Technology development plane.

Since researchers have to use that which is known to help them fathom that which is unknown, they rely on instruments, other scientific apparatus, their current knowledge of the physical world, calculation, and inductive and deductive thinking to inch their way into new discoveries. During their attempts to solve the main problem, they encounter associated problems in their ability to measure certain characteristics and must take a detour to push their associated technology far enough along its own plane to bring it to a point of application where it is of use on the primary problem. Because of this constant interaction of known and unknown, it is impossible to make a flat statement that "this work is basic research," and "that work is development" except perhaps in the sense of the main technology plane of interest.

What is needed is to recognize that different technologies follow different planes and that where one is depends on what one is looking for. While the end use of the main scientific effort of a basic research project may be knowledge for its own sake or a testing of a current scientific paradigm, the supporting technologies will have progressed along their own planes until they are mature enough to be relied upon to prove or disprove the main thesis. A diagram of the concept of intersecting Technology Development Planes is shown in Figure 20.

Given that the main purpose of the project is to prove or disprove the existence of the Higgs Boson, there is a massive association of machines, technologies, techniques, and people all forming their own social structure (25) and all in various stages of their own development. To press on with the search for the fundamental building blocks of nature requires sophisticated devices like the Large Hadron Collider. (It seems enigmatic that to investigate smaller and smaller phenomena, scientists need larger and larger types of apparatus.) But to bring the Large Hadron Collider to an operating state, they needed to develop improved detectors. These devices must all be able to be trusted or their weaknesses known and compensated for. Components of any large apparatus may consist of commercial devices that may be far along in their own individual planes. But, applying those units to the detection of unique physical events may stretch their intended capabilities such that some reversion to a previous point on the plane is required to redevelop the commercial device or material to be more sensitive in the area of interest. The controls to be applied to the development and demonstration of any device's capabilities are contingent upon the effect that device's function has upon the main effort, or to the safe operation of the detector, or even pure economics. Such assessments may be influenced by factors quite different from those affecting the use of the data from the main effort. That is, while the redevelopment of a safety relief valve or a glove box for handling irradiated materials is driven by the need to protect the research personnel, the main outcome of the primary scientific effort may have no direct bearing on anything for the foreseeable future and, by itself, would require much less verification to be usable.

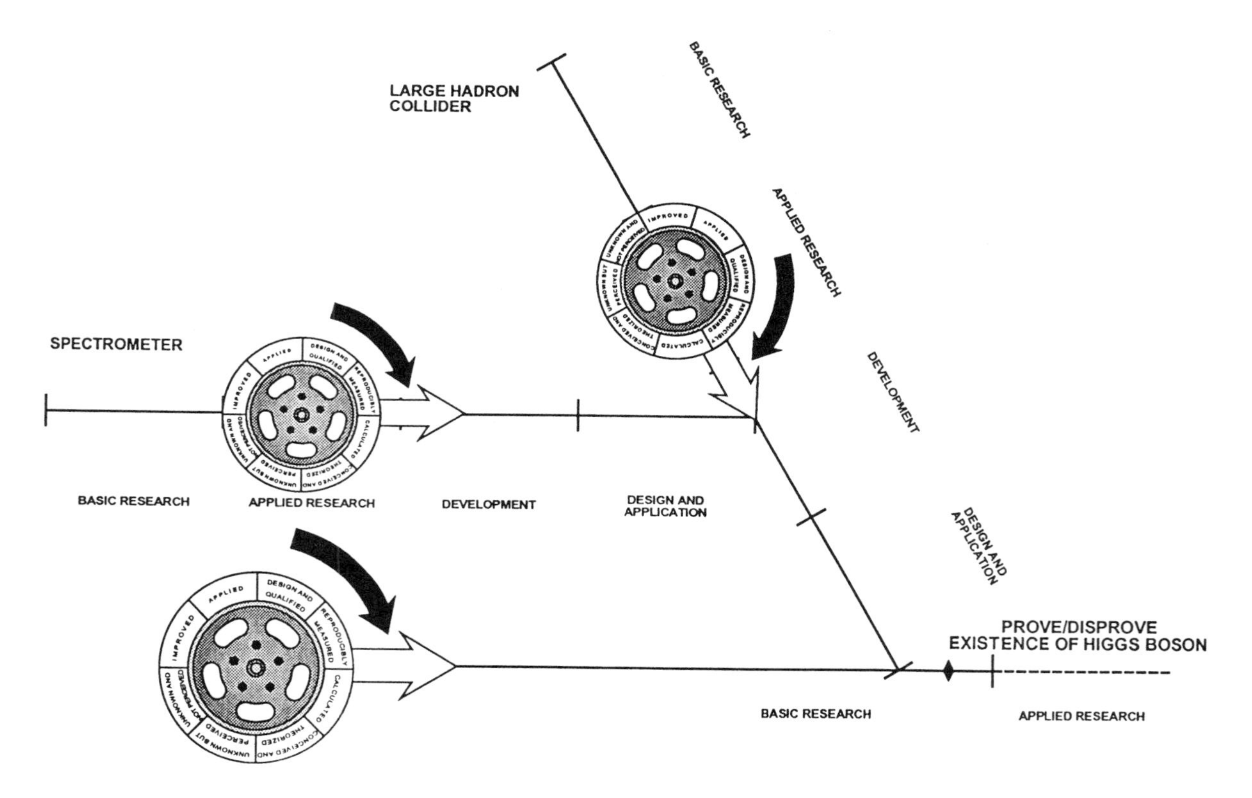

Figure 20 Intersecting technology development planes.

Thus, the detailed planning for any project must take into account several facets of operation and may require not one but several risk assessments (however informal) as to what technology development planes are involved and their intersections and, therefore, what good scientific and engineering practices are applicable.

3. Project Quality Assurance Plans

The QA plan is developed in conjunction with the project technical plan to selectively impose verification activities and processing controls. Typically, such plans are only required in response to a customer's imposition of a quality specification on the proposed work. However, some advantage has been gained in nonregulated research by having a checklist incorporated within a guidelines of good engineering and scientific practices. The checklist would prompt the project leader to consider such things as notifying the Safety Administrator if hazardous chemicals were to be used and to obtain concurrence with decisions regarding the use of incremental peer reviews prior to release of the final report. Figure 21 is an example of this type of plan.

QA plans may be narrative or of the checklist variety. Figure 22 illustrates a plan using a checklist format with specific implementing procedures noted adjacent to each of the applicable QA elements to be imposed on the project. The implementing procedures provide the specific detailed activities to be followed when performing the QA actions required for this project. Additional short notes can be entered in the Remarks column and supplemental sheets are only added when necessary for additional clarification or to make special provisions not otherwise covered in the implementing procedures. This format has the advantage of rapid preparation and, where implementing procedures are used, allows the project leader to know immediately which procedures he should have on hand for that project. The final page of this plan includes a checklist of all the quality records expected to be generated for this project. The checklist can be used by the project leader to verify that his files are complete. This particular illustration assumes that the user has a QA manual, a QA plan, and detailed implementing procedures. It further assumes the QA manual is a policy manual and does not give sufficient direction to permit detailed implementation.

As the research project grows larger, more elements are identified as being critical to the data. The design and construction of a large high-energy physics test apparatus, such as the Superconducting Super Collider or the Continuous Electron Beam Accelerator Facility, are examples of projects that warranted a full program status with full-time QA personnel assigned. These QA programs would appear very similar to a traditional QA program for any high-technology product line. Verification activities, however, would not concentrate only on measuring the physical aspects of the test apparatus to assure compliance with the design. Verification would include the conceptual design and the data resulting from the experimental use of the apparatus.

✔

- ❑ WORK PLAN
- ❑ FUNCTIONS AND RESPONSIBILITIES DOCUMENTED
- ❑ PLANT ENGINEERING NOTIFICATION
- ❑ JOB SAFE PRACTICE
- ❑ RADIATION WORK PERMIT
- ❑ SPECIAL PERSONNEL TRAINING/QUALIFICATIONS
- ❑ FORMAL DESIGN REVIEW
- ❑ INDEPENDENT REVIEW OF CALCULATIONS
- ❑ SOFTWARE RISK ASSESSMENT
- ❑ FORMAL DRAWINGS
- ❑ APPROVED SUPPLIERS
- ❑ EQUIPMENT WITH SPECIAL PERFORMANCE CRITERIA
- ❑ RECEIPT INSPECTION OF PURCHASED MATERIALS
- ❑ CERTIFIED TEST REPORTS
- ❑ CERTIFICATION OF CONFORMANCE
- ❑ RECEIPT INSPECTION OF CUSTOMER-SUPPLIED MATERIAL
- ❑ CUSTOMER APPROVAL OF MATERIAL SUBSTITUTIONS
- ❑ INDEPENDENT INSPECTION REQUIREMENTS
- ❑ MAINTAIN LIST OF INSTRUMENT SERIAL NUMBERS
- ❑ SPECIAL PROJECT RECORD RETENTION REQUIREMENTS
- ❑ SPECIAL TEST PROCEDURES REQUIRED
- ❑ PRELIMINARY DATA TRANSMITTAL *(MANAGEMENT APPROVAL REQUIRED)*
- ❑ FINAL REPORT
- ❑ RETRIEVABLE INFORMATION TO CORPORATE INFORMATION CENTER
- ❑ PROJECT LOGBOOK AND RECORDS TO SECTION OR PROJECT FILES
- ❑ IDENTIFICATION OF ORAL PRESENTATION MATERIAL

Figure 21 Standard practice checklist.

RC 375-1
REV. 2 2/87

QUALITY ASSURANCE PLAN
PAGE 1 OF 3

CUSTOMER BWNS
QA PROJECT NO. 93004
CUSTOMER CONTRACT NO. 3470
REVISION 0 DATE 8/4/93
R&D PROJECT NO. 47351

PROJECT PTS Curves on Compact Tensions DATE August 3, 1993

IN ACCORDANCE WITH CUSTOMER SPECIFICATION 09-1427-00 DATED 10/27/75

PREPARED BY:
QA ADMINISTRATOR R. K. Gill 8-4-93

APPROVED BY:
PROJECT LEADER C. S. Wade 8/4/93

APPROVED BY:
QA MANAGER G. W. Roberts 8-4-93

APPROVED BY:
SUPERVISOR B. P. Miglin

THE SECTIONS OF THE R&D QUALITY ASSURANCE MANUAL DESIGNATED BELOW ☒ (AND IMPLEMENTING PROCEDURES REFERENCED IN THE MANUAL) ARE APPLICABLE TO THIS PROJECT.

SECTION		REMARKS
1.0 INTRODUCTION	☒	
2.0 QA PROGRAM	☒	MANUAL REVISION 1/20/88 RDD-1702-03 (2/12/87)
3.0 DESIGN CONTROL PTP	x	RDD-1704-02 (12/9/87)
DESIGN REVIEW	☐	
INDEPENDENT TECHNICAL REVIEW	☐	
CALCULATIONS	☐	
COMP. PROGRAMS	☐	
4.0 PROCUREMENT DOCUMENT CONTROL (QA REVIEW)	☒	RDD-1705-01 (9/05/86) RDD-1211-01 (4/01/93)
5.0 INSTRUCTIONS, PROCEDURES & DRAWINGS		
DRAWINGS	☐	
ROUTE SHEETS	☐	
INSPECTION CHECKLISTS	☐	
ADMIN. PROCEDURES	☒	
TECHNICAL PROCEDURES	☒	RDD-1706-01 (2/10/87)
6.0 DOCUMENT CONTROL		
ADMIN. PROCEDURES	☒	
DRAWINGS	☐	
INSPECTION CHECKLISTS	☐	
PROPOSAL	☐	
PROJECT TECHNICAL PLAN	☒	RDD-1704-02 (12/09/87)
QA MANUAL	☒	RDD-1702-02 (6/30/93)
QA PLANS	☒	RDD-1702-03 (2/12/87)
ROUTE SHEETS	☐	
FINAL REPORT	☒	RDD-1801-11 (6/01/89) RDD-1801-10 (11/09/89)
TECHNICAL PROCEDURES	☒	RDD-1706-01 (2/10/87)
RELEASE OF DATA	☒	RDD-1718-04 (4/26/93)

Figure 22 Checklist style quality assurance plan.

REFERENCES

1. M. Bodnarczuk, "Peer Review, Basic Research, and Engineering; Defining a Role for QA Professionals in Basic Research Environments," Sixteenth Annual National Energy Division Conference, ASQC Energy Division, Fort Lauderdale, FL, 1989
2. T. B. Kinni, "What's QFD. Quality Function Deployment Quietly Celebrates Its First Decade in U.S.," *Industry Week*, November, 1993
3. H. J. Harrington, *The Improvement Process*, McGraw-Hill, New York, 1987
4. J. M. Juran, *Quality Control Handbook*, McGraw-Hill, New York, 1974
5. G. W. Roberts, Quality Assurance in R & D, *Mechanical Engineering*, Vol. 100, No. 9, p. 41 (1978)
6. Atomic Energy Commission, Division of Reactor Development and Technology, "Quality Assurance Program Requirements," RDT F2-2
7. *Code of Federal Regulations 10 CFR 50, Appendix B*, Quality Assurance Criteria for Nuclear Power Plants and Fuel Reprocessing Plants, U.S. Government Printing Office, Washington, D.C., 1970
8. ANSI N45.2—1971, *Quality Assurance Program Requirements for Nuclear Power Plants*, American Society of Mechanical Engineers, New York, 1971
9. ASME NQA-1-1989, *Quality Assurance Program Requirements for Nuclear Facilities*, American Society of Mechanical Engineers, New York, 1989
10. ASTM C1009–89, "Standard Guidelines for Establishing a QA Program for Analytical Chemistry Laboratories Within the Nuclear Industry," *Annual Book of ASTM Standards*, American Society for Testing and Materials, Philadelphia, 1989
11. DOE Order 5700.6C, *Quality Assurance*, U.S. Department of Energy, Washington, D.C., 1991
12. DOE-ER-STD-6001-92, *Implementation Guide for Quality Assurance Programs for Basic and Applied Research*, U.S. Department of Energy, Washington D.C., 1992
13. R. W. Peach, *The ISO 9000 Handbook*, CEEM Information Services, Fairfax, VA, 1992
14. ISO 9004 (1987), *Quality Management and Quality System Elements—Guidelines*, ANSI Version ANSI/ASQC Q94 (1987), American Society for Quality Control, Milwaukee, WI, 1987
15. ISO 9001 (1987), *Quality Systems: Model for Quality Assurance in Design/Development, Production, Installation and Servicing*, ANSI Version ANSI/ASQC Q91 (1987), American Society for Quality Control, Milwaukee, WI, 1987
16. ISO 9002 (1987), *Quality Systems: Model for Quality Assurance in Production and Installation*, ANSI Version ANSI/ASQC Q92 (1987), American Society for Quality Control, Milwaukee, WI, 1987
17. ISO 9003 (1987), *Quality Systems, Model for Quality Assurance in Final Inspection and Test*, ANSI Version ANSI/ASQC Q93 (1987), American Society for Quality Control, Milwaukee, WI, 1987
18. ISO 9000 (1987), *Quality Management and Quality Assurance Standards: Guidelines for Selection and Use*, ANSI Version ANSI/ASQC Q90 (1987), American Society for Quality Control, Milwaukee, WI, 1987
19. J. J. Dronkers, "Report on the Activities of the ASME NQA Committee Working Group on Quality Assurance Requirements for Research and Development," Eigh-

teenth Annual National Energy Division Conference, American Society for Quality Control, Danvers, MA, 1991

20. Draft White Paper, ''A Description of Difficulties Encountered When Applying NQA Standards to Research and Development Activities,'' ASME Committee on Nuclear Quality Assurance, Working Group on Quality Assurance Requirements for Research and Development, American Society of Mechanical Engineers, New York
21. D. Marquardt et al., ''Vision 2000: The Strategy for the ISO 9000 Series Standards in the '90s,'' *Quality Progress*, Vol. 24, No. 5, p. 25 (1991)
22. J. M. Juran, ''The Quality Trilogy,'' *Quality Progress*, August, 1986
23. Open letter, ''To University of California President Gardner and Lawrence Livermore National Laboratory Directory Nuckolls,'' dated April 13, 1990, Lawrence Livermore National Laboratory, Livermore CA
24. R. R. Geoffrion, R. K. Gill, G. W. Roberts, *ASQC Quality Assurance Guidelines for Research and Development*, American Society for Quality Control, Milwaukee, WI, 1994
25. M. Bodnarczuk, *The Social Structure of Experimental Strings at Fermilab*; A Physics and Detector Driven Model, Fermilab-Pub- 91/63, Fermi National Accelerator Laboratory, Batavia, IL, 1990

6

DESIGN OF EXPERIMENTS

Robert H. Lochner
Consultant in Quality Improvement
Milwaukee, Wisconsin

"The facts." An overused expression. Everyone assumes he knows, but no one actually knows.

Hitoshi Kume (7)

I. INTRODUCTION

To many people experimentation is synonymous with research and development. Clearly, without experiments there would be no research and development. In this chapter *statistically designed* experiments are presented. These experiments allow us to gain a surprising amount of information in a relatively few experimental trials. Statistically designed experiments come in many forms and levels of complexity. In this chapter we will look at a few basic experimental designs which are representative of the designs that have been found to be particularly effective in quality improvement. There will also be references for more advanced experimental designs.

"Experiment" is sometimes defined as any act of observing. That is a little too broad for quality improvement, since in order to determine if quality or performance has improved, it is generally necessary to be able to measure the improvement. A better definition is *an experiment is a sequence of trials or tests performed under controlled conditions which produces measurable outcomes.*

Prior to performing an experiment the researcher identifies factors (temperature, time, thickness, amount of training, percent of solvent, etc.) which may affect one or more measurable characteristic of interest (yield, strength, processing time, thickness, etc.). The experiment is then performed to evaluate the effects the factors have on the characteristics of interest and also discover possible relationships among the factors. The goal is to use this new understanding to improve products and processes.

Not all experiments are created equal. A frequently used rule regarding "scientific experimentation" is to change one factor at a time. Although this is often perceived as a conservative approach, it is usually a risky and inefficient way of gathering information. It is generally not possible to detect *interactions* among factors with change-one-factor-at-a-time experiments. (If the effect one factor has on a product or process characteristic depends on the value or level of a second factor, the two factors are said to interact. Discovering and measuring the strengths of interaction effects is an important use of statistically designed experiments.) The experimental designs to be considered in this chapter are *orthogonal*, meaning that the effects of various factors on the product or process characteristic being measured can be estimated without being biased by the effects of other factors. [See Lochner and Matar (8) or Box, Hunter and Hunter (3) for a more precise definition of orthogonal designs.] Changing-one-factor-at-a-time experiments involving two or more experimental factors are almost never orthogonal.

Academic research is done to discover new relationships among variables or to develop models to describe natural or social phenomenon. Within the context of Total Quality Management, experiments are done to improve products or processes in one or more of the following ways:

Optimize the average of a process or product characteristic. For example reduce average set-up time, increase average tensile strength, or have average batch weights as close to target as possible. Most classical experimental designs were developed with this type of application in mind.

Minimize the variability in a product or process characteristic. There is little benefit in reducing the *average* set-up time or average tensile strength if variation is so large that set-up times cannot be reliably scheduled or if tensile strengths of some parts are far below the average. Although reduction in variability has long been recognized as critical for quality improvement, traditional approaches to analyzing experimental data inhibited use of experimental designs in reducing variation.

Minimize the effects of uncontrollable variability on product or process characteristics. Genichi Taguchi, a prominent Japanese quality engineer, has lead the way in this use of experimental designs. It is often more practical and less costly to reduce the *effects* of variation than to reduce the variation itself. For example, if ambient temperature adversely affects a process, it may be cheaper to redesign the process to reduce the effects of room temperature than to air condition the facility. Similarly, it may be better to reinforce and cushion an instrument case than to print "handle with care" on all six sides.

Taguchi believes the most important use of experimental designs is in reducing the effects of uncontrollable variation, which he calls *noise factors*. A process which is relatively insensitive to noise is said to be *robust*. There are three types of noise factors: *internal*, *external* and *unit-to-unit*. Internal noise refers to product and process wear or aging which occurs with time or use. External noise refers to uncontrollable variation outside the process or product which can affect performance. Ambient temperature, air pollutants, vibration and variation in incoming supply voltages are external noises. Unit-to-unit variation is the inevitable variability in products made to the same specifications under "identical" conditions.

Experiments can be applied in three distinct ways in quality improvement:

Explore the relationship among product/process characteristics. Which process factors have the strongest effect on product reliability? Which factors control the amount of variation in plating thickness or amount of waste? Does the effect of temperature on a process change as relative humidity changes? Experiments are typically used for exploration early in a design or improvement project.

Estimate the strength of the effects of important factors. If we increase pressure 5 percent, by how much will the yield be increased? If we set the pH at 9.2, how long will the reaction take to complete?

Confirm the conclusions reached from earlier studies and experiments. Based on earlier experiments, appropriate levels for process/product characteristics were determined. A follow-up experiment should always be done under the recommended conditions to confirm that the recommendations are based on repeatable results.

Most quality improvement or product/process design projects include all three of these activities. Depending on the situation, a different type of experimental design will be used each time.

II. PLANNING AN EXPERIMENT

Statistically designed experiments can save an organization thousands of dollars in reduced development time, increased productivity and more reliable products and processes at lower costs. But use of this methodology requires careful planning. A first step in designing an experiment, indeed the first step in any improvement project, should be to write a mission statement for the experiment or project. (See Chapter 5.) This brief paragraph should state what exactly is to be accomplished by the project. It should state a specific problem or opportunity to be addressed, but should not state an expected solution. For example, "Reduce the number of product defects" is too broad. "Reduce the number of visible scratches appearing on the face of the product" would be appropriate. "Reduce the number of visible scratches appearing on the face of the product by training employees in proper handling procedures" would not generally be appropriate unless data were available indicating that improper employee handling was a leading cause of scratches.

The mission statement should be accompanied by:

1. A statement of project scope. (All visible scratches or just those on the painted surfaces? All products or just one model?)
2. Reasons why the project was selected (high scrap costs, late delivery, or unhappy customers, for example). Supporting data should be included if at all possible.
3. Measures of performance and quality with respect to critical product/process characteristics. That is, how is quality, or poor quality, measured. (What is a "visible" scratch? How is adhesive force measured?) What are the current tolerances for these characteristics?
4. List of individuals who will be involved in the project, either as team leaders, team members or resource personnel. Also, what resources will be available, in terms of time, dollars and facilities.

The team to be involved in a project should, at one of their first meetings, review the mission statement and associated material to assure there is understanding and buy-in among the team members. They should also establish tentative project milestones which include completion dates and lead persons for critical process activities such as:

- Flow chart the process
- Gather and analyze existing data on the process/product
- Develop a cause-and-effect diagram
- Select the experimental design and experimental factors
- Analyze the experimental data
- Recommend changes/improvements based on the data
- Carry out a confirmatory study.

III. PRE-EXPERIMENT DATA ANALYSIS

An important first step in process improvement, whether it involves experiments or not, is to flow chart the process. Even if the process is felt to be trivial and well understood, there are clear benefits to developing a flow chart. First, it assures that everyone involved understands what the process is, where it starts and ends, and who the suppliers and customers of the process are. The flow chart helps an individual or group in looking for potential root causes of problems or excessive variation (if the solution was easy or obvious, the problem would have been solved already).

Next, all available data on the process/product characteristic or problem should be collected and analyzed. Brainstorming using a process flow chart is a good way to develop the list of available data. Evaluate the data over time—is the process stable or are there cycles, trends or spikes in process measurements? If the process is varying wildly and in an unpredictable way, an experiment will probably give unreliable information (''garbage in—garbage out.'') If the quality characteristics being measured have tolerances assigned to them, determine from the data if the process can consistently hold these tolerances. If excessive variability causes the process to be frequently out of tolerance (over 1 percent of the time), it might be best to do some initial experiments to identify how to reduce the variability. There is a tendency to use expert opinion in place of data analysis during this step. The urge must be strongly resisted. Kume (7) is a good source for basic data analysis tools.

''Is/Is Not Analysis'' can be used to help identify key factors affecting process performance including lack of stability (6). This involves asking when, where and for whom does the problem occur, and when, where and for whom does the problem NOT occur. Or when, where and for whom is the problem more/less severe? If the problem happens only when it rains, or only at the Chicago plant, or more often on Mondays and Fridays than on Tuesdays, this may be very helpful information. Possible explanations for these patterns should be formulated and tested.

An essential tool in designing an experiment is the cause-and-effect diagram. One should be prepared after the flow chart is completed but before experimental factors are selected. To draw a cause-and-effect diagram first draw a horizontal arrow across the center of a pad of paper. Write the problem described in the project mission statement in a box at the right end of the arrow. Next write the major categories of possible causes above and below the arrow and connect them with smaller arrows. Using brainstorming, add other causes, subcauses, and so on, to the diagram. Connect them with lines at appropriate places. A completed cause-and-effect diagram for the problem ''popcorn doesn't pop'' is given in Figure 1.

IV. SELECTING EXPERIMENTAL FACTORS

Once it has been determined that the process is stable enough to produce meaningful measurements of the quality characteristics of interest, prepare a list of those

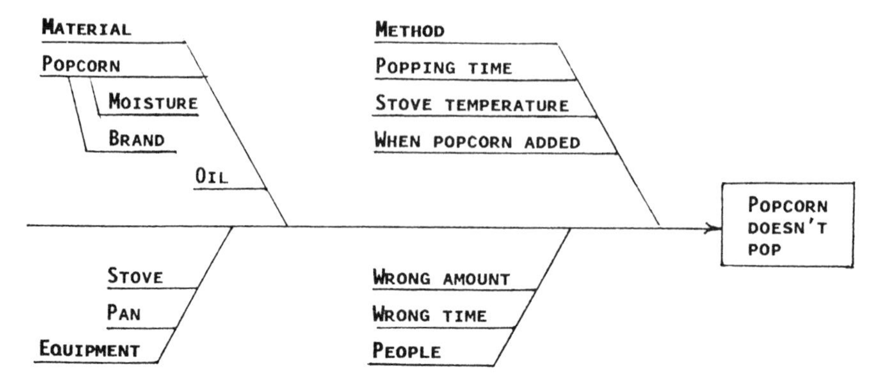

Figure 1 Cause-and-effect diagram for why popcorn doesn't pop.

factors which are most likely to affect these characteristics. Flow charts, cause-and-effect diagrams, is/is not analyses, process data and the expert knowledge of people familiar with the process should be used when developing the factor list. If process variability is a serious problem, it might be best to carry out an experiment to identify the sources of variation. (See Section XII, "Using Experiments to Reduce Variability.") Reducing process variability can be a challenging task since the experimental data may be unreliable and the root cause of the variation may be an unknown or unsuspected factor. When performing experiments to reduce variability the researcher will need to obtain repeated measurements at each combination of factor levels used in the experiment, control as many factors as possible, and monitor or measure all factors which cannot be controlled during the experiment. Experimental data should always be plotted in the time order it was collected as a check for "lurking" variables which produce unexpected patterns in the data. Such a plot can be particularly useful when looking for causes of excessive variability.

Each factor suspected of possibly affecting the characteristic of interest should be classified by *type*, *status*, and *use in the experiment*.

Type: Is the factor controllable or is it a noise factor?

Status: In terms of the experiment is the factor (1) controllable (it can be set at different levels by the experimenter), (2) uncontrollable but measurable, or (3) uncontrollable and unmeasurable. (Note: a noise factor can sometimes be controlled during an experiment. For example, by putting ice around a pipe to reduce temperature or doing some trials on days with high humidity and other trials on days with low humidity to "control" humidity.)

Use in the experiment: During an experiment each factor identified as possibly having an effect on the product or process characteristics of interest is put into one of the following four categories: (1) set at two or more predetermined levels, (2) held fixed throughout the experiment, (3) allowed to vary but be measured, or (4) allowed to vary and not measured. Usually the factors which are felt to be most important become the *experimental factors* (category 1). As many as possible of the remaining factors are held fixed during the experiment (category 2). Usually factors are put in categories 3 or 4 only if there is no possibility of controlling them, since uncontrolled variability in important factors can result in misleading experimental data. For factors which cannot be held fixed but can be measured, the observed values of the factors should be recorded for each experimental trial. As part of the analysis of the experimental data the possibility of relationships between an uncontrolled factor and the experimental factors should be explored by treating the uncontrolled factor as a response variable. In addition, the residuals (arithmetic differences between observed values for characteristics of interest and the estimated expected values based on the experimental data and model) should be calculated. The residuals should then be plotted against the observed values of the uncontrolled factor to determine if the uncontrolled factor could explain some of the variability not accounted for by the experimental factors.

Occasionally an experimenter will intentionally let some noise factors vary in order to observe the resulting variability in the process. This happens most often in experiments which have as their goal the reduction of variability.

For each experimental trial, measurements are made of one or more product or process characteristics of interest. These characteristics are called *response variables*. The purpose of the experiment is to identify which experimental factors affect which response variables, and to estimate the form and strength of the relationship.

Selecting the factors to be set at different levels during the experiment and determining which combinations of levels of these factors to be included in the experiment is at the heart of design of experiments. This issue, and consideration of how to analyze the data from an experiment, will occupy much of the remainder of this chapter.

V. EXPERIMENTS WITH ONE EXPERIMENTAL FACTOR

The simplest experiments involve varying one factor while holding all other factors fixed. For example, use three different batch temperatures while holding other factors such as batch formulation, size of batch and mixing time fixed. Usually several trials are done at each level of the factor which is being varied. This provides the experimenter with the data needed to estimate the amount of variation in the response variables for each of the test levels of the factor. The more replication, the more reliable the results.

A. Sample Average

Suppose an experimenter wants to compare surface hardness of two different alloys. The response variable could be the depth of an indentation (in mm) left by a probe pressed into the material with a given force. The experiment could be concerned with the *average* depth of the indentations, the *percentage* which had indentations exceeding 1.5 mm, or the *variability* of the indentation depths.

Suppose the experiment consists of testing ten specimens of each alloy, producing the data in Table 1. Note that the data from the two samples overlap, so we cannot say that one alloy will *always* have larger response values than the other. But we can ask whether one alloy has larger values *on the average*. Average responses are calculated by adding up the measurements in a sample and dividing the sum by the sample size. Averages are typically denoted by the letter Y, or some other letter, with a bar over it. For our data set:

$$\bar{y}_A = (1.4 + 1.6 + 1.3 + 1.4 + 1.5 + 1.6 + 1.4 + 1.3 + 1.5 + 1.4)/10 = 14.4/10 = 1.44$$

$$\bar{y}_B = (1.5 + 1.7 + 1.3 + 1.4 + 1.5 + 1.4 + 1.6 + 1.5 + 1.6 + 1.5)/10 = 15.0/10 = 1.50$$

(It is customary to round off calculated values to as many significant digits as appeared in the raw data used in the calculations. However, in statistical analyses it is appropriate to carry one or two additional decimal places in intermediate calculations, with the understanding that appropriate rounding will be done for final reports.)

B. Sample Standard Deviation

The average penetration depth for alloy A is less than that of alloy B, suggesting that alloy A is stronger. But is the difference because A is stronger, or is it just due to random variation in testing and piece-to-piece strength? This issue may be resolvable by estimating the variability present in the measurement process. The commonly used measure of variability is the *standard deviation*. The standard deviation of a random sample of measurements is obtained as follows:

1. Calculate the sample mean, $\bar{y}$.
2. Subtract $\bar{y}$ from each measurement in the sample.

Table 1 Penetration Depths from Alloy Strength Study

Alloy A	Alloy B
1.4, 1.6, 1.3, 1.4, 1.5	1.5, 1.7, 1.3, 1.4, 1.5
1.6, 1.4, 1.3, 1.5, 1.4	1.4, 1.6, 1.5, 1.6, 1.5

3. Square the differences obtained in step 2.
4. Add the squared differences obtained in step 3 and divide the sum by the sample size minus one. This statistic is called the sample variance and is denoted by s^2.
5. Obtain the square root of s^2. This is the sample standard deviation and is denoted by s.

Example: Calculation of Standard Deviation

The penetration depth measurements in Table 1 for alloy B are listed in column 1 of Table 2. The differences of step 2 above are in column 2 of Table 2. The squared differences are in column 3. The sample variance (step 4) is

$$s_B^2 = 0.12/(10-1) = 0.0133$$

and the sample standard deviation (step 5) is

$$s_B = \sqrt{0.0133} = 0.115$$

Using the above five steps with the measurements from alloy A gives a standard deviation of

$$s_A = \sqrt{0.104/9} = 0.107$$

How is the standard deviation interpreted? First, the more the sample values vary from each other, the more they will vary from the mean, and so the larger s will be. In the above example it was found that $s_A = 0.107$ and $s_B = 0.115$. Figure 2 has histograms of the two samples. Note that the alloy B sample, with the larger standard deviation, is also more spread out.

Table 2 Calculation of Sample Standard Deviation

y	$(y - \bar{y})$	$(y - \bar{y})^2$
1.5	0.0	0.00
1.7	0.2	0.04
1.3	−0.2	0.04
1.4	−0.1	0.01
1.5	0.0	0.00
1.4	−0.1	0.01
1.6	0.1	0.01
1.5	0.0	0.00
1.6	0.1	0.01
1.5	0.0	0.00
	Sum =	0.12

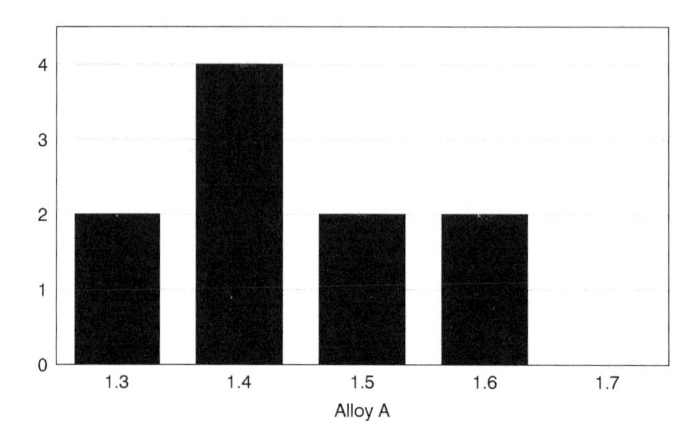

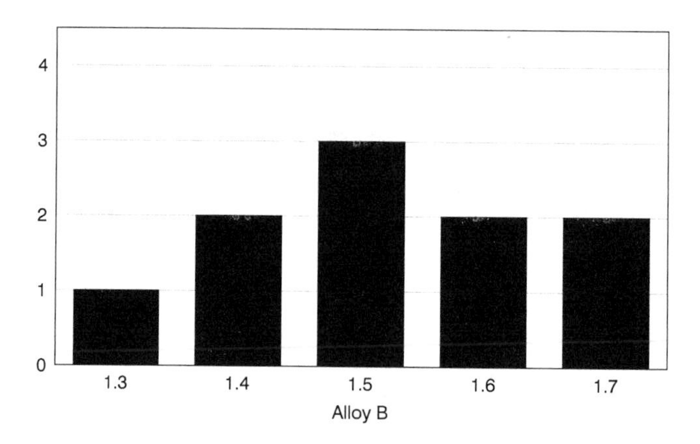

Figure 2 Histogram of hardness test data.

There is another use we can make of the standard deviation. If the data has an approximately bell-shaped, or *normal, distribution* form (see Figure 3), and if the sample size is reasonably large (at least 20) we can expect, approximately:

1. 68 percent of all measurements to lie between plus or minus 1 standard deviation from the mean,
2. 95 percent of all measurements to lie between plus or minus 2 standard deviations from the mean, and

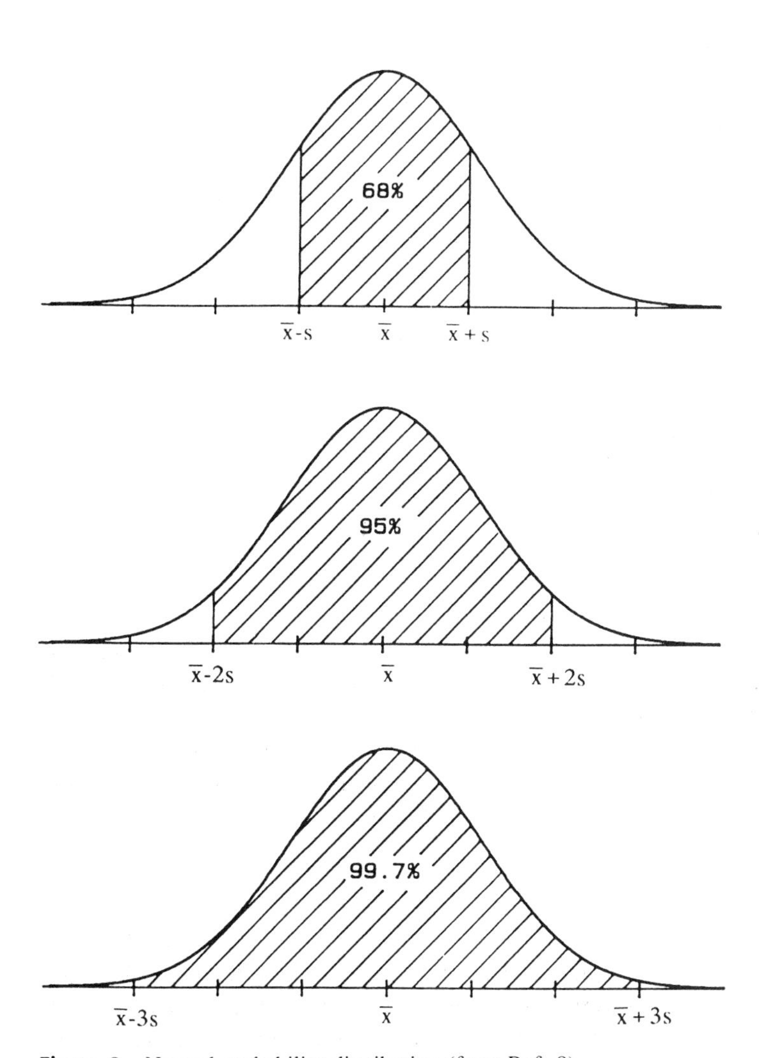

Figure 3 Normal probability distribution (from Ref. 8).

3. 99 percent of all measurements to lie between plus or minus 3 standard deviations from the mean.

We assume here that the sample mean and standard deviation are based on a large sample from a stable process which produces measurements having a normal distribution. If the process being sampled is not stable—that is, if the distribution of measurements changes over time—then it would be inappropriate to predict tomorrow's results based on today's data.

Example: Interpreting a Standard Deviation

In the previous example, $\bar{y}_B = 1.5$ and $s_B = 0.115$. If the sample measurements are from a process which produces normally distributed values, then:

1. About 68 percent of all the measurements will lie between $1.5 - 0.115 = 1.385$ and $1.5 + 0.115 = 1.615$
2. About 95 percent of all the measurements will lie between $1.5 - (2 \times 0.115) = 1.27$ and $1.5 + (2 \times 0.115) = 1.73$,
3. About 99 percent of all the measurements will lie between $1.5 - (3 \times 0.115) = 1.155$ and $1.5 + (3 \times 0.115) = 1.845$

(Note: These three statements are for illustrative purposes only and should not be taken too seriously since the sample size used to calculate $\bar{y}_B$ and s_B are so small. Usually samples of twenty or thirty are preferred since they give more reliable estimates and provide a clearer picture of whether the data appear to have come from a bell-shaped distribution.)

C. Comparing Two Means

When evaluating possible differences between two processes, it is not enough to simply ask which sample has the larger sample average. Random variation can lead to one sample mean being larger than the other simply by chance. But there should be some magnitude of difference between the two means for which we can safely say that the differences are real. There are several ways of comparing averages from two processes. For example, take two independent random samples, one from each process, of sizes n_A and n_B, each at least twenty. If the two processes have the same average value, then

$$Z = \frac{\bar{y}_A - \bar{y}_B}{\sqrt{s_A^2/n_A + s_B^2/n_B}}$$

is approximately a normal random variable with mean zero and standard deviation equal to one. If the two processes have equal mean values, the probability that the absolute value of this *test statistic* will exceed 1.645 is 0.10. The probability it will exceed 1.96 in absolute value is 0.05. The probability it will exceed 2.58 in absolute value is 0.01. So if Z is, say, 1.8 then we decide the two process means are not equal since the difference between the two sample means is "statistically significant at the 0.10 significance level." That is, we observed a value of the test statistic which had a less than 10 percent chance of occurring if the two processes had equal averages. Note that if $Z = 1.8$ the difference between the two sample means is *not* significant at the 5 percent significance level since 1.8 lies between -1.96 and 1.96.

In academic research the significance level to be used in a study is usually selected prior to performing the experiment. In data analyses done in businesses the experimenter will more often state the smallest level at which the test statistic is significant. If the test statistic is not significant at the 0.10 significance level, the experimenter simply says that the test statistic is not significant.

The values given above at which the test statistic became significant (1.645, 1.96 and 2.58) are called *critical values* for the test statistic. The values given are appropriate if the experimenter wants to determine whether the two sample means differ significantly without regard to which mean is larger. However, if the question is "do the two processes have the same mean or does the first process have a larger mean?" then the process expected to have the larger mean is designated as $\bar{y}_A$ in the formula for the test statistic Z and the test statistic is deemed to be statistically significant if Z exceeds 1.28 (0.10 significance level), 1.645 (0.05 significance level) or 2.33 (0.01 significance level).

The validity of the above test method is based on the "Central Limit Theorem," which says that even if individual measurements do not follow a bell-shaped (normal) distribution, the *means* of random samples will still be approximately normally distributed when sample sizes are large. For discussion on other statistical tests for equality of two process means, including small sample situations for normal processes, see Box, Hunter and Hunter (3) or any basic statistics text. For a discussion on appropriate tests of process means for small samples from non-normal processes see Ott (10) or any text on nonparametric statistics.

Example: Two-Sample Test of Means

Two different methods for filtering a solution have been suggested. The first method is tried on twenty batches. The average filtration time was 47.6 minutes with a standard deviation of 3.4 minutes. The second method was used with twenty-six batches, producing an average filtration time of 44.0 minutes with a standard deviation of 5.7 minutes. We have $n_A = 20$, $\bar{y}_A = 47.6$, $s_A = 3.4$, $n_B = 26$, $\bar{y}_B = 44.0$ and $s_B = 5.7$. Then

$$Z = \frac{47.6 - 44.0}{\sqrt{3.4^2/20 + 5.7^2/26}} = \frac{3.6}{\sqrt{1.8276}} = 2.66$$

Since 2.66 is greater than 2.58, we can say that the *mean values for the two processes are significantly different from each other at the 0.01 level of statistical significance*.

Comparing More than Two Means

Whole books have been written on how to compare more than two averages. The most frequently used approach to analysis involves a rather complicated but mathematically satisfying class of statistical methods called *analysis of variance* (ANOVA). This methodology formed the basis for much of the theory of experimental design developed in this century. ANOVA is not used as extensively in quality improvement projects as in some other areas of research, however. One reason for this is that ANOVA assumes that the standard deviation remains constant, even though averages may change from one process or time period to the next. Experience with processes in a variety of industries shows that this is not a very realistic assumption. In fact, for quality improvement the focus is often more on reducing process variation than on changing a process average value. Another difficulty with ANOVA is that the test statistic typically used will tell you when there are significant differences among process average values, but will not tell you which mean differs from which. To discover that requires another set of statistical methods called multiple comparisons, which takes us beyond the scope of this book. There are several good references on this subject (3, 9).

Most statistical packages now on the market offer a variety of multiple comparisons techniques. Normal probability plots, to be discussed later in this chapter, are an easy graphical alternative to ANOVA. Another alternative for the researcher who only occasionally finds the need to compare three or more average values is to use the two-sample method on each pair of means. There is one caution in using this approach. When testing at, say, the 5 percent significance level, 5 percent of the time the statistic will indicate there is statistical significance when none is really present. That means that if say ten comparisons of means are made and there are no real differences in averages among the processes, the chance of making the correct decision for all ten based on the test statistic is equal to $(0.95)^{10} = 0.60$. So there is a 60 percent chance of correctly finding no significant differences, and a *40 percent chance of incorrectly concluding that at least one pair of means is significantly different*.

VI. MULTIFACTOR TWO-LEVEL EXPERIMENTS

So far in this chapter we have considered experiments which analyze the effects of one factor (method of filtering, for example) on average response. A test statistic for comparing the average effect of two different levels of a factor was discussed in the section entitled "Comparing Two Means." Then the discussion was expanded to include an experiment involving several distinct levels of a single factor ("Comparing More than Two Means"). The experimental designs to be considered in the remainder of this chapter are used to analyze the effects of two or more factors. We will focus on experimental designs in which each factor is tested at two different levels (two methods of filtering, for example, not three or four). This may seem rather restrictive, but many experimenters working to improve processes have found that two-level experiments are usually the most effective. They provide maximum information in a minimum number of experimental trials. Unlike three- or four-level experiments, two-level experiments cannot detect nonlinearity in functional relationships. For example, the relationship between pump speed (rpm) and pressure produced may appear as a curve when plotted, but since a two-level experiment would only use two pump speeds, that situation would not be detected. However, since two-level experiments require so few trials it is often possible to do several two-level experiments at less cost than one three- or four-level experiment. The data from the series of two-level experiments can be used to estimate the nonlinear relationship. In the next section an experimental design for use with three experimental factors will be presented. Experiments involving two factors at two or more levels each are discussed in references 3, 8, 9, and 10.

VII. THREE-FACTOR TWO-LEVEL EXPERIMENTS

The effects of three factors on a response variable can be evaluated in as few as four experimental trials. However, a more useful experimental design includes all possible combinations of levels, so there are $2 \times 2 \times 2 = 2^3 = 8$ trials in the experiment. The two levels used for a factor may represent two selected values of a continuous factor (temperature, pressure, time, thickness, etc.) or two different discrete possibilities (machine A or B, operator Amanda or Brook, new method or old method, water in cooling sleeve turned on or off, acid dip prior to applying coating or no acid dip, two or three coats of finish). Although these controllable factors can be continuous or discrete, the process characteristics which are functions of these factors and which will be observed during the experiment (the responses) should be continuous variables. There are exceptions to this. For example, if a large sample is used at each combination of the experimental factors, the percent of the sample which is ''defective'' can be used as a response variable. More advanced methods and special techniques permit use of discrete and multidimensional variables as response variables. But this goes beyond the scope of this chapter, which was written with the intention of presenting some of the basic statistical tools for experiments.

The trials which comprise a two-level eight-run design are listed in Figure 4. We will study this design rather closely since many important characteristics of good experimental designs can be illustrated with it. According to Figure 4, the first trial would be run with all factors at level 1. In the second trial, factors A and B would be set at level 1 and factor C at its level 2, etc. The level designations of 1 and 2 are arbitrary and need not imply that level 1 has a numerically lower value than level 2.

Trial Number	Levels of Factors A	B	C	Response
1	1	1	1	y_1
2	1	1	2	y_2
3	1	2	1	y_3
4	1	2	2	y_4
5	2	1	1	y_5
6	2	1	2	y_6
7	2	2	1	y_7
8	2	2	2	y_8

Figure 4 Design matrix for a three-factor, eight-run experiment.

Before an experiment is performed, the order in which the trials will be run should be determined randomly. This can be done by writing the numbers 1 through 8 on eight slips of paper, putting the slips in a bowl and taking them out randomly one at a time. Run the eight experimental trials in the order in which the slips were drawn.

It is important for the integrity of an experiment that the trials be run in random order whenever possible. If the trials are run in the order listed in Figure 4, unsuspected factors which change with time may distort the analysis and result in misleading conclusions. For example, suppose one of the chemicals used in an experiment deteriorated over the course of the experiment or suppose unmeasured environmental conditions such as humidity or dust increased. Such changes might cause the observed responses to be significantly different in the second half of the experiment than in the first. Since factor A was set at level 1 during the first four trials (using standard order) and level 2 during the second four trials, this difference in response might be incorrectly interpreted as being due to the change in level of factor A. Randomizing the order in which the trials are performed minimizes the chance of this sort of thing happening.

The *effect* of a factor on a response variable is the change in the response when the factor goes from its level 1 to its level 2. With the above eight-run design we can estimate the effect of each factor by finding the average value for the response variable at level 1 of the factor, and also at level 2 of the factor, and then taking the arithmetic difference between these two average values.

For example, in order to estimate the effect of factor A on the average value of the response variable, we first add together the four observed responses at level 1 of factor A (y_1, y_2, y_3, y_4 in Figure 4). We then divide the sum by 4 to obtain the average response at level 1 of factor A. This average is denoted by $\bar{A}_1$. In a similar manner we obtain the average of the four response values at level 2 of A ($[y_5 + y_6 + y_7 + y_8]/4$), and denote it by $\bar{A}_2$. The estimated *effect* of A on the average response is then:

$$\text{Effect of A} = (\text{Average at level 2 of A}) - (\text{Average at level 1 of A})$$
$$\bar{A}_2 - \bar{A}_1.$$

The effects of factors B and C on the response variable can be similarly estimated:

$$\text{Effect of B} = (\text{Average at level 2 of B}) - (\text{Average at level 1 of B})$$
$$\text{Effect of C} = (\text{Average at level 2 of C}) - (\text{Average at level 1 of C})$$

The experimental design in Figure 4 has the property of being *orthogonal*, which means the effects of each of the factors can be estimated without fear that the estimates are being distorted by effects of other factors. We know this experiment is orthogonal because for each pair of factors, each combination of factor levels is used in an equal number of trials. For example, in Figure 4 we see that factors A and B are both set at level 1 for two trials, A is at level 1 and B at level 2 for two trials, A at level 2 and B at level 1 for two trials, and both A and B at level 2 for two trials. For further discussion on orthogonality of experimental designs see References 2, 5, 8 and 9.

Example: Increasing the Yield of a Nitration Process

[This example is based on an experiment analyzed on Davies (5, pp. 258–259).] A laboratory study was undertaken to increase the yield of a batch nitration process. The product formed by this process is used as a base material for a variety of dyestuffs and medicinal products. The study considered the following three factors: time of addition of nitric acid, time of stirring and effect of "heel" (residual material in a mixing pan left over from the previous batch). The effect of heel is important in that it can be time-consuming to remove all residual material after each batch, and it would be more economical to allow some heel to remain if it did not adversely affect yield.

Each factor was run at two levels during the experiment. The levels used for the factors are listed in Figure 5. Every combination of factor levels in Figure 5

Factor	Level 1	Level 2
A: Time of addition	2 hours	7 hours
B: Time of stirring	0.5 hour	4 hours
C: Heel	no heel	with heel

Figure 5 Factor levels for nitration experiment (from Ref. 5).

was used in the experiment. This means that $2 \times 2 \times 2 = 8$ experimental trials listed in Figure 4 were performed. These trials are also listed in Table 3. The factor combinations used are in columns 3 to 5. The eight trials were carried out in the random order indicated in the first column of Table 3. The resulting yields, as percent of theoretical, are given in the last column.

Based on the data in Table 3, the average response (yield) at level 1 of time of addition is

$$\frac{y_1 + y_2 + y_3 + y_4}{4} = \frac{87.2 + 86.7 + 82.0 + 83.4}{4} = \frac{339.3}{4} = 84.82$$

Similarly the average response at level 2 of time of addition is

$$\frac{y_5 + y_6 + y_7 + y_8}{4} = \frac{88.4 + 89.2 + 83.0 + 83.7}{4} = \frac{344.3}{4} = 86.07$$

Then

$$\begin{aligned}\text{Effect of time of addition} &= \text{(Average at level 2)} \\ &\quad -\text{(Average at level 1)} \\ &= 86.07 - 84.82 = 1.25\end{aligned}$$

Similarly,

$$\begin{aligned}\text{Effect of time of stirring} &= \text{(Average at level 2)} \\ &\quad -\text{(Average at level 1)} \\ &= 332.1/4 - 351.5/4 = -4.85\end{aligned}$$

Table 3 Nitration Experimental Design and Observed Responses

Random order	Standard order	Time of addition	Time of stirring	Heel	Yield
6	1	2	0.5	no	87.2
7	2	2	0.5	with	86.7
4	3	2	4	no	82.0
3	4	2	4	with	83.4
8	5	7	0.5	no	88.4
1	6	7	0.5	with	89.2
5	7	7	4	no	83.0
2	8	7	4	with	83.7

and

$$\text{Effect of heel} = (\text{Average at level 2}) - (\text{Average at level 1}) = 343.0/4 - 340.6/4 = 0.6$$

Since the average response is higher for the second time of addition than for the first level, this estimated effect is greater than zero. For time of stirring, on the other hand, the estimated effect is less than zero, indicating that the average response is higher at level 1 than at level 2. The estimated effect of heel (0.6) is so small relative to the other estimated effects that it should probably be viewed as due to random variation rather than as a "real" heel effect. The data suggests that in order to increase yields, time of addition should be set at 7 hours (level 2) and time of stirring set at 0.5 hours (level 1). Since heel apparently has no appreciable effect on yield, it should be set at the most economical level (probably level 2). There should, of course, be a subsequent confirmatory experiment in which the factors are set at their recommended levels and the actual yields observed. It might also be desirable to do additional experiments for time of addition between 2 and 7 hours to see if yields comparable to that observed for 7 hours can be achieved in less time.

VIII. REPLICATION OF EXPERIMENTS

If there is excessive variability in the process being studied, estimates of factor effects may be far from the true process average values. The effects of high variability on experimental results can often be reduced by performing an experiment more than once—that is, by "replicating" the experiment. There are at least three benefits of replicating experiments:

1. Average values have less variability than individual measurements, so our calculated averages will tend to be closer to "true" values.
2. Without replications, a single erroneous or unusual sample value can distort the whole analysis.
3. Data from replicated experiments can be used to estimate the level of variability in the process for each combination of factor levels used in the experiment. This information can then be used to identify which factors affect process variability and select combinations of factor levels which will reduce variability.

There are two basic ways to replicate experiments:

1. Take repeated measurements during each experimental trial.
2. Take one measurement during each experimental trial, and then repeat the process several times.

The first option has the advantage that it is often relatively inexpensive to take repeated measurements once the experimental trial is set up. This approach provides a good estimate of inherent process variability when all critical process factors are being held fixed. The second option is usually more time-consuming and expensive. But if the process is unstable and tends to drift over time, this second approach will provide a more realistic estimate of true process variability. Sometimes an experimenter using the second option will call each replication a "block" and will treat block as an experimental factor having as many levels as there are replications. Plotting block averages over time graphically illustrates stability of the process average.

IX. FACTOR INTERACTIONS

Very often the way in which a factor affects a response variable is independent of the levels of the other factors included in the experiment. But sometimes factors do influence each other. For example, a protective coating may be much more effective at low temperatures than at high temperatures. When the effect of one factor is influenced by the level of one or more other factors, we say there is an "interaction effect" among the factors. The estimates for factor effects defined earlier (which are often called *main factor effects*) do not measure interactions—this must be done separately. It is often possible to estimate interaction effects without performing additional experimental trials. For example, data from an eight-run experiment can be used to estimate the main effects of three factors plus all interactions among the factors.

The (main) effect of a factor upon a response variable is equal to the average value of the response variable at level 2 of the factor minus the average value at level 1 of the factor. The estimated *interaction* effect between two factors is calculated in much the same way. To calculate the interaction effect between factors A and B, say, the effect of factor A *at level 1 of factor B* is calculated. Then the effect of A *at level 2 of factor B* is calculated. Next the *difference* between these two effects is obtained. The interaction effect between A and B is, by convention, half the difference between these two effects. If "factor A" and "factor B" are interchanged in the calculations described in this paragraph, the resulting value of the interaction effect would be the same. Interaction effects are illustrated in Figure 6. Figure 7 shows the calculation of an interaction effect for the nitration process example.

When factors interact, it is critical that the interaction be considered when making recommendations. In the second illustration of Figure 6, for example, recommendations on how to adjust factor B to control the average response should take into account the level of factor A since we see that changing B from level 1 to level 2 changes the response by only 3 units if A is at level 1, but

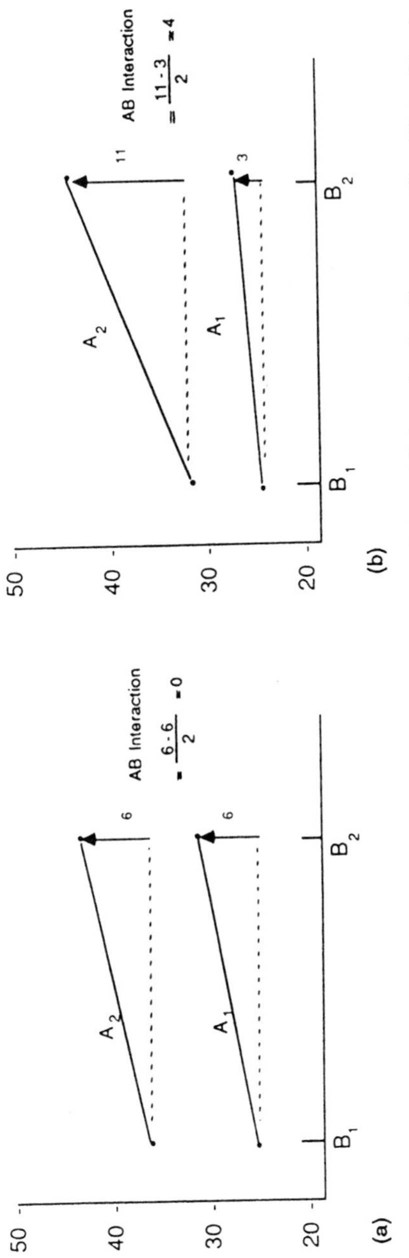

Figure 6 Interaction effects (from Ref. 8, reprinted by permission of the publisher, Quality Resources, One Water Street, White Plains, NY 10601).

TWO-FACTOR INTERACTION TABLE

Factors			
Add Time	Stir Time	Observed Values	Average
1. 2	1. 0.5	87.2 86.7	87.0
1. 2	2. 4	82.0, 83.4	82.7
2. 7	1. 0.5	88.4, 89.2	88.8
2. 7	2. 4	83.0, 83.7	83.4

TWO-FACTOR INTERACTION PLOT

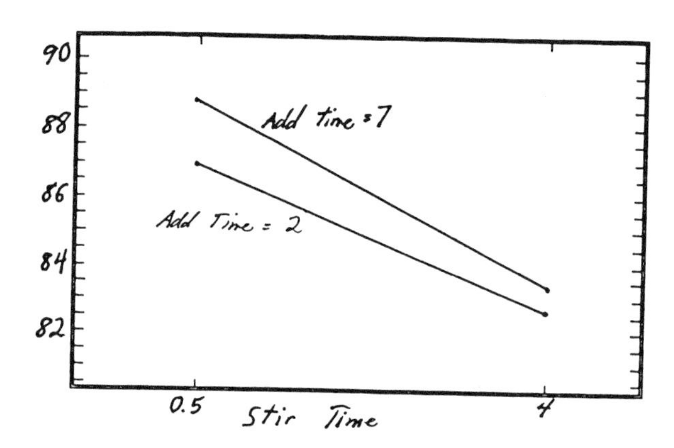

Figure 7 Interaction effects—calculation.

changes the response by 11 units if A is at level 2. The two lines on an interaction plot can have opposite slopes. So it is possible that, for example, when A is at level 1 the response is maximized when B is at level 1, but when A is at level 2 the response is maximized when B is at level 2.

With an eight-run experiment involving three factors, three main effects (A, B, and C) and three two-factor interaction effects (AB, AC, and BC) can be estimated. In addition a three-factor interaction (ABC) can also be estimated. The three-factor interaction effect measures the degree to which the interaction between any two factors is influenced by the level of the third factor.

Figure 8 is a *response table*, which was developed by Randy Culp of G.E. Medical Systems Division and used by Lochner and Matar (8) to simplify calcula-

Random Order Trial Number	Standard Order Trial Number	Response Observed Values y	A		B		C		AB		AC		BC		ABC	
			1	2	1	2	1	2	1	2	1	2	1	2	1	2
	1															
	2															
	3															
	4															
	5															
	6															
	7															
	8															
TOTAL																
NUMBER OF VALUES																
AVERAGE																
EFFECT																

Figure 8 Blank response table (from Ref. 8, reprinted by permission of the publisher, Quality Resources, One Water Street, White Plains, NY 10601).

tion of estimated effects. In the first column of the response table the order in which the trials were run is recorded. The second column gives the trial numbers in "standard order," which is the order listed in Figure 4. The response values are recorded in the third column of the response table. The remaining columns are used to calculate the various main and interaction effects. Each response value is copied into all the blank spaces to its right. Then column values are added and averaged. For example, in the fourth and fifth columns we see that y_1, y_2, y_3, and y_4 are used to calculate $\bar{A}_1$ and y_5, y_6, y_7, and y_8 are used to calculate $\bar{A}_2$. Differences between averages at levels 1 and 2 are calculated for each main and interaction effect and recorded in the last row of the response table. [It is not obvious that the columns in the response table assigned to interactions will in fact produce the desired estimates, but a little algebraic manipulation will show they are correct (8).] A completed response table for the illustrative example on increasing the yield of a nitration process is given in Figure 9. Notice, for example, that the response averages at levels 1 and 2 of the factor *time of addition* are 84.82 and 86.07, respectively. The *time of addition factor effect* is then $86.07 - 84.82 = 1.25$. The "grand average," $\bar{y}$, is calculated at the bottom of column 3 to be $683.6/8 = 85.45$.

Random Order Trial Number	Standard Order Trial Number	Response Observed Values y	A: Time of Addition		B: Time of Stirring		C: Heel		AB		AC		BC		ABC	
			2 hrs.	7 hrs.	½ hr.	4 hrs.	No	With								
			1	2	1	2	1	2	1	2	1	2	1	2	1	2
6	1	87.2	87.2		87.2		87.2			87.2		87.2		87.2	87.2	
7	2	86.7	86.7		86.7			86.7		86.7	86.7		86.7			86.7
4	3	82.0	82.0			82.0	82.0		82.0			82.0	82.0			82.0
3	4	83.4	83.4			83.4		83.4	83.4		83.4			83.4	83.4	
8	5	88.4		88.4	88.4		88.4		88.4		88.4			88.4		88.4
1	6	89.2		89.2	89.2			89.2	89.2			89.2	89.2		89.2	
5	7	83.0		83.0		83.0	83.0			83.0	83.0		83.0		83.0	
2	8	83.7		83.7		83.7		83.7		83.7		83.7		83.7		83.7
TOTAL		683.6	339.3	344.3	351.5	332.1	340.6	343.0	343.0	340.6	341.5	342.1	340.9	342.7	342.8	340.8
NUMBER OF VALUES		8	4	4	4	4	4	4	4	4	4	4	4	4	4	4
AVERAGE		85.45	84.825	86.075	87.875	83.025	85.15	85.75	85.75	85.15	85.375	85.525	85.225	85.675	85.7	85.2
EFFECT			1.25		−4.85		0.6		−0.6		0.15		0.45		−0.5	

Figure 9 Completed response table for nitrogen example.

A. Plotting Estimated Effects

It is generally helpful to prepare a graphical presentation of the effects estimated from experimental data. Figure 10 is a graph of the estimates calculated in Figure 9. Note in Figure 10 that:

1. A horizontal line is drawn to indicate the "grand average."
2. For each main and interaction effect a line is drawn between the average at level 1 and the average at level 2. The arrowhead is at the point representing level 2. The length of each line is equal to the magnitude of the effect being represented. Each line is symmetric about the grand average line.

These graphical representations help in determining which estimated effects are large enough to indicate the presence of a real effect rather than just random variation.

X. NORMAL PROBABILITY PLOTS

Although Figure 10 can be helpful in assessing which estimated effects are "real" and which are random variation, often such subjective evaluations are

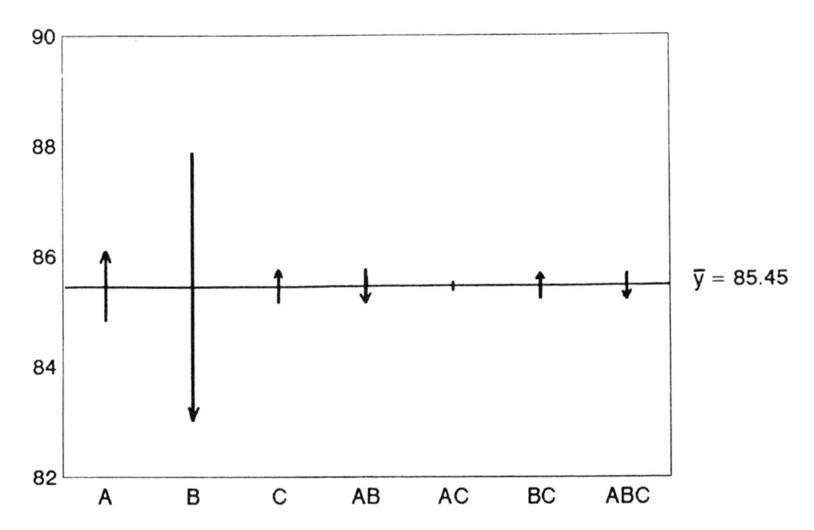

Figure 10 Plot of estimated effects for nitrogen example.

unsatisfactory. Statistical hypothesis testing, particularly analysis of variance, has traditionally been used to decide when statistical measures from experimental designs indicate significant effects. Section V.C on comparing two means, used a test of hypothesis to assess when a difference between two averages was statistically significant. But in recent years resistance has developed among both practitioners and theoretical statisticians to use of traditional hypothesis testing for business and industrial applications. The most frequently heard complaints about traditional hypothesis tests are:

1. Virtually any hypothesis can be rejected with a large sample, but it is almost impossible to find statistically significant differences with small samples.
2. Many test statistics, especially those used with design of experiments, assume normally distributed populations with fixed standard deviations for measurements. This is clearly false for many industrial experiments.

An alternative to some statistical hypothesis tests is the normal probability plot. This plot is not as dependent on assumptions of constant standard deviation or of normally distributed measurements. For an eight-run experiment there are seven effects estimated, if the interaction effects are included. Each of these effects is the difference between two averages. The Central Limit Theorem of statistics says that the sample averages for large samples are approximately normally distributed. Actually, for even small samples of four or five measurements it is common to see a normal distribution for sample averages. In terms of the Central Limit Theorem the seven effects estimated in an eight-run experiment can each be considered as an average of eight values. So we usually expect the averages to be approximately normally distributed. Further, if the factors included in the experiment had no real effect on the response values, then the seven estimated effects should represent an independent random sample from a single normal population. When a random sample from a normal distribution is plotted on special graph paper called normal probability paper (see Figure 11) the points display approximately a straight-line pattern. If some factor or interaction has a large effect on the response values, the absolute value of the estimated effect will be large and the corresponding point on the normal probability plot will not follow the straight-line pattern of other estimated effects. Use of normal probability plots was first proposed by Cuthbert Daniel and is discussed in Daniel (4).

Example: Normal Probability Plot

Continuing the nitration process example the seven estimated effects, as calculated at the bottom of Figure 9, are:

$$1.25,\ -4.85,\ 0.6,\ -0.6,\ 0.15,\ 0.45,\ -0.5$$

These estimated effects are recorded *in increasing order* above the normal probability plot in Figure 12. The points plotted in Figure 12 have these estimated

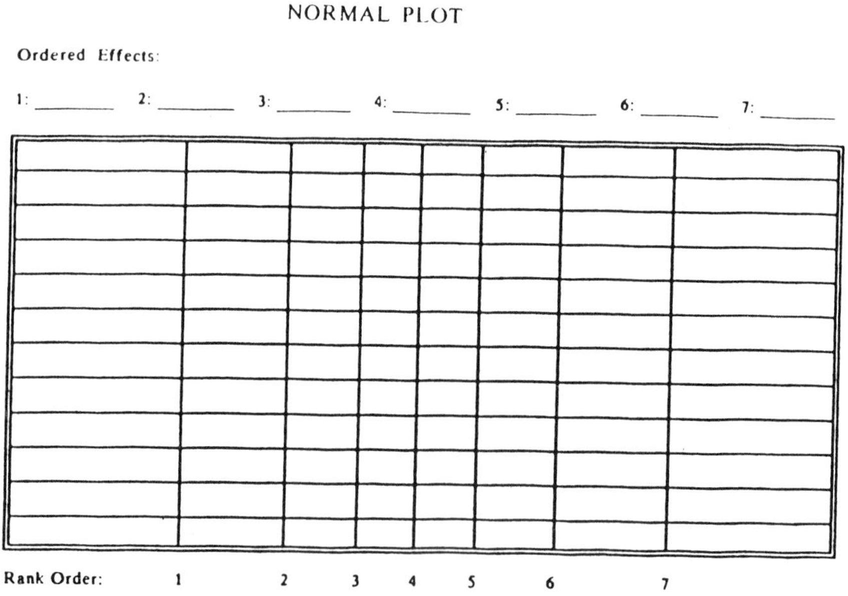

NORMAL PLOT

Ordered Effects:

1: ______ 2: ______ 3: ______ 4: ______ 5: ______ 6: ______ 7: ______

Rank Order: 1 2 3 4 5 6 7

Figure 11 Normal probability paper.

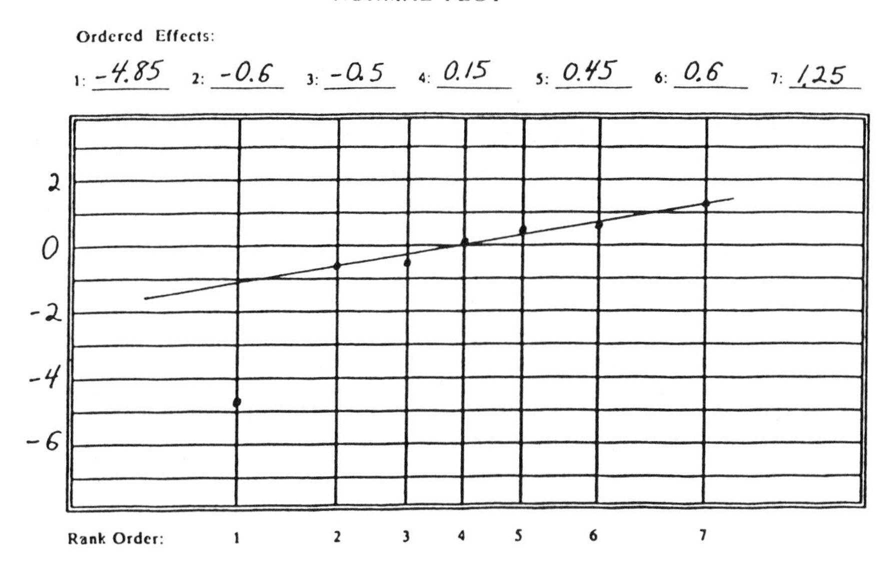

Figure 12 Normal probability plot for nitrogen example.

effects values for their vertical coordinates. The horizontal coordinates are scaled so that estimates which reflect only random variation will follow a straight-line pattern. Figure 12 suggests that only the effect due to time of stirring is real.

Three cautions should be made regarding interpreting normal probability plots. First, there is a loss of inferential precision, in that we cannot say a point deviates significantly from a fitted straight line at some specific level of significance. Instead, we are reduced to statements such as "the point *appears* to depart from the straight-line pattern established by the points for estimated effects of smaller absolute value." Second, fitting a line to, say, four points may give a very different line than one fitted to five or six of the points, and it may not be clear which line to use. Third, if an estimated effect is *not* clearly away from the line, this is *not* sufficient evidence to conclude with certainty that the effect is not real. For the example we have been considering, time of addition may well have a real effect on yield even though the effect isn't large enough to show up on the normal probability plot. These problems with interpretation make more traditional methods such as analysis of variance look rather attractive. But we need to keep in mind that methods which give us more precise answers ("cannot reject the hypothesis of no effect at the 10 percent significance level," for example) depend on assumptions which are frequently untrue in the real world, and are intended for use with one-shot experiments rather than with the ongoing, iterative approach to process improvement used today.

XI. EIGHT-RUN EXPERIMENTS WITH FOUR OR MORE FACTORS

Suppose an experimenter wanted to assess the possible effects of four factors. If each factor is tested at two levels, and all possible combinations of factors are considered, $2^4 = 16$ experimental trials would be required. Five factors would require $2^5 = 32$ experimental trials. Experiments of this type are frequently done. But if only some of the factor combinations are used, a considerable reduction in number of trials is possible. The factor level combinations for three factors in eight runs was given in Figure 4. These factor combinations were also represented in the response table in Figure 8. Figure 13 expresses the levels used in calculating the main and interaction effects as given in the response table.

The seven columns in Figure 13 are *orthogonal* to each other since if any two columns are selected, when one column is at level 1, the other column is at level 1 half the time and is at level 2 half the time. Similarly, when one column is at level 2, the other column is again at level 1 half the time and at level 2 half the time. So far we have used columns 1, 2, and 3 for determining experimental levels for the three design factors (compare columns 1 to 3 of Figure 13 with the A, B, and C columns in Figure 4). But other columns could be used as well. For example, in order to assess the effects of four factors in eight experimental trials, assign the fourth factor to column 7 in Figure 13. Then the design matrix would be as appears in Figure 14.

Analyzing four factors in eight runs involves a loss in ability to estimate effects. In particular, the last column of the response table formerly estimated

Trial	Level in Column							
Number	1	2	3	4	5	6	7	Response
1	1	1	1	2	2	2	1	y_1
2	1	1	2	2	1	1	2	y_2
3	1	2	1	1	2	1	2	y_3
4	1	2	2	1	1	2	1	y_4
5	2	1	1	1	1	2	2	y_5
6	2	1	2	1	2	1	1	y_6
7	2	2	1	2	1	1	1	y_7
8	2	2	2	2	2	2	2	y_8

Figure 13 Design matrix for an eight-run experiment.

Trial Number	Factor (Column) A (1)	B (2)	C (3)	D (7)
1	1	1	1	1
2	1	1	2	2
3	1	2	1	2
4	1	2	2	1
5	2	1	1	2
6	2	1	2	1
7	2	2	1	1
8	2	2	2	2

Figure 14 Design matrix for a four-factor, eight-run experiment.

the ABC interaction effect. Now it also estimates the D main effect. The D main effect and ABC interaction effect are said to be *confounded*. That is, the last column in the response table will estimate the combined effect of both D and the ABC interaction. Similarly, the fourth column of the response table formerly estimated just the AB interaction effect. But that effect is now confounded with the CD interaction. Other confoundings include AC with BD, and BC with AD. Confounding of effects can sometimes present serious problems with interpretation of data from an experiment, but often does not. See Box, Hunter and Hunter (3) and Lochner and Matar (8) for further discussion on this point.

In setting up an eight-run experiment for four factors, it is best to "put" the fourth factor in column 7 since this minimizes the potential damage due to confounding. If five or six factors are to be used in an eight-run experiment, the factors are better assigned to columns 4, 5, or 6. It is possible to include seven factors in an eight-run experiment by assigning a factor to each column in Figure 13. This is an example of a *saturated* design—this is as many orthogonal factors as you can squeeze into eight runs. With five to seven factors we encounter heavy confounding in the experimental design. For example if factor E is put in column 5, then the main effect E is confounded with the AC two-factor interaction effect. The effect calculated in column 5 of the response table estimates the sum of the E main effect and AC interaction effect, and possibly some other interaction effects as well. Further, if E is confounded with AC, then A is confounded with CE and C is confounded with AE. This opens up some serious possibilities for misinterpretation of data. There are several ways of dealing with possible confoundings. One is to examine previous data or rely on expert opinion to assess which factors are capable of potentially interacting. Another is to decide that interaction between two factors is unlikely unless at least one has a large estimated main effect. Still another is to approach the heavily confounded experiment as a "screening experiment" where the focus is on identifying which factors merit further investigation—a small main effect means the factor will be dropped for now from further consideration. Another possibility is to look upon all experiments as part of a process of continual learning and improvement. Then if a particular experiment produces confusing results, a second experiment can be run to clarify the issue. See Box, Hunter and Hunter (3) and Lochner and Matar (8) for further discussion on selecting follow-up experiments, particularly "fold-over" experiments.

Example: Reducing Voids in Die Castings

[This example is based on a project carried out at the Ford Motor Company's Rawsonville plant and reported by Duane E. Becknell (1).] A die casting which is part of the throttle body for a Ford 5.0L engine was subject to gas porosity voids. Although the actual scrap rate due to this defect was low, approximately 30 percent of the units had some visible voids. This was felt to be unacceptable and a team was organized to improve the situation. A scatter diagram of visual voids was done to identify where most of the casting voids occurred. Next a cause-and-effect diagram was prepared by the team to help them focus on root causes. Seven process parameters were identified as being the most likely candidates affecting occurrence of voids. The seven factors, and their experimental levels, are listed in Figure 15.

A saturated eight-run experiment was performed to identify which of the seven factors were most important, and at what levels these factors should be set in order to reduce the incidence of voids in the castings. The response variable used in the experiment was the percent of a sample of castings exhibiting any visual

Factor	Level 1	Level 2
A. Metal cleanliness	present	degas
B. Shot size	3 inch	2 7/16 inch
C. Intensification	1000 psi	1500 psi
D. Spray pattern	1.125 oz	0.875 oz
E. Profile velocity	63 ips	121 ips
F. Die temperature	250° F	350° F
G. Metal temperature	1280° F	1220° F

Figure 15 Experimental levels for die castings experiment (from Ref. 1).

gas porosity. The design matrix is given by Figure 13, where factors A through G were placed in columns 1 through 7, respectively.

The completed response table is in Figure 16. A graph of the estimated effects is given in Figure 17. Based on the experiment the project team recommended that:

B: Shot size be set at Level 2, 2⁷⁄₁₆ inches.
C: Intensification be set at Level 1, 1000 psi.
D: Spray pattern be set at Level 2, 0.875 oz.
E: Profile velocity be set at Level 2, 121 ips.

The project resulted in a 73 percent reduction in voids, at an estimated annual cost savings of over $200,000. The project took two weeks and required an estimated 120 person-hours.

XII. USING EXPERIMENTS TO REDUCE VARIABILITY

The discussion and examples in this chapter have focused on process/product average values. But the two-level experiments described here can also be used

Random Order Trial Number	Standard Order Trial Number	Response Observed Values y	A: Metal Clealiness		B: Shot Size		C: Intensi-fication		D: Spray Pattern		E: Profile Velocity		F: Die temperature		G: Metal Temperature	
			present	degas	3"	2 7/16"	1000	1500	1.125	0.875	63	121	250	350	1280	1220
			1	2	1	2	1	2	1	2	1	2	1	2	1	2
	1	14.2	14.2		14.2		14.2			14.2		14.2		14.2	14.2	
	2	39.3	39.3		39.3			39.3		39.3	39.3		39.3			39.3
	3	9.3	9.3			9.3	9.3		9.3			9.3	9.3			9.3
	4	27.0	27.0			27.0		27.0	27.0		27.0			27.0	27.0	
	5	44.5		44.5	44.5		44.5		44.5		44.5			44.5		44.5
	6	37.2		37.2	37.2			37.2	37.2			37.2	37.2		37.2	
	7	14.5		14.5		14.5	14.5			14.5	14.5		14.5		14.5	
	8	15.8		15.8		15.8		15.8		15.8		15.8		15.8		15.8
TOTAL		201.8	89.8	112.0	135.2	66.6	82.5	119.3	118.0	83.8	125.3	76.5	100.3	101.5	92.9	108.9
NUMBER OF VALUES		8	4	4	4	4	4	4	4	4	4	4	4	4	4	4
AVERAGE		25.22	22.45	28.00	33.80	16.65	20.62	29.82	29.50	20.95	31.32	19.12	25.07	25.37	23.22	27.22
EFFECT			5.55		−17.15		9.20		−8.55		−12.20		0.30		4.00	

Figure 16 Response table for die casting example.

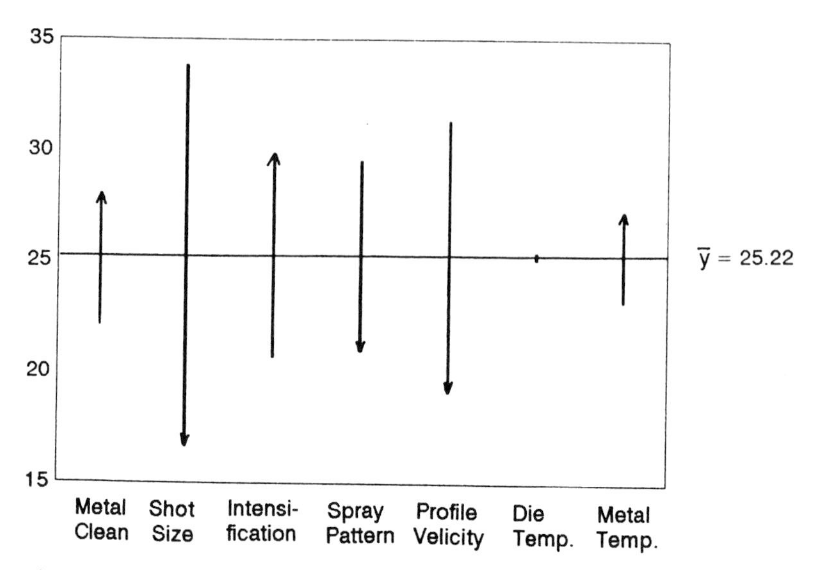

Figure 17 Graph of estimated effects for die casting example.

to attack problems related to variability. Instead of using observed measurements or averages of observed measurements, a function of the sample standard deviation can be the response variable. This means that for each experimental "trial" there must be multiple measurements so that a sample standard deviation can be calculated for each combination of factor levels used in the experiment. The *logarithms* of sample standard deviations are entered as the observed response values from the experiment (column 3 in the response table in Figure 8). *Logarithms* of the standard deviations are used because sample standard deviations are usually not normally distributed (the Central Limit Theorem does not apply to them), but for processes which produce approximately normally distributed measurements the logarithms of sample standard deviation, to any base, will be approximately normal. The completed response table and normal probability plot obtained using the logarithms of the sample standard deviations are interpreted in the same way as described in previous examples.

With processes which have excessive variability relative to specified requirements or tolerances, it is generally advisable to use statistically designed experiments to reduce the variability before using experiments to change average values.

If an experimenter wants to compare just two standard deviations, a special test statistic involving the ratio of the two sample standard deviations is used as the test statistic. See Box, Hunter and Hunter (3), Ott (10), or any basic statistics text.

XIII. BLOCKING EXPERIMENTS

Suppose an experimenter wants to perform an experiment involving more trials than can be done in a day, or with a given batch, or at a given location. To perform the experiment over several days could jeopardize the integrity of the experiment since factors beyond the control of the experimenter can change from one day or batch to the next. One way to resolve this problem is to treat days or batches as different levels of an experimental factor. Each day or batch is referred to as a block. Block factors can be treated the same as any other factor, although it is usually assumed that block factors do not interact with other factors. See Box, Hunter and Hunter (3) and Lochner and Matar (8) for further discussion.

XIV. MORE ADVANCED EXPERIMENTAL DESIGNS

In this chapter we focused on two-level eight-run experimental designs. There are many other experimental designs also available—for example, two-level designs in four, twelve, sixteen, and more trials. Lochner and Matar (8) discuss some of these designs at an elementary level and give response tables for four-, eight- and sixteen-run experiments. Box, Hunter and Hunter (3) discuss a broader array of such designs. Experimental designs involving factors at three, four, or more levels are also sometimes called for. Some three- and four-level designs are given in Lochner and Matar (8). A broader selection of experimental designs can be found in Box, Hunter and Hunter (3), Davies (5), and Montgomery (9). Also see Ott (10).

Better results can usually be obtained with a sequence of two-level designs than with a single higher-level design. An experiment involving five factors at three levels each, for example, will involve at least eighteen trials and will not provide any clear information on factor interactions. An eight-run two-level experiment, on the other hand, requires half as many trials and does give information on factor interactions.

A major drawback to two-level experimental designs is that they provide no information on nonlinearity of factor effects. One effective approach to this situation is to first use two-level experiments to identify the important factors and any interactions. Then additional experimental points are added as a second block to produce a combined experiment with factors at three levels. The data from the combined experiment can be used to estimate nonlinearity. For more specific information on this approach see Box and Draper's (2) discussions on "composite designs."

A more serious problem occurs when an experiment is unbalanced or there is missing data. "Unbalanced" refers to the situation where some combinations of factor levels are used more often than others. Unbalanced designs usually occur when the trials must be grouped into blocks which are too small to allow proper balance within the blocks or some experimental data must be discarded as invalid due to problems encountered during the experiment. "Missing data" refers to a situation where some combinations of factor levels do not appear at all in the final experiment. This may be due to experiments being terminated before scheduled completion, certain combinations of factors being unreasonable or unsafe to perform, certain combinations being specified by the client, or experimental data being invalid due to problems encountered during the experiment. Davies (5) and Montgomery (9) discuss how to handle such situations. When faced with unbalanced designs or missing data situations, the best approach is to seek the assistance of a competent statistician, preferably before the experiment is carried out.

XV. SUMMARY

This chapter provides an introduction to the tools available for designing experiments and analyzing experimental data. These methods can be used to improve product and process quality, increase productivity, reduce variability, and resolve issues related to interactions among factors. The chapter is not intended as a complete guide to experimental design. Extensions exist for each of the methods which have been presented, and references for additional reading have been provided.

A critical element for success in research is adequate control over systems and processes which are part of the research activity. Chapter 7 will address equipment setup, calibration, test procedure control, documentation and other issues which affect the validity and usefulness of experiments.

REFERENCES

1. Becknell, D., "Die Cast 5.0L Throttle Body Porosity Study," *Fifth Supplier Symposium on Taguchi Methods*, American Supplier Institute, 1987, pp. 301–312
2. Box, G. E. P. and Draper, N. R., *Empirical Model-Building and Response Surfaces*, Wiley, 1987
3. Box, G. E. P., Hunter, W. G. and Hunter, J. S., *Statistics for Experimenters*, Wiley, 1978
4. Daniel, C., *Applications of Statistics to Industrial Experimentation*, Wiley, 1976
5. Davies, O. L., *Design and Analysis of Industrial Experiments*, Oliver and Boyd, London, and Hafner, New York, 1963
6. Kepner, C. H. and Tregoe, B. B., *The Rational Manager*, McGraw-Hill, 1965
7. Kume, H., *Statistical Methods for Quality Improvement*, The Association for Overseas Technical Scholarship, Tokyo, 1985
8. Lochner, R. H. and Matar, J. E., *Designing for Quality: An Introduction to the Best of Taguchi and Western Methods of Statistical Experimental Design*, Quality Resources, 1990
9. Montgomery, D. C., *Design and Analysis of Experiments* 3rd ed., 1991
10. Ott, Lyman, *An Introduction to Statistical Methods and Data Analysis*, 2nd ed., Duxbury, 1984

7

PROJECT EXECUTION

George W. Roberts and Maureen J. Psaila-Dombrowski
Babcock & Wilcox
Alliance, Ohio

I. APPLICATION OF TOOLS

In this chapter, we will discuss some of the tools that may be employed to aid the researcher during the execution of a project. Obviously, not all tools are appropriate for every situation, but the reader is encouraged to consider each as the occasion arises.

A. Communication

From the perspective of project management, the most important tool is communication. The project leader will spend a great deal of time expressing needs and providing instructions to the research team and its suppliers. More time may be spent dealing with the team's customer, usually a sponsor, providing reports of technical and financial progress. The project team is supplied by the support sections at the research center and inherent with the position of an internal customer are some responsibilities. A customer's needs have to be clearly communicated to enable the suppliers to "do their thing." This may involve working with the plant facilities people to obtain proper test cells, plumbing, and utilities. Facility details are usually thoroughly reviewed during the preparation of a large proposal, and the plant engineer has some knowledge of what is needed, but the small job may pop up as an unpleasant surprise. Kerzner comments on the problem of informal project management as practiced in research and development (R & D):

A common mistake made during the planning phase in the informal type of project management is that the functional managers are not included in the planning of the project, even when the project involves work performed in their department. Time, costs, and procedures may be established for the functional department by the project manager without input from the functional manager. The functional manager can be difficult to work with, once he learns that someone else has planned his involvement and that he is supposed to adhere to these plans. (1)

B. Safety

Project personnel should all be familiar with and supportive of safety requirements germane to the work. Specifics such as personal protective equipment, equipment operation notices, hazardous chemical use, storage, and disposal, job safe practices, and special work permits can all have an impact on project performance and costs. For some business units, workman's compensation claims (or the lack thereof) have meant the difference between quarterly profit and loss. Safety concerns need to be understood beforehand.

C. Work Scope Changes

To improve budget performance, the project leader should insist on written authorizations for work scope changes and should issue concomitant authorizations to team members and suppliers. Regular project reviews should be scheduled (and documented) with appropriate internal management and the customer (sponsor). Monthly reports including schedule and budget performance should be issued to everyone with a reasonable need to know. The project leader should be prepared to stop the project if it becomes difficult or impossible to obtain approvals or needed information.

D. Experimental Configuration

Many of the basic needs of a project involve the ability to control the construction of a test article in order to verify or document characteristics important to experimental results. Inherent in this is the ability to recall/restore the experimental configuration. In any experiment, there is a possibility of losing track of how the experiment was set up. During endless trials and retrials, various equipment hookups are tried, different arrangements and orientations have to be checked out, and materials within a test section are used and substituted until just the right combination is found. It is easy during the intensity of preparing for a test run to misplace notes and measurement results of every change to the facility. Multiply the usual levels of excitement with the numbers of different workers and disciplines to bring a large test apparatus to readiness for taking data and there is ample opportunity to misplace information that will be sorely needed when reducing the data or writing the final report.

Monitoring several sets of metallurgical specimens through a variety of heat treatments and exposure to a variety of environmental conditions creates an enormous problem. Each specimen must be uniquely numbered and the individual processes followed closely to prevent mixing them. The experience researcher knows well the need for careful documentation and control of test arrangements and materials, regardless of the particular methods selected. Some prefer laboratory notebooks, others separate procedures or process forms, while others prefer the computerized approach. Most likely there will be a combination of all these and others.

E. Calculations

Calculations are used throughout the research process and are vitally important to substantiate the work being done. Typical discipline for calculations says that they should be identified by the project number and title, and should include date performed, originator's name, and a cross-reference to the experimental apparatus or the system or test phase to which the calculations apply. Nuclear standards require that calculations should be entered in a manner that is legible and in a form suitable for reproduction. They should be sufficiently detailed as to the purpose, method, assumption, design input, references, and units, such that the person technically qualified in the subject can review and understand the calculation and verify the adequacy of the results without contacting the originator (2). This detail should typically include:

1. Identification of the objective of the calculation
2. Identification of the design inputs and their sources
3. Documentation of assumptions and identification of those assumptions that must be verified as the calculation proceeds
4. Functional performance requirements: What is the range of the parameters to be investigated?
5. Quantitative values necessary for defining fabrication, installation, or testing
6. Results of the analysis: Do the results conform to the objectives established in the project technical plan?
7. Conclusions: How do the results compare to the predicted values?

F. Drawings

There is a need to establish some standardized techniques for generating and revising drawings. This will ensure that drawings receive proper reviews and are distributed to all who need them. Some sort of centralized drawing control may be desirable, regardless of whether one organization does all the design or whether the project leader does some drafting himself and has some done by others at the research center. Once the drawing has been completed, it is efficient to use a centralized function to ensure that drawing numbers are assigned and that appropriate drawing reviews have been performed. These reviews should include a verification of compliance with state and local building codes, and a review to ensure in-process verification of critical attributes.

Test section designs are highly fluid, particularly during the early phase of the project, and there should be provisions for rapid changes to drawings used for test section construction.

G. Materials

Any discussion of the construction or erection of a test section immediately leads into the area of material control. It is in that area that the most apparent departure from standard quality control techniques is observed. The term *apparent* is used because upon closer examination the differences aren't nearly as significant as one may assume. For a construction project such as the construction of a pressure vessel to an ASME code (3), the material is generally all collected into central areas and appropriately identified by tags or markings. Issuance of material from stock is closely controlled with markings properly transferred from the base stock item to the piece about to be machined or used. Often material control areas are segregated with security fences. There is documentation showing the use of the items issued on a given project or to a given person. This type of centralized stock control with limited access and controlled issuance is a vital part of material control, particularly for "code" construction. But, in a small research project, the person most likely to be involved with the material is the project leader. Since he is the one with the overall responsibility for the success of the research project, he might prefer to maintain his own controlled area for his project materials and keep them segregated from other projects. The end result is that throughout the research center there may be several storage areas maintained by the various technical sections and project leaders to assure that their materials are separate and available when needed. Adequate controls can still be applied to these decentralized project stores. Project leaders can maintain their own identification tags and markings on the project materials and prevent them from being mixed in with unidentified stock. Standard quality control tags, such as "hold for inspection" tags, green "accepted" tags, and red "discrepancy" tags can be conveniently used. If the project is to be constructed by a support group, such as the machine shop or welding ship, it may be prudent to establish some central storage areas. Then, when the support groups has to draw some material out of general stock for a project, that material will be available and appropriately protected. This type of system is more nearly similar to the ASME Boiler and Pressure Vessel Code material control systems.

The project leader should maintain a record of materials or components that affect the validity of the test results. The record should indicate if certified test reports or certificates of conformance are required. Any additional testing by the research center to verify material identity should be specified. Copies of material certificates should be included in the project files. If the material is for a project or a test facility which is liable to be at least semipermanent at the research center, copies of those material test reports should be on file. Material substitutions should be judged by the effect they have on the test results. State and local building codes may heavily influence whether material can be substituted or not. From a purely research standpoint, the adequacy of material substitutions may be determined by the project leader. If the material was specified in the project technical plan, the project leader should obtain approval for material substitutions from the same party(s) that approved the project technical plan.

H. Special Instructions

Special instructions such as those required for cleaning or handling test articles should be identified by the project leader in the contract proposal or the project technical plan. The project leader should either issue a technical procedure detailing these requirements or coordinate with the design engineer to have the requirements defined by drawing. The technical procedures would include provisions for recording inspection of the work, if such inspection is required. If drawings are used, the associated route sheets and inspection checklists document accomplishment of the action and the inspection.

One of the primary activities of an R&D Division is the development of processing techniques which may not yet be defined by any set code or standard. The refinement of an existing process may be such as to advance the state-of-the-art above what is considered a production standard normally regulated by existing codes. This may be the reason for the project in the first place or it may be an ancillary process used to support the project (for example, welding of a pipeline system for experiments in flow characterizations). To the extent that special processing techniques are needed to support fabrication or to perform an experiment or test, the process technique should be documented in a formal technical procedure and approved by the project leader. Qualification of the equipment to be used should be directed by the technical procedure or by standard calibration techniques.

I. Personnel Qualifications

If personnel qualifications are required, such as for welding, appropriate qualification records should be maintained reflecting the examination results, the training records, and the maintenance of proficiencies. In those areas where the special process is to support the fabrication of a test section, existing codes may be acceptable. Those qualification processes with their documentation and testing systems developed for production work [i.e., ASME Section IX, Welding and Brazing (3)] should be quite adequate for the construction. However, if the special process is a part of the research itself, it will be up to the project leader to ensure that appropriate qualification criteria and testing methods are developed and documented. In many areas, the special process might be by laboratory personnel who are directly engaged in the research aspect of that process and who, by the nature of their work, may be qualified well above standards imposed on the production facilities. These qualifications should be documented and maintained at least in the personnel department, if not by the project leader. A word of caution: There may be some tendency by technical personnel to want to deviate from standardized qualification techniques or process control techniques because of their extensive experience in that field or the experimental nature of the work. But if the process controls or the qualification requirements are mandated by a code, then the code must be rigidly adhered to. The temptation may be to assume that because technical sections know how to write codes, they have the capability and automatic authority to deviate from those codes. This attitude must be carefully avoided.

J. Nonconformances

Nonconformances detected during the fabrication of test articles should be recorded. The article itself should be tagged with some significant identifier, such as a red tag, and segregated whenever possible. The project leader should then review the nonconformance to determine any adverse effect to the test results. The project leader should approve actions taken to correct the nonconformance since they may affect subsequent test results. The project leader must always be cognizant of nonconformance decisions to determine whether such decisions will cause a deviation from any of the customer requirements or any commitments made in the proposal or project technical plan. If a deviation will exist, the project leader has the responsibility to obtain approval from the customer before he approves the action. Obviously, nonconformances requiring supplier corrective action should be coordinated with Purchasing.

If there is a formal quality assurance or quality control organization, its proper role would be to verify that the appropriate corrective action has been implemented to fix the nonconformance and to assess whether long-term corrective action is warranted. This also permits the collecting of defect data and long-term analysis over various research projects so undesirable trends or chronic violations can be detected and corrected.

Forms such as nonconformance reports or corrective action reports are often used to report problems. It may be desirable to separate the forms used to report manufacturing defects related to fabricating the test article from systems compliance problems detected during audits in order to permit more specialized and rapid data collection and analysis. The basics of recording the discrepancy, identifying the action to be taken, and follow-up to verify implementation apply to any form used. If an incident or condition has occurred which could jeopardize the attainment of quality objectives, an appropriate report should be made to the cognizant project leader, group supervisor, or upper-level management.

II. PROJECT CYCLE

A typical interrelationship between project activities and support systems is shown in Figure 1. Here the project leader has responded to the customer's needs and developed an experimental concept which takes into account technical and quality criteria established by the customer, and standard systems used by the research center. Having received approval for the technical and quality assurance (QA) plans, the project leader is, as usual, the main impetus for accomplishing the task. In this case, the first support system called upon is detailed design of a test section. The technical and QA requirements are transmitted to a design group that prepares the test section design documentation including fabrication drawings and a bill of materials. This design is independently checked within the designs group for compliance with drafting standards and state and local building codes. The design package is then routed back to the project leader. The design is compared with the technical and QA plans to assure that test section operating conditions will be as agreed upon with the customer. At this point, portions of the design may require an independent check; for example, if they had not been specified by the technical plan and, therefore, not approved directly by the project leader's customer. (Calculations made by the project leader prior to giving the input to the designs group should have been checked if they would have affected the design group's approach.)

The next major support system is procurement, whereby the project leader, having determined from the test section design package which items are "make" and which are "buy," initiates requisitions to purchasing.

The "QA" diamond indicates a supplier evaluation and approval action. The project leader then performs or arranges for inspection of incoming materials. Inspection records, tags, and nonconformance reports are standard quality forms common to any QA operation—the difference is in *who* is doing the assurance. Whether it is more economical to have a full-time inspector depends on the individual research center.

The third major support activity is construction of the test section to technical requirements specified on the fabrication drawings with inspection recorded by checklists or route sheets as necessary. The calibration system is shown here for the first time supporting the project through instrument and gage control. Whether the calibration is adequate depends on the tolerances to be measured, the capability of the gage or instrument, and needs of the project leader. The project leader must consider these factors in selecting the proper instrument or gage as well as in reviewing the standard calibration procedure. All measuring devices have their limitations. Conversely, unnecessary cost and delay may result if an exotic

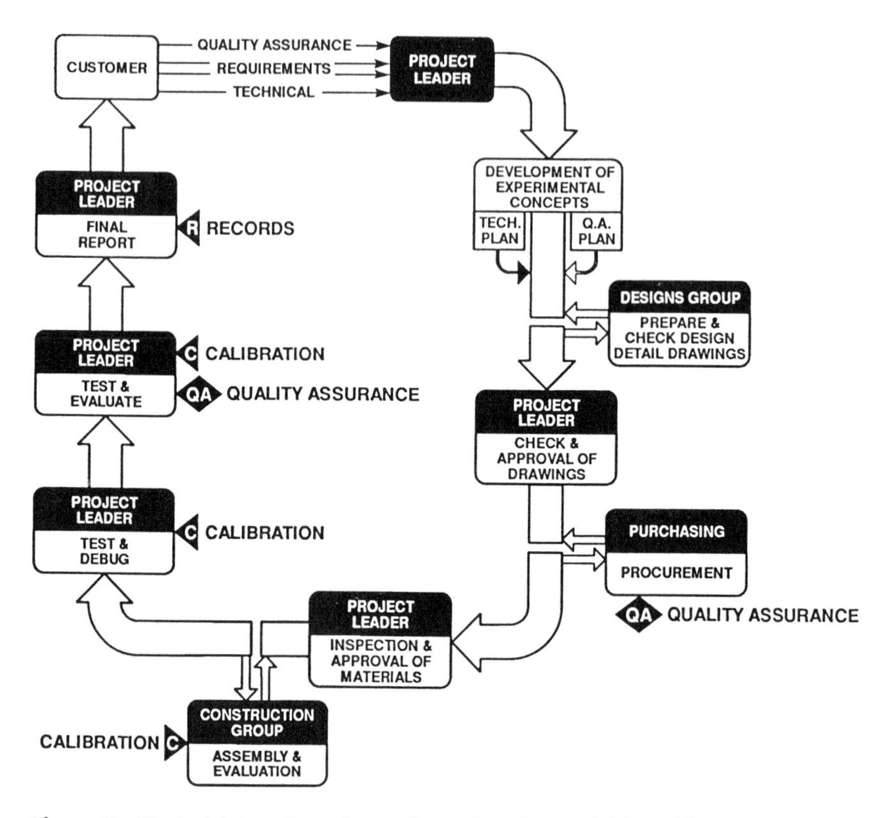

Figure 1 Typical interaction of experimental project activities with support systems (from Ref. 4).

measuring device is used for a simple, fairly loose-tolerance (but necessary) measurement. After erection of the test section, the project leader may then proceed to check out the apparatus prior to taking research data. Additional calibrations may be necessary, and if a long series of tests are planned, a recall and recalibration system may be needed. The QA diamond here indicates that at some point a decision may be made to perform an audit to verify that all quality systems are in place and effective. ''Records'' indicate data collection, storage, and retrieval of backup records supporting the final research report.

While the preceding example does not show all of the interactions to be expected on a given project, it does indicate the overall relationship between different groups and systems with the main thrust by project personnel.

III. PROCUREMENT

A. Evaluating Potential Suppliers

The R & D center may have access to a corporate Approved Suppliers' List or various lists developed by other operating divisions if the research center is a part of a large corporation. Using prior developed lists saves a lot of money in evaluating and developing unknown suppliers to meet your particular needs. Placing purchases through central purchasing allows for combining needs of several facilities and the advantage of bulk rates and more leverage on a supplier to meet your special needs. The disadvantage is that R & D centers may only need a small piece here and there, and cannot wait the long lead times required for large-quantity orders. The smaller orders are placed with local retailers who can turn the order out within 24 to 48 hours. However, all too often, the local retailer will not have material certifications or will charge an exorbitant price to provide them. Then, what may be received is a certification prepared by a distributor which may purport to be a copy of the original analysis. Some construction codes do not permit such copies and require the analysis results to be contained on the letterhead stationery of the organization that actually performed the analysis.

A goal of the vendor evaluation and selection process for any research center is to find those suppliers who can provide special-order items on fairly short notice, and who are willing to provide the proper certifications, all for a reasonable price—no small accomplishment indeed. When developing new suppliers, the research center should give consideration to establishing centralized supplier history files. The purchasing organization often has such files, but just as often, they are limited to prices and achievement of delivery schedules. This is excellent information in evaluating a supplier's performance, but it doesn't go far enough. Quality problems should also be included in the centralized files and their associated costs of investigation and resolution should be used together with the quoted price as a basis for evaluation and placement of a supplier on an Approved Suppliers' List.

Project leaders should be encouraged to add their experiences with their suppliers into the central history files. During the course of a research project, a project leader may have several interactions with a supplier of specialized test equipment. The results of those interactions, both positive and negative, should become a part of the research center's data bank. All too often, the results of these transactions are kept in the projects leader's personal files and remain unknown to anyone else at the research center. Since supplier capabilities and limitations may vary from one product line to another, particularly where different physical plants are involved, centralized evaluation files based on the supplier's overall performance will allow the comparison of different experiences with the same suppliers by different project leaders, and highlight the relative strengths and weaknesses within the same supplier structure. Some suggested criteria for evaluation and selection of suppliers are:

1. The supplier should have a good record of supplying acceptable items. The research center may use supplier history data obtained for other sources, such as corporate Approved Suppliers' Lists or other division's supplier quality history. The history should be for similar items, however, so the comparison remains valid. If the purchase is of a complex or high-cost item, the supplier should be evaluated by the technical section project leader and QA to determine the capability of the supplier to perform.
2. For some procurement, the research center may wish to perform an audit or, alternatively, to accept the supplier on the basis of a certificate of accreditation; for example, American Society of Mechanical Engineers' (ASME) Quality Systems Certificate for Materials; Certification to an International Standards Organization quality systems standard such as ISO 9001 (5); National Voluntary Laboratory Accreditation; or an evaluation performed by industry evaluation groups.
3. For commercial or off-the-shelf items, a receiving inspection may be sufficient to determine compliance with the procurement document requirements. This is somewhat risky, however, since you have to wait until you receive the product in hand before you know whether the supplier is any good. An alternate consideration, particularly for expensive items, would be to perform a source inspection at the supplier's facility prior to permitting the supplier to ship the item to the research center. If this is to be done, the supplier should be notified well in advance and given the specifics as to when the research center wishes to inspect the product, whether it's during the course of the fabrication process or at the completion of the fabrication process just prior to shipment. The advantage of this, of course, is that if there is something wrong with the item, the supplier has it at his facility and taking up *his* space instead of the research center's storage space. Additionally, the materials and personnel are available immediately to start correcting the item if correction is possible.

When setting up the system for procurement control at the research center, much effort must be expended in coordination meetings and discussions with all affected parties to assure that everyone understands the requirements of the system. This means getting together those persons who are primarily involved in the requisitioning process, the project leaders, the designers, and the purchasing agents, and going over the detail of the procurement control system item by item with appropriate examples, overheads, viewgraphs, or whatever is necessary to make sure everyone understands the system and its salient features. This is a key responsibility of a QA organization and, unless this understanding is achieved from the very outset, problems will crop up during the implementation phase beyond those that can normally be expected when instituting any new system.

B. Specification of Requirements

The supplier should be told specifically what is wanted, ideally *before* he is asked to ship the order. Although this sounds self-evident at first, in many cases purchasers are in such a great rush to receive their material, they will place the order over the phone and state that the purchase order will follow later. The supplier assumes he has all the information, proceeds to ship the order (usually taking the item from stock), but finds out when he receives the actual purchase order that some specific requirements have been added. Then the time saved by the verbal purchase order and quick shipment is lost because the material is held up for resolution of the discrepancy. Often, purchase orders are quite vague about what they're asking the supplier to provide. Casual statements such as "certification required" do not tell the supplier very much. What he needs to know is what *kind* of certification—does it involve a chemical analysis or physical test, or a certificate of conformance to a particular specification?

Purchase orders should be carefully reviewed to ensure they specify the material or item by appropriate ASME or ASTM (6) designation or catalog number, material grade or size, performance rating, pressure rating, or other functional or physical characteristics that are necessary. Quantity, length and weight, and identification of the test reports required should be spelled out carefully. When citing the procurement to a specific code, be sure to include the revisions or addenda of that code. The same rigor applies when purchasing calibrated equipment or calibration services. Just because the research center may be operating to some calibration specification, such as MIL-STD-45662A (7), do not assume that suppliers who calibrate measurement standards are also operating to the same specification. Nor should it be assumed that a certification of traceability to the National Bureau of Standards is any more valid for a calibration service than a certificate of conformance is for materials. For some levels of activity, these paper certifications are sufficient to establish some degree of confidence. But for highly critical experiments with high cost or safety-related consequences, a more thorough investigation of the supplier's capability behind the certificate is warranted.

Specific requirements should be identified to define the quality expected of a supplier of calibration services, since these services so directly affect the quality of the data at the research center. Such requirements should include a request for the supplier to list the calibration equipment used at the supplier's facility by serial number and a statement to the effect that those items have a verified uncertainty with respect to known naturally occurring phenomenon or artifacts maintained by the National Institute for Science and Technology. It should then be possible at a later date for the research center to verify by audit that those supplier standards were properly maintained. "Traceability" for a research center requires traceability of the *data*, not just an equipment check against a standard claimed to be four or ten times more accurate. Actual numbers must be on file to substantiate a numerical uncertainty statement. In those instances where units are being sent back to a supplier for recalibration, a check should be made by the supplier "as received" before any repairs of adjustments are made. A precalibration report furnished to the research center with the data points measured enables the research center to determine whether the equipment drifted out of calibration during use. The effect of that condition on any previous data that was taken by the instrument can then be assessed. If no repairs or adjustments are made to the unit, this precalibration check can then be used as the final calibration report. If there are repairs or adjustments made to the unit, the supplier then needs to furnish a final calibration report showing the final readings taken with the instrument. Finally, the supplier's procedures for the calibration should be referenced on the calibration report. The research center then has the option of reviewing those procedures later if they did not do so before the supplier was placed on the Approved Suppliers' List. Although many commercial calibration labs use standard calibration manuals supplied by their corporate headquarters, there may be a tendency on the part of the technicians in the local areas to modify the calibration steps to suit themselves. These modifications are not always condoned or authorized by the supplier's management. For that reason, it is prudent to require the supplier to use a written procedure which has evidence of management approval and to reference that procedure on the calibration report.

The formal specification of these requirements on a purchase order may come as a shock to many suppliers who have not been in business in the military, aerospace, or nuclear markets. It is an important function of a survey team to explain these requirements to prospective suppliers so they can understand their customer's needs and the impact of the requirements on the supplier's own operation.

Any organization anticipating the procurement of research should be concerned with the quality of the service. Like many processes, the quality to be designed and built into the deliverable end item may not be measurable at the time of delivery. The basis for achieving quality must have been laid down from the beginning and adhered to throughout the entire project. Quality is not applied retroactively by auditing the paper at the end of the work.

It is extremely important that there be a meeting of the minds between customer and supplier as to what is going to be accomplished. Too often, the customer project leader and the supplier project leader will become so completely engrossed in following the project technical details that they forget to pay attention to business. The money runs out before the test is complete and both sides wonder what happened. Then the finger pointing begins.

A statement of what is to be done including any verification steps and documentation requirements should be developed at the very beginning and adhered to throughout the project by both parties with written agreements as to what changes are needed as the project progresses and what the cost and schedule impact will be. These details can be contained in a project technical plan and QA plan and can be as brief or as lengthy as necessary to adequately cover the task.

Depending on the amount of control to be exercised by the customer and whether a supplier has a QA function, the requirements and commitments for verification and documentation of test articles and results can be included in a separate QA plan or written in with the project technical plan.

The decision to impose independent technical review of the test results and calculations depends on the reason for subcontracting the research in the first place. If the customer has the technical expertise, a sufficient review may be performed by the customer, provided there is access to the raw data taken and the customer can witness the data collection process. As with all quality-related activities, the more remote the customer is from the ''action,'' the more reliance must be placed on a supplier's internal independent QA function.

C. Procurement Documents

A key element of the procurement cycle is the procurement document. It may be called a purchase requisition or a purchase order, or, in some cases, a combination of forms all having the purpose of specifying the product or service to be purchased by the research center, along with appropriate technical and quality requirements to be met by the supplier. Many of the materials to be purchased for a project for R & D may not actually be critical to the quality or validity of the project data, but there are a sufficient number that are critical to warrant a formal system for review of procurement documents. This review may be performed by the QA organization if one exists, since it is familiar with the overall aspects of supplier evaluation and control and knowledgeable about the consequences of not adequately specifying material type, composition, markings, applicable codes, etc. At the time the project technical plan and the QA plan are developed, the project leader should have a general idea of which procurements will be critical to the results. If possible, those procurements should be identified for later reference when the purchase requisitions actually are generated. Purchase requisitions should be identified to the job identification number to allow a reviewer to trace back to the basic technical plan and QA plan requirements.

D. QA Review

A QA review would include verifying the inclusion of basic technical requirements, provisions for access to the supplier's facility for a source inspection or audit, documentation submittal and retention requirements, general requirements for establishment of a QA program at the supplier's facility, provisions for extending applicable requirements to lower tier suppliers, and requirements for material certifications or certificates of conformance. The review would also verify whether the supplier had been formally evaluated and approved, either by the R & D center, another company division, or a third party such as an accredited agency for the International Standards Organization (ISO) or the American Society of Mechanical Engineers (ASME). These requirements are similar to those imposed on production-type operations. The difference from an R & D standpoint is that the criteria depends on whether the procurement affects the quality or validity of the data.

Of all the procurement that can have an effect on research, the procurement of data-taking devices is of the most significant importance. Therefore, evaluation and control of suppliers of instrumentation becomes a significant issue. Since the normal output of the research center is data, the experimentation, data collection, and analysis functions of the research center are areas of critical importance. The procurement documents for instrumentation and gages for the research center should be reviewed.

IV. DOCUMENTATION

The reports published by a research and development (R & D) facility are the principal evidence of the quality of its performance and ability. Since personnel at the R & D center are judged by the contents of these reports, their writing ability must match their research ability in the laboratory. The production of effective reports, therefore, is a major part of professional performance.

Informal "letter" reports are typically used to communicate within a corporate structure and are intended to promptly transmit detailed results of a project to technical counterparts in business units. The format of a letter report is not rigid, but the data, operating procedures, equipment descriptions, etc., should be in sufficient detail to be understood by present and future technical specialists on the subject. Illustrations and tables are appropriate. Conclusions and recommendations may be omitted or partially covered, depending on progress of the work. A lengthy report probably indicates the need to subdivide. A summary is usually necessary at the beginning of the report, particularly when the report contains more than three pages of narrative.

"Formal" reports are intended to communicate the significant results, conclusions, and recommendations of an R & D project, or major part thereof, and are written for a broad spectrum of readers, particularly middle and upper management. They should be prepared annually, or at the completion of a major work phase, whichever occurs sooner. They include comprehensive summaries of the projects.

An introduction, results, discussion, conclusions, and recommendations are usually applicable to every formal report. Additional sections and/or arrangements are optional. Exhibits should be designed to communicate general ideas, and graphs and curves need not be readable with high precision. Appendices (perhaps in a separate volume of limited distribution) are preferred for voluminous tables of data and results. All pertinent ''letter'' reports prepared during the project should be included as part of the appendices.

Most beginning writers tend to order their material in a chronological or problem-solving sequence, just as they organized the research project itself. Unfortunately, this arrangement emphasizes the methods and equipment used, rather than the implications of the research. It is preferable, therefore, to alter the sequence with emphasis on giving management what they want most to know. In this case, the first sections would be introduction, results, discussion, conclusions and recommendations, followed by the remaining supporting sections and appendices. This outline requires more writing skill, since the details of apparatus, procedure, and analysis have not yet been introduced. However, once learned, the continuity of thought from problem through objectives, results, conclusions, and recommendations is preserved without interruption.

Project records should include any information, memos, meeting minutes, laboratory documentation, or special references that have a bearing on the direction the project has taken, the data acquired, or the interpretation of results. These would normally include the following: work orders and customer communications, purchasing documents, test article configuration, laboratory notebooks, and reports. To this can be added the specific records committed to by the project leader to meet quality assurance requirements and that are listed in an appropriate section of the quality assurance plan (see Chapter 5). As a general guideline, if a commitment was made by the research center to a customer or from a supplier to the research center, that commitment should be documented in the project files and compliance with that commitment should also be documented.

It is very helpful if a unique job or project number is assigned and used to mark each of the records. This allows them to be quickly collected and sorted for future reference. For large, long-term test programs involving many independent test phases, separate QA plans may be necessary to adequately identify the needed controls. If this is the case, it is recommended that the job traceability number be likewise modified so each QA plan has a unique number. Since the QA plan specifies the project records, the project phases can be accurately documented. Numbers are often assigned to formal research reports, but the reports are usually written at the completion of the project. Marking of records should begin when the records are generated. The retention of records is of concern because there is always the question, how much should be kept and for how long? Nuclear requirements specify retention for the life of the plant, which is usually assumed to be forty years, but could go on much longer depending on the time and effort required for complete decommissioning of the facility. ASME NQA-1 (2) describes typical lifetime records, among which are thirteen types directly involved with the design. Since the results of R & D are most often used to affect the design of hardware or systems, these document retention requirements should be looked at carefully. The application of R & D to the manufacturing installation, inspection, and operation is not intended to be downplayed, however, and research reports having a significant bearing on these areas must also be considered for long-term retention.

Not all R & D activities have the same regulatory environment as nuclear. But the potential for damage to the public and resultant litigation against the designer, manufacturer, or a research organization is at least as great in the non-nuclear field because of the general complacence toward these seemingly safer items offered in the marketplace. Automobiles have been shown to be more dangerous than nuclear power plants, and expensive recalls are almost commonplace, but there are few, if any, standards or guidelines for R & D. The background and thought put into nuclear standards are extensive and much of the rationale is applicable to any product line.

Generally speaking, research records should be retained as long as any of the vital records for a research center. If the data pertains to a specific product, process, or facility, the records should be retained as long as that entity is in use. Results of basic research should be retained as long as the research organizational entity exists.

IV. CASE STUDY: STRESS CORROSION CRACKING SUSCEPTIBILITY OF DIFFERENT HEATS OF ALLOY XYZ

A customer requested a series of tests be performed on different heat treatments of Alloy XYZ to evaluate and optimize the material for its resistance to stress corrosion cracking (SCC) in high-temperature caustic environments. The heat treatments were confidential so all test materials were supplied by the customer and the material was arbitrarily labeled A, B, and C to distinguish between the different heat treatments. Constant extension rate tests were performed on smooth tensile specimens and precracked compact fracture specimens. This provided information on susceptibility to SCC initiation and SCC growth. Tests were performed in an autoclave containing 10 percent sodium hydroxide (NaOH) solutions maintained at a test temperature of 300°C. In addition, the effect of 50 ppb of certain solution contaminants (lead, sulfate, and chloride) was also examined.

The solubility of lead in caustic solution was found to be dependent on the test solution oxygen content. The solutions had to be made up under deaerated conditions. This required that a mixing cell be built and a new procedure be written for this type of mixing process. In addition, the use of lead in an autoclave environment, the possibility of lead escaping into the test facility due to a pressure boundary breach, and the disposal of the lead-contaminated solution had to be addressed.

Because of the number of specimens to be tested and the time schedule imposed on the project by the customer, a new test fixture was to be designed and built to test multiple specimens simultaneously. Prior to initiating the research project, a preliminary design of the test fixture was generated and a budgetary estimate of the machining costs obtained. In order to save costs, the fixture was designed so that it could accept either smooth tensile specimens or compact fracture specimens simply by using different specimen grips. A series of pretests of the new equipment was also necessary to ensure that the load and displacement rate applied by the new test fixture was uniform. The cost of these additional tests was discussed with the customer.

The customer insisted that the frictional resistance of the pull rods in the pressure boundary sealing fittings be minimized because this factor is difficult to account for. A new sealing assembly was designed to minimize this frictional resistance. Because these fixtures represent the pressure boundary of the autoclave, their design had to meet the ASME Code (3).

After the scope, requirements, and cost of the project were agreed upon, a project technical plan and schedule were written and provided to the customer for approval. The initial project schedule is given in Table 1. Reporting and meeting schedules were set up to ensure that the customer was kept informed of project status.

A. Overall Project Description

The project was scheduled to take 18 months to complete and consisted of four phases (Table 1). Phase 1 involved the design and manufacturing of the various fixturing needed to perform the test, writing any technical procedures that were

Table 1 Initial Project Schedule

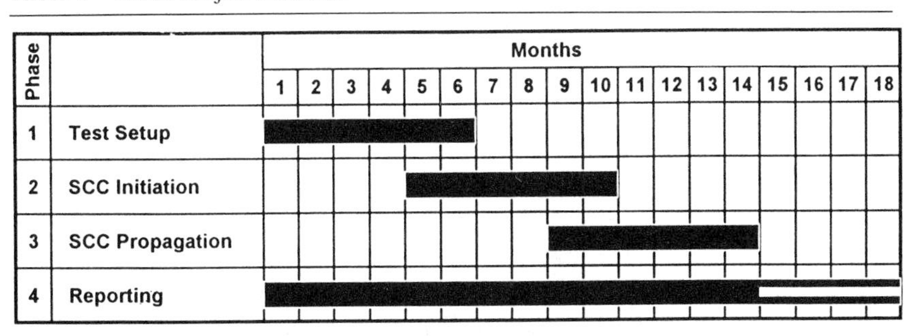

required, and examination of any environmental issues that arose from the use of lead. In Phase 2, the resistance of the alloys to stress corrosion crack initiation using smooth tensile specimens was evaluated. Phase 3 involved testing using compact fracture specimens to evaluate the resistance of the test materials to propagation of stress corrosion cracks. Phase 4 covered reporting to the customer, which included monthly and final reporting and scheduled meetings with the customer.

B. Phase 1: Test Setup

An initial meeting of the primary group of engineers and technicians involved in the project was held. Work and responsibility was assigned. Separate assignments included the design and machining of the multiple specimen test fixture, the design and construction of a new pull rod sealing assembly (including approval as a pressure boundary), specimen machining and precracking, solution preparation (including environmental issues concerning lead testing and solution disposal), and cleaning and preparation of two test facilities. The details of each assignment are outlined below.

Design and Machining of Test Fixture:

1. Preliminary design of test fixture (sketch completed during costing of project and preliminary estimate made).
2. Preliminary design review meeting (attendance included project leader, person assigned responsibility for this portion of the project, machine shop manager, technician who will be using the test fixture, and the designer)
3. Finalize design.
4. Final design review (same attendance as preliminary design review).
5. Obtain materials for machining.
6. Machine fixture, pull rods and specimen grips.
7. Do postmachine check on critical dimensions.
8. Perform preliminary tests on materials of known K_{ISCC}.

Design and Manufacturing of Pull Rod Sealing Assembly

1. Preliminary design (sketch completed during costing of project and preliminary estimate made).
2. Preliminary design review meeting (design review team included project leader, person assigned responsibility for this portion of the project, machine shop manager, technician who will be using the test fixture, engineer responsible for the evaluation of the pressure boundary conformance to the ASME code, and the designer).

3. Perform calculations to ensure that the design meets ASME code.
4. Finalize design.
5. Review final design (same attendance as preliminary design review).
6. Obtain materials.
7. Machine and assemble grips.

Specimen Machining, Precracking and Equipment Evaluation:

1. Obtain drawing of the specimens to be used in the test.
2. Obtain material from customer.
3. Have specimens machined.
4. Assign specimens a unique specimen number.
5. Measure and record critical specimens dimensions.
6. Precrack specimens according to standard laboratory procedure.
7. Clean specimens and put in desiccator.
8. Calibrate test equipment.

Solution Preparation:

1. Find material safety data sheets (MSDS) for chemicals used in the project and place copies in appropriate locations (or make available to personnel involved in testing).
2. Perform and document calculation to determine the level of vaporous lead that would exist if pressure boundary was breached during testing. Ensure that these levels are below action levels in MSDS sheets. If not, determine what safety procedures must be taken during lead testing.
3. Draft technical procedure for mixing contaminated solutions in a dearated manner.
4. Review technical procedure for mixing of test solutions.
5. Issue final technical procedure.
6. Construct mixing apparatus.

During this initial phase of the test program, a number of difficulties arose. The first problem came from machining of the test fixture. Even though a detailed cost estimate for machining and constructing the test fixture had been obtained before when the cost of the project was agreed upon, the money allotted was quickly spent before the test fixture was completed. This increased rate of expenditure was due to the combination of a number of events: The hourly charge-out rates for the machine shop increased, materials cost more than expected, and there was a complication in the machining process (equipment failure which resulted in a less efficient process being used for machining certain parts). Because of the imminent danger of overexpenditure, the machining process was halted and an estimate for completion of the work was obtained. This increase in cost estimate was given to the customer for evaluation and was approved (in writing). Test fixture machining was reinitiated and completed within the new budget. Schedule was not significantly affected. The post-machining measurements of critical dimensions showed that all specifications for machining had been met. A preliminary test was performed using alloy ABCD specimens in sea water. These specimens were used because their behavior is well understood and documented (also compact fracture test specimens were available, hence, reducing project costs). The preliminary tests indicated that the test fixture was performing adequately.

The pull rod sealing assembly design and manufacture proceeded relatively uneventfully. There was a complication in the design—the engineer responsible for ensuring that the design conformed to the ASME code was concerned about using a certain plastic in the pull rod seal. An initial meeting occurred between the engineer responsible for this part of Phase 1, the design engineer, the engineer responsible for calculations involving ASME code, and the technician involved with the project. After considerable discussion, it was decided that a new plastic could be located that would be better suited for the environment to which it would be exposed. In addition, a slight modification to the design was discussed that would make the fitting easier to assemble. A new plastic was found and the design modified to accommodate this change. The design review team met again to review the final design (particularly the design change and the new choice of plastic). There was no additional cost to the project. Since there was no indication of increased cost or a delay in schedule, neither the overall project leader nor the customer were involved in the negotiation process. The project leader was kept informed of progress with weekly memos. The final design was approved as a pressure boundary.

Specimen machining proceeded without any complications. The thirty smooth bar tensile specimens were received ahead of schedule. The specimens were stamped with a unique specimen number and measurements of critical specimens taken and recorded in the test log notebook. However, when the compact fracture specimens came back from the machine shop, five out of thirty specimens did not meet the drawing specifications. A critical length of the specimen (Figure 2) was smaller than required and the specimens did not meet the fracture mechanics criteria required for evaluation of K_{ISCC}. These specimens had to be remachined. Because of scheduling conflicts in the machine shop, the length of Phase 1 had to increased by three weeks. Since testing of the compact fracture specimens did not start until Week 8, this did not delay the overall test program. The customer was notified of the problem and an extension of Phase 1 was requested and received. A new project schedule was issued. Because of the underexpenditure in machining the smooth specimens, there was no increase in overall project costs. All of the test equipment to be used in the project was evaluated and

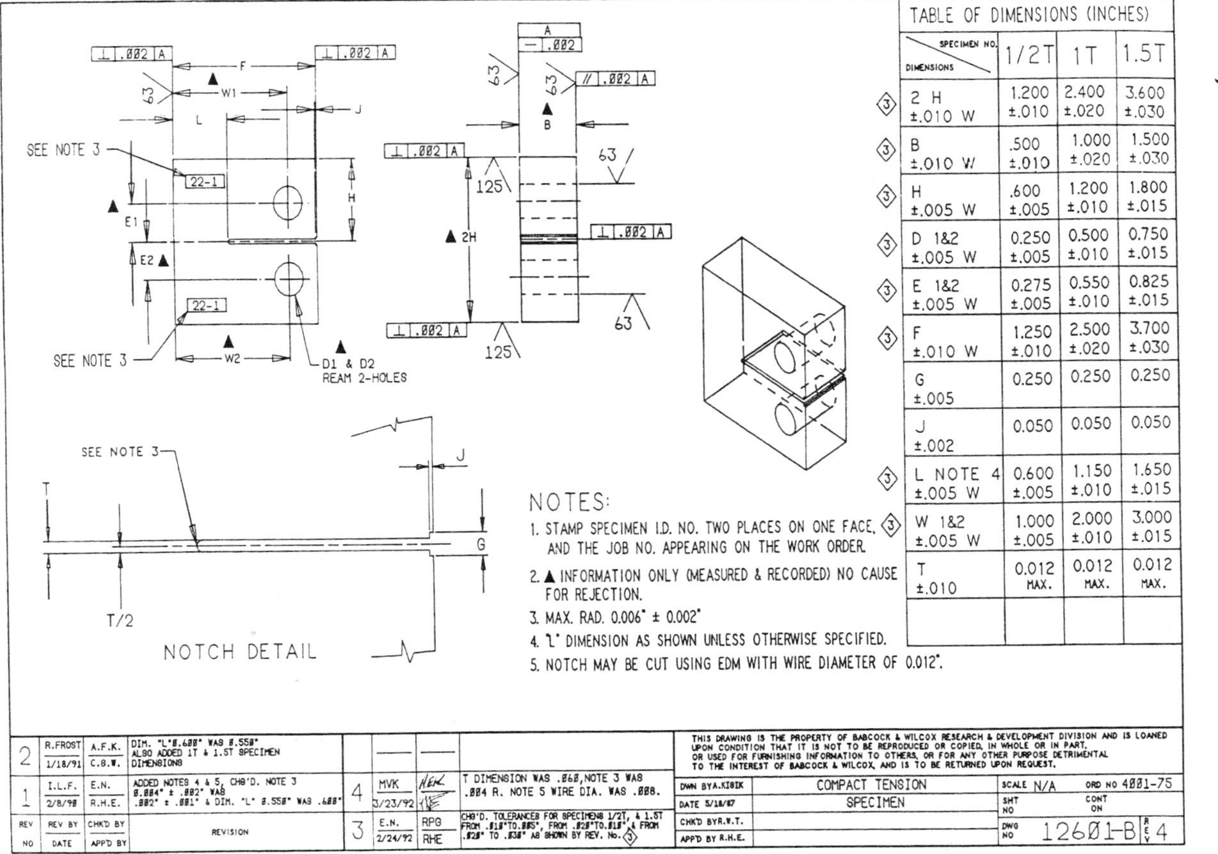

Figure 2 Fabrication drawing for compact tension specimen.

calibrated without difficulty. The test facility was set up and cleaned; a schematic is shown in Figure 3.

MSDS sheets for all chemicals to be used were obtained and evaluated. As anticipated, the only chemical that was going to be a problem was lead. After thorough evaluation of the MSDS sheet, it was decided that the worst case scenario would be a pressure boundary breach of the autoclave during testing and vaporizing and release of the test solution into the lab. To investigate the implications of this scenario, a calculation was performed to evaluate the maximum concentration of lead that would be in the laboratory in this event (Figure 4). The calculation procedure was discussed with the site environmental officer. The maximum amount of lead in solution and the volume of the laboratory were used in the calculation. The maximum amount of volatilized lead was well below the limits set in the MSDS sheet. The calculation was reviewed by the laboratory group leader.

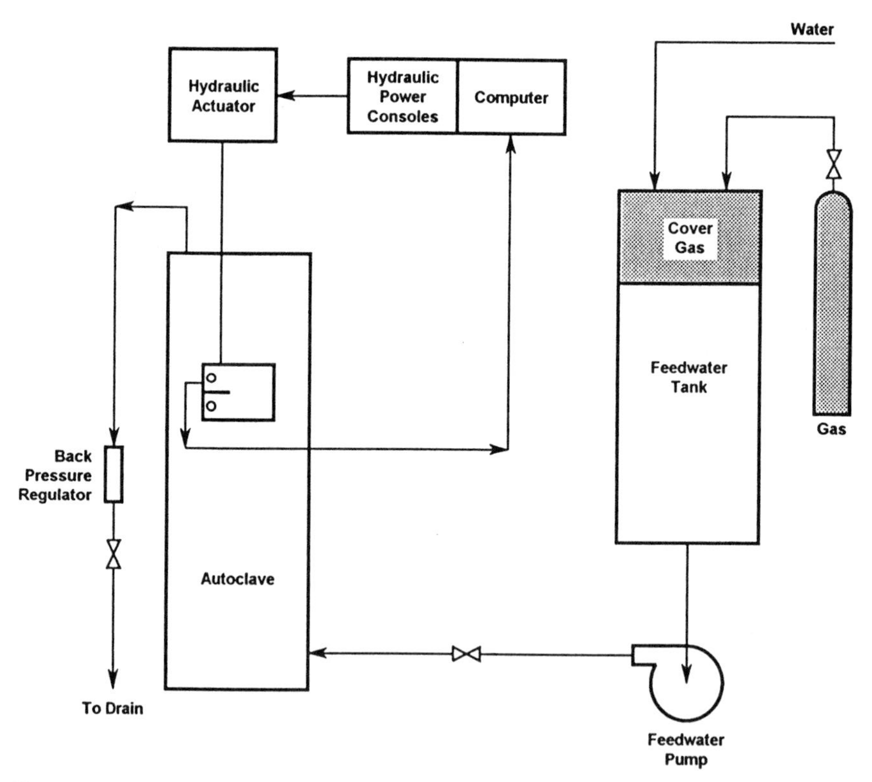

Figure 3 Schematic of a typical once-through refreshed autoclave system.

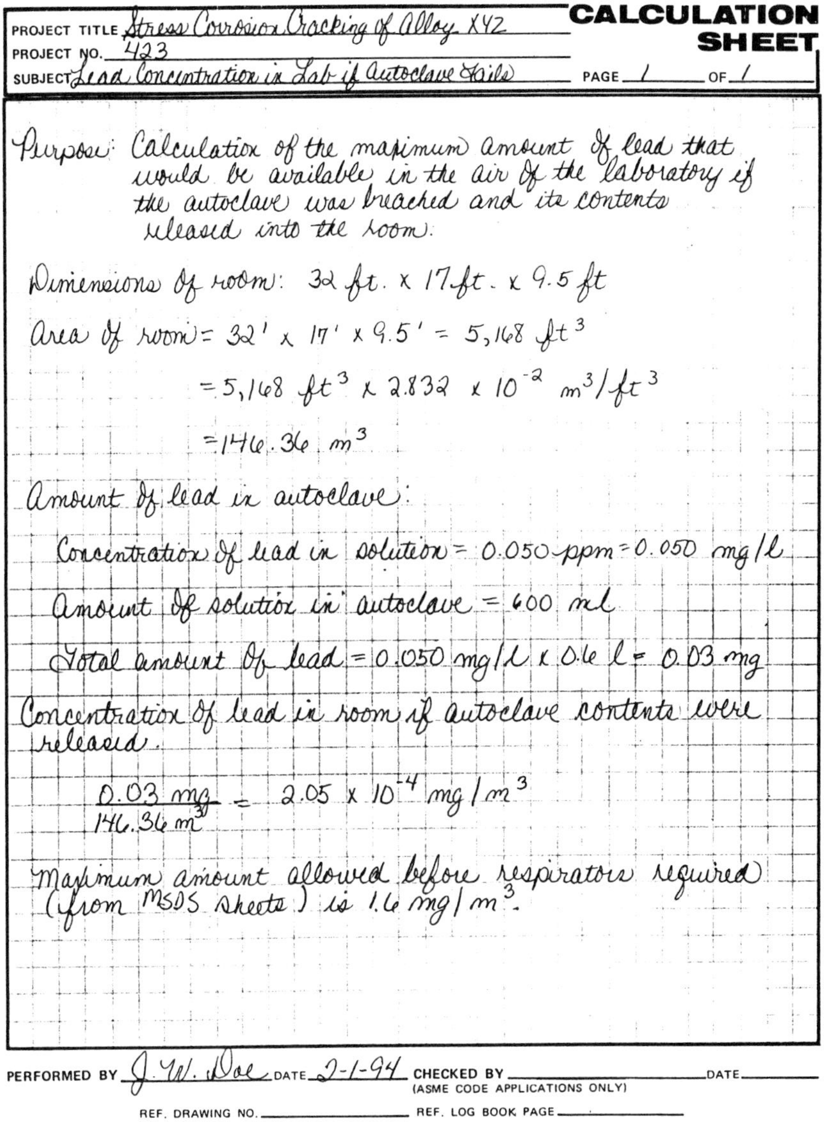

CALCULATION SHEET

PROJECT TITLE Stress Corrosion Cracking of Alloy X42
PROJECT NO. 423
SUBJECT Lead Concentration in Lab if Autoclave Fails
PAGE 1 OF 1

Purpose: Calculation of the maximum amount of lead that would be available in the air of the laboratory if the autoclave was breached and its contents released into the room.

Dimensions of room: 32 ft. x 17 ft. x 9.5 ft

Area of room = 32′ x 17′ x 9.5′ = 5,168 ft^3

= 5,168 ft^3 x 2.832×10^{-2} m^3/ft^3

= 146.36 m^3

Amount of lead in autoclave:

Concentration of lead in solution = 0.050 ppm = 0.050 mg/l

Amount of solution in autoclave = 600 ml

Total amount of lead = 0.050 mg/l x 0.6 l = 0.03 mg

Concentration of lead in room if autoclave contents were released.

$$\frac{0.03 \text{ mg}}{146.36 \text{ m}^3} = 2.05 \times 10^{-4} \text{ mg/m}^3$$

Maximum amount allowed before respirators required (from MSDS sheets) is 1.6 mg/m^3.

PERFORMED BY J. W. Doe DATE 2-1-94 CHECKED BY ______ DATE ______
(ASME CODE APPLICATIONS ONLY)

REF. DRAWING NO. ______ REF. LOG BOOK PAGE ______

Figure 4 Calculation of potential release of lead.

C. Phase 2: SCC Initiation

Four tests were conducted using six smooth bar specimens (two of each heat of the test material) in each test. The test specimen numbers are given in Table 2. Note that the test matrix in the table contains the unique specimen number that was assigned to each specimen; the first portion of the number refers to the project number 423 and A# is the designated specimen number from heat treatment A. These tests were performed in 10 percent NaOH without contaminants and with 50 ppb lead, chloride, and sulfate respectively. The tests performed in the 10 percent NaOH with 50 ppb lead proceeded without any problems. During the test performed in the 10 percent NaOH with 50 ppb chloride, the load cell failed on one of the pull rods. The test was continued until all specimens had failed so that data was only lost on one specimen. However, the load history of the specimen attached to the failed load cell was not known. This information was not critical to the test since the displacement history of the load frame was known and each specimen had a duplicate.

The customer was notified of the problem and wanted to repeat the test. The cost and schedule delay of the repeat were assessed. Because of the time constraints involved, it was not possible to obtain a load cell from the vendor normally used. Other vendors were sought but none of them were approved vendors for the research facility. The customer was provided with the following options: (1) wait for load cells from an approved vendor, (2) obtain approval status for a new vendor, or (3) accept responsibility for the use of nonapproved test equipment. The customer opted to use a nonapproved vendor and supplied the project leader with a letter accepting responsibility. The increase in project

Table 2 Test Matrix with Specimen Numbers

H.T.	SCC INITIATION STUDIES				SCC GROWTH STUDIES			
	None[1]	Pb[1]	SO4[1]	Cl[1]	None[1]	Pb[1]	SO4[1]	Cl[1]
A	423A1 423A2	423A3 423A4	423A5 423A6	423A7 423A8	423A9 423A10	423A11 423A12	423A13 423A14	423A15 423A16
B	423B1 423B2	423B3 423B4	423B5 423B6	423B7 423B8	423B9 423B10	423B11 423B12	423B13 423B14	423B15 423B16
C	423C1 423C2	423C3 423C4	423C5 423C6	423C7 423C8	423C9 423C10	423C11 423C12	423C13 423C14	423C15 423C16

costs and schedule were discussed with customer and approval to proceed was obtained. The schedule was extended two months for Phase 2 and the resulting project schedule is shown in Table 3. The new load cell was received and calibrated. The test was repeated and further testing proceeded without incident.

The results were well received by the customer. The results of the two specimens of each heat treatment tested together in the same solution were extremely close, hence the accuracy was good. It was clearly demonstrated that all the test heat treatments performed similarly in 10 percent NaOH without contaminants and with lead and chloride, but that heat treatment B performed did not perform as well as heat treatment A and C in sulfate as shown in Figure 5. The specimens from material B failed much sooner and at a lower stress than for A and C. In addition, examination of the gauge length of the specimens indicated that specimens from material B had many more SCC initiation sites than those from material A and C. All failures were intergranular SCC. Since in most applications, structures failure time by SCC is dominated by SCC initiation time, a structure made out of material B would fail sooner unless its crack growth rate was significantly slower.

D. Phase 3: SCC Growth

Four tests were conducted using six compact fracture specimens (two of each test material). As in the case of the SCC initiation tests, these tests were performed in 10 percent NaOH without contaminants and with 50 ppb lead, chloride, and sulfate respectively. The testing proceeded without any difficulty. However, the posttest calibration of the test equipment showed that one range of the controller did not meet specifications. An out-of-calibration report was issued. The implications of the size of the error for test results were evaluated by the project leader. The calibration error was small and only in one range of the instrument. Since

Table 3 Final Project Schedule

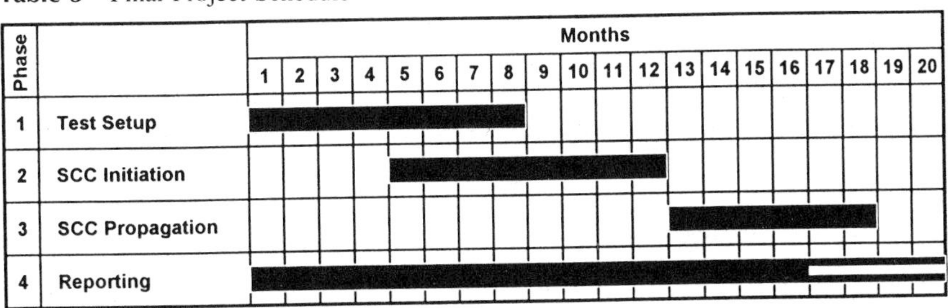

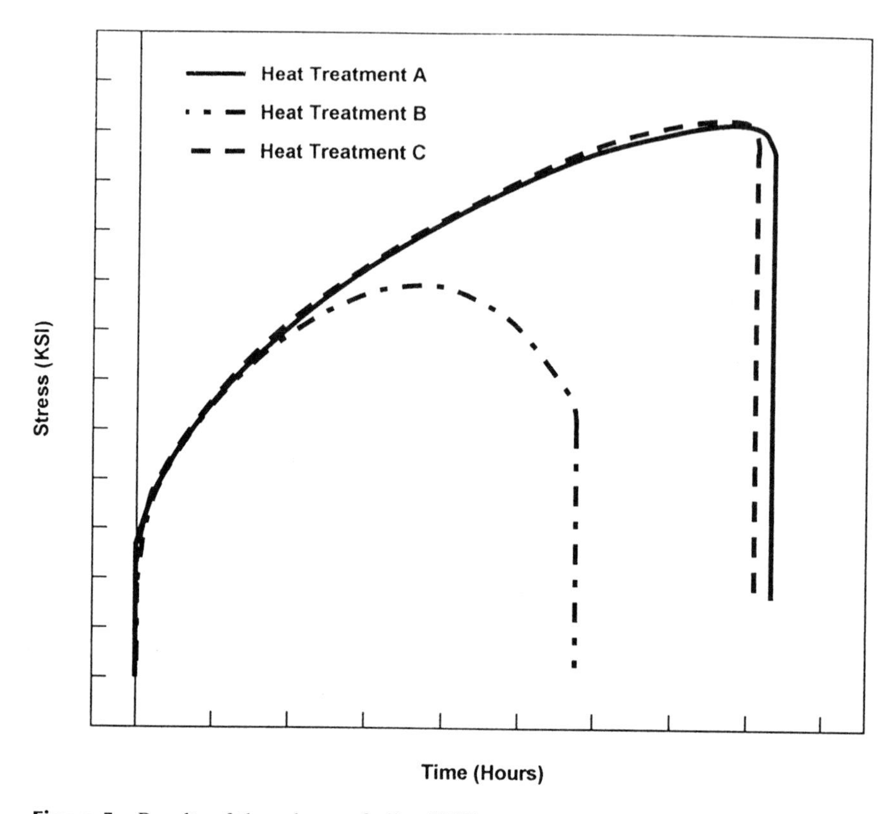

Figure 5 Results of three heats of alloy XYZ.

the range was not the one used at the end of each test, the error did not affect the final K_{ISCC} obtained. The customer was informed and accepted the test data.

The measured crack growth rates for the three test materials were very similar in each test solution case, but test material A had a slightly slower crack growth rate (6 percent). It was not clear if this was statistically significant. Hence, the materials could not be differentiated by crack growth rate alone. However, in conjunction with the SCC initiation tests, either test material A or C would be adequate. Test material A would be chosen from those two because of the slight difference in SCC growth rate.

E. Phase 4: Reporting

Monthly reports were issued throughout the duration of the twenty-month project. Each monthly report was issued within ten days of the end of the month. A draft annual report was issued thirty days after the end of the first year for customer approval. A final annual report was issued sixty days after the end of the first year. The draft final project report was issued thirty days after the end of the project and an approved report was issued ninety days after the end of the project because of delays with the approval process.

F. Case Study Summary

Experimental testing can be extremely difficult even without the added burden of paperwork, time, and costs associated with performing the project as described in this book. Nonetheless, the procedures described do ensure that the project proceeds in a manner acceptable to the customer. These procedures provide the researcher the paths and mechanisms for communication and delegation with the least chance of misunderstanding. In the current environment where priority is given to cost cutting and customer satisfaction, researchers find themselves performing a balancing act between the two forces. To remain competitive, the lowest possible costs are proposed to the customer, often neglecting documentation and cross-checking and perhaps cutting some corners. This can lead to experimental nightmares and unhappy customers. The key is to have enough in the project for reasonable project management and documentation and to keep the customer as informed as possible.

REFERENCES

1. H. Kerzner, *Project Management: A Systems Approach to Planning Scheduling and Controlling*, Van Nostrand Reinhold, New York, 1989
2. ASME NQA-1-1989, *Quality Assurance Program Requirements for Nuclear Facilities*, American Society of Mechanical Engineers, New York, 1989
3. *ASME Boiler and Pressure Vessel Code*, The American Society of Mechanical Engineers, New York, 1992.
4. G. W. Roberts, *Quality Assurance in Research and Development*, Marcel Dekker, New York, 1983
5. ISO 9001 (1987), *Quality Systems—Model for Quality Assurance in Design/Development, Production, Installation and Servicing*, ANSI Version ANSI/ASQC Q91 (1987), American Society for Quality Control, Milwaukee, WI, 1987
6. *Annual Book of ASTM Standards*, American Society for Testing and Materials, Philadelphia, PA.
7. Department of Defense MIL-STD-45662, *Calibration System Requirements*, U.S. Government Printing Office, Washington, D.C., 1980

8

MEASUREMENTS

Michael T. Childerson
Babcock & Wilcox
Alliance, Ohio

Rolf B. F. Schumacher
COAST Quality Metrology Systems, Inc.
San Clemente, California

In Chapter 6, the planning of an experiment was discussed. Statistically designed experiments were shown to provide a wealth of information in a relatively few experimental trials. An experiment was defined to be a sequence of trials or tests which produce measurable outcomes. In this chapter, the focus is on the measurements used in an experiment.

When investigators are using experimentally determined data in an analytical solution, or comparing the results of a mathematical model with experimental data, they will eventually ask "How good is the data?" Indeed, this question should be asked long before an experimental apparatus is constructed, and data are taken. If the solution to a problem must be known to within 5 percent, it makes no sense to spend time and money to conduct a test only to find that the likely amount of error in the results is considerably more than 5 percent. The planning of measurements is a critical activity in the experiment. Estimates of experimental errors are a part of this planning activity.

Measurement "uncertainty" is an estimate of experimental error. The concept of uncertainty is used to describe the "degree of goodness" of a measurement or an experimental result. In the performance of an experiment, activities are included to insure that the estimates of measurement uncertainty are appropriate for the acquired data. These activities are termed "data qualification." In this chapter, the concepts of measurement uncertainty and data qualification are described.

I. DATA QUALITY PLANNING

As described in Chapter 6, an experiment should be planned beforehand to establish an agreement as to what is to be studied and what characteristics are to be varied. Determination of the type and quality of data needed should involve key users of the data as well as those responsible for activities affecting data quality.

The planning should identify the data quality objectives, and how to achieve them. The overall test plan is driven by the design of the experiment to be performed. This experimental design does not always benefit from the same rigor as a formal review, such as design reviews performed to verify the physical design of the test facility, the prototype test article, or the production hardware. But if the overall test program is planned and documented in a technical plan and subjected to a competent review, there would be added confidence that the program is technically sound. This should be done prior to embarking upon an expensive test program.

A. Data Quality Objectives

The objective of measurement planning is to ensure one gathers useful, accurate, and reproducible data to the specifications in the project technical plan, or the governing project scope of work. The measurements critical to the success of the test must be identified for focus. Note that measurements do not have to be close tolerance to be "critical." The tolerance is defined by the acceptable uncertainty, which then sets the requirements for calibration.

Sound data quality planning can lead to reduced testing costs through early identification of problem measurements and techniques. For example, a level of redundancy in critical measurements can help point out faulty instruments prior to performance of an expensive test.

Finally, one must be capable of identifying data quality to the end users of the data. If, for example, the data will be used to assess the performance of a tool such as a computer code, the code analyst must be aware of the confidence that can be entrusted in the measurements. If a measurement is questioned in a legal proceeding, its "shadow of doubt" must be established just as with any other evidence. The experiment should include the necessary support measurements to enable one to make a clear statement of measurement uncertainty.

B. Meeting the Objectives

1. Measurement Planning

In planning the experiment, one should select measurements and qualification activities to enhance the capability to assess:

- Instrument performance
- Experimenter's adherence to the test procedure or test objectives
- Performance of control systems used in the test
- Overall test data quality.

It is highly recommended that one select a data acquisition system that lends itself to ''painless'' implementation of measurement system validation. The computer hardware and software should be flexible so that the user is encouraged to create tools to implement quick and useful measurement system checks.

One can meet the data quality objectives by incorporating redundancy in critical measurements. For example, try to include independent measurements of mass flow across test section boundaries for comparison. Grouping of redundant or related instruments provides for the use of Statistical Process Control (SPC) methods as a potentially valuable tool for tracking historical performance of a group of instruments. SPC is discussed further later in this chapter.

Examine the signal path for key instruments, and look for opportunities to gain confidence in the hardware components. For example, periodic checking of the data acquisition system analog-to-digital conversion equipment is highly recommended. In an experiment using a reference junction for thermocouple compensation, periodically monitor the reference junction temperature to increase confidence in the thermocouple-based temperature measurements. Better yet, include a recording of this reference junction temperature in the test electronic data base if possible, and you have a convenient way of verifying its performance throughout a test.

Perform pre-test checks on the measurement set at a frequency that provides acceptable confidence in instrument performance. The checks may be performed hourly, daily, or weekly. The frequency depends on the risk in acquiring data of unknown quality, and the cost of doing checks.

Perform post-test checks on the measurement set for each acquired data set. Repeat the pre-test checks, but expand the checks to include performance of any control systems used in the test. For example, if an automatic controller is used to simulate a particular test section boundary condition, verify that the simulation was acceptable by comparing the measurement with the ''desired'' value. The ''desired'' value may be based on the control equations. This type of check may include electric power versus time, or perhaps test section coolant flow versus local pressure. These checks provide confidence to proceed through the test matrix when the same systems are used repeatedly. More on these checks will be provided in Section III.

2. *Test Control*

Once the overall test plan has been agreed upon, specific tests described by the plan are defined and controlled by individual test procedures. These may include standardized tests, such as those published by the American Society for Testing and Materials (ASTM) and the American Society of Mechanical Engineers (ASME). These can be identified in the project files by number and revision, since they are readily available for future reference. The key experimental results will be obtained from other procedures developed especially for each test. Some may be very open-ended and must be developed as the test proceeds. The usual practice in this case would be to write a test procedure with only those steps and acceptance criteria that are known beforehand. As much of the detail of the test as is known is specified at the beginning of the test run, and then amended as necessary as the experiment proceeds. Obviously, in a situation where the phenomena are being characterized, there are no acceptance criteria. This is not a performance or qualification test to be passed or failed.

Test procedures should consider prerequisites, such as special operator training or certification requirements and test conditions that should be verified before the test begins. If tests are sequenced, the procedure should require verification of completion of prior tests before proceeding with the current tests. If there is a long series of tests to be performed to different test procedures, a route sheet or test matrix is beneficial for directing progression through the testing process, much as a shop route sheet (traveler) directs manufacturing processes.

A listing of typical test equipment to be used with the required range and accuracy should be included in the procedure, but the listing of actual equipment serial numbers should be avoided since, if units must be changed for any reason, the procedure must also be revised. It is important, however, to know what specific units were actually used during the test. A block diagram of the test setup is very useful. This information can be logged on data sheets or in the laboratory notebooks. If duplicate units are used, it is important to specify which serial numbers were used in which locations. If, during the recalibration of the test equipment after the test, an out-of-calibration condition is noted, the effect of that condition on the test results can then be assessed.

3. Documentation and Review of Results

The data should be recorded in some prescribed format on data sheets or in a laboratory notebook. Often, test results are processed directly from the test station to a computer data acquisition system and emerge as partially or completely reduced data. Supplemental computer codes may be used to complete the data analysis process. Just as calculations ought to be independently verified and test instruments periodically calibrated, the computer data acquisition system should be checked both as an instrument for calibration of the signal detection amplification and processing and as a computer with appropriate software verification and validation (see Chapter 9).

The results of experimental testing are evaluated by the project leader or project engineers under the direction of the project leader. Just as the calculations for designing the test apparatus should be verified, so should the calculations for analyzing and interpreting the data being reviewed. The emphasis, aside from mathematical correctness, is on whether the experimental objectives were met and whether the effects and interactions were interpreted correctly. This type of technical review is practiced in all types of research centers and is discussed in more detail in Chapter 10.

II. MEASUREMENT ASSURANCE*

The purpose of the following paragraphs is to present a brief discussion of the techniques involved in measurement process control. It is not possible to include all of the technical background material with which one should be familiar to set up and operate a measurement process control program. Instead, this section is intended to expose the reader briefly to the key concepts and urge the reader to consult additional reference materials for detailed information on these topics.

The crucial aspect of measurement process controls is that one has some way of monitoring in real time the quality of the measurements and detecting any sudden or gradual deterioration of the quality, either arising from an increase in the random error of the process or an increase in the systematic error of the process. The other aspect of measurement process controls is that one have a good quantitative estimate of the errors (or uncertainty) in the measurements made in the process. If both of these conditions are achieved, then one has measurement assurance of the process.

A. Evaluation of Uncertainties in a Measurement Process

In a measurement process, one normally compares an instrument or device whose properties are to be determined with some other instrument or device whose properties are known (a reference standard). This comparison process inevitably introduces uncertainties. So, one cannot determine with complete certainty the exact value of the item to be measured even if the reference standard's value is known exactly (which it is not). Furthermore, most reference standards and calibrated items are not perfectly stable. Their properties may change with time. For this reason, one of the most important aspects of any measurement process is the specification of the extent to which the value or values determined during the measurement process may be in doubt, i.e., the development of a valid uncertainty statement.

*From Rolf B. F. Schumacher, COAST Quality Metrology Systems, Inc., text from draft of ANSI/ASQC M1/M2 Appendix A.

The uncertainties associated with any measurement process fall into one of two categories, those which can be estimated by statistical methods and those that cannot be so estimated. They are frequently called uncertainties due to random and systematic errors, or uncertainties due to Type A and Type B errors. The total uncertainty must reflect both kinds of uncertainties, and thus it is obtained from an appropriate combination of best estimates of these two types of uncertainties.

In the application of the more traditional approach to measurements involving "accuracy ratios," one typically finds requirements to the effect that any instrument or standard used to calibrate another device must have an "accuracy" ten times better than that of the device being calibrated. Accuracy may be defined as the closeness of agreement between a measured value and the true value [1]. As an example of the accuracy ratio method, a device having a +/−0.5 percent tolerance might be calibrated using a standard having an accuracy of +/−0.05 percent or better, and it in turn might be calibrated using an item having an accuracy of +/−0.005 percent or better as determined by the standards laboratory. In such a process, one makes no attempt to determine precisely the actual uncertainties introduced at each step. It is assumed that because of the built-in "safety factor" of ten at each level, the measurement hierarchy will produce devices of adequate accuracy.

In most cases, experience shows that the accuracy ratio method works quite well, provided devices having the necessary accuracy are available and provided the "waste" of information is tolerable. By contrast, the measurement process control approach seeks to quantify the actual degradation due to uncertainty at each step in a chain of measurements. Uncertainty may be defined as an estimate of the error (or inaccuracy) based on some degree of confidence. The measurement process control approach also involves regular monitoring of the measurement process where deemed necessary and the introduction of feedback so that any measurement problems that may not be apparent when using the accuracy ratio approach will readily be identified as soon as they appear so that they may be corrected.

The whole purpose of a measurement process is to compare the properties of one reference item to an unknown item. It is obvious that one must be able to state quantitatively an upper bound to the possible uncertainties associated with this comparison if the results of such a comparison are to be applied with a high level of confidence. A measurement that has no statement of uncertainty associated with it is generally much less useful than one that has such a statement. In the case of measurements based on accuracy ratios, one does not know exactly what the uncertainty is, although one can usually make worst-case estimates. For many purposes such estimates are adequate, and in any case a measurement based on accuracy ratios is vastly superior to no calibration whatsoever. For particularly critical measurements where one needs to know the extent to which the measurement may be trusted, there is no substitute for a valid uncertainty statement.

1. Random Errors

Measurement uncertainties which can be estimated by statistical methods are referred to by various designations, including uncertainties due to random error or Type A uncertainties. Random error can be defined as a component of the error of measurement which, in the course of a number of measurements of the same item, varies in an unpredictable way. It is related to the precision or degree of reproducibility of a measurement process. It is a measure of the variability of the process. The concept is very easy to grasp if one thinks of a simple standard such as a mass standard. In measuring an unknown mass using a high precision balance and a set of known weights, one may make several determinations and average the mass values obtained. For high-precision weighing, one would not expect that all values obtained would be exactly the same down to the last digit; some scatter in the data would be expected. Suppose that one makes six mass determinations on one day and averages the readings. The size of the deviations of each measurement from the average is an indication of the magnitude of random error or imprecision. Some multiple of the standard deviation of the measurement data is the usual measure of the limits of random error, also being referred to, in this case, as "within-group" random error.

If one were to make an additional six mass determinations the following day, one would expect to observe again a certain amount of scatter in the data. The standard deviation of the readings for the second day may or may not agree with the amount of scatter in the data seen during the first day.

The average for the second day may also differ from that of the previous day. If the difference of the averages of the different days cannot be explained solely by the variability due to within-group random error, then another source of variability, a "between-group" random error, should be suspected to exist.

There are many sources of variability in most measurement processes, and so one may speak of day-to-day variability, month-to-month variability, operator-to-operator variability, instrument-to-instrument variability, etc. The point is that random error is not just a property of an instrument or standard, it is a property of the measurement process.

In order to quantify the variability of a measurement process, one must make repeated measurements of the same item or items. When the measurement of one item is based on a number of "observations," say three, four, or ten, the "within-group" standard deviation of the process can be determined from many such measurements taken over an appreciable period of time on a large number of different items. But normally, any one group of a limited number of observations can only be taken over a short period. The result is a measure of the short-term variability of the process. But the longer-term variability, the "between-group'' variability, remains indeterminable by a simple pooling of the within-group standard deviations of many of such groups. In fact, ignoring the between-group variability is perhaps the most common reason for underestimating measurement uncertainties. Measurement assurance approaches are usually necessary to quantify and control between-group random errors.

Frequently, as illustrated in Figure 1, the between-group random error may be much larger than the within-group random error so that five to ten observations may not yield much more information regarding the measurement uncertainty than one, two, three, or four observations. Thus, small within-group variabilities often deceive the unwary into believing that the measurement uncertainty is much less than it really is.

To obtain a measure of the within-group variability and the between-group variability, one has to measure a stable item repeatedly over a long period. A ''check standard'' is such an item. Thus, by using a check standard, one can obtain a good estimate of the total variability of the measurement process. At the same time one assumes that no unexpected systematic error offsets the process beyond the normally experienced limits of between-group variability. The available process control tools can show whether this assumption is warranted. The extent to which one gets the same answer each time the check standard is remeasured is an indication of the variability of the measurement process.

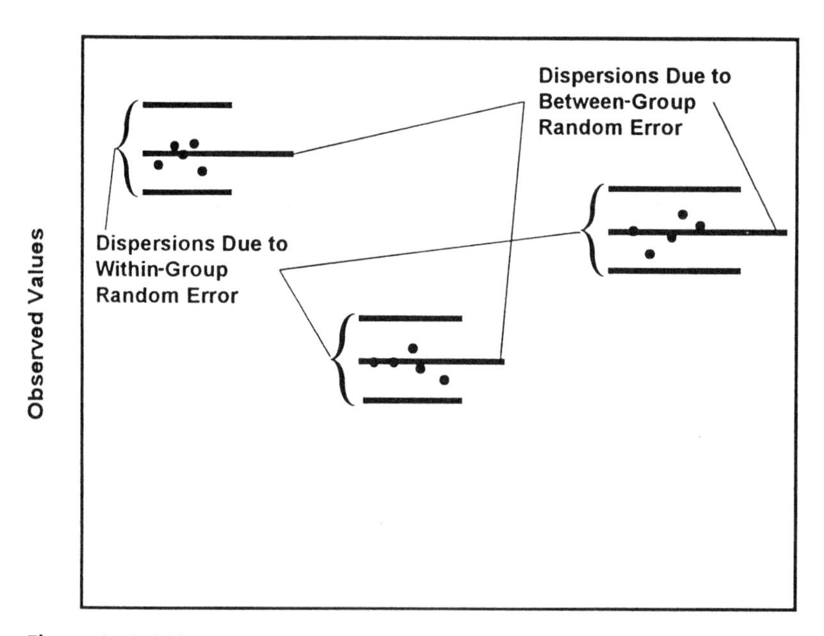

Figure 1 Within-group and between-group random error.

The mean value of a large number of measurements of the check standard may be compared with the value of the check standard obtained by a calibration performed with a measurement process which is independent of the process the check standard helps to control. In this way, an indication of the bias of the measurement process may be obtained.

This assumes, of course, that the measurements on the check standard are performed with the same process, i.e., in exactly the same manner as the measurements of the unknown items being normally measured with that process. If the measurements of the check standards are made with painstaking care, but the actual calibrations are performed carelessly, or with different techniques, then the data from the regular measurements on the check standards are not representative of the actual measurement process.

In later sections, the subject of control charts as a tool for determining and monitoring the random error of a measurement process is discussed.

Shewhart [2] has called the process which includes all sources of variability the "cause system." The metrologist is well advised to decide which sources of variability should be included in the cause system and write a procedure which ensures that the included sources are represented in the data sufficiently that they do not appear as assignable causes. For instance, different technicians may cause different variabilities, depending on their skill and care in making measurements.

2. *Systematic Errors*

The concept of measurement uncertainties which cannot be estimated by statistical methods is on the surface quite straightforward. A systematic error can be defined as a component of the measurement which, in the course of a number of measurements of the same item, remains constant or varies in a predictable way. It is simply an unwanted offset or bias that is present in the measurement process. The sign and magnitude of this error may be unknown.

In another method of determining and reporting measurement uncertainties, such uncertainties as must be quantified by means other than statistical methods are called Type B uncertainties. Such uncertainty components may be either fixed in sign and magnitude or varying as a function of some other influence. Often they vary as a function of time. Because they are offsets or biases, one obviously must answer the question "offset or biased relative to what?" That is, one must specify the reference base to which the measurements are referred. For example, the reference base for mass is maintained by the International Bureau of Weights and Measures, BIPM (i.e., the International Prototype kilogram). The reference base is usually specified by law to be a standard maintained by the national standards laboratory of the country in which the measurements are made. Thus, one is concerned with the possible offset of a particular measurement process relative to nationally maintained standards.

Consider an example of a systematic error: a laboratory calibrates gage blocks by using a mechanical comparator to compare unknown blocks to a working set of blocks which is in turn periodically compared to a primary reference set of blocks calibrated by the national standards laboratory. This laboratory calibrates its gage blocks at 20°C. If the calibration laboratory maintains its environment at some temperature close to but slightly different from 20°C, and the temperature coefficients of the blocks are unknown, then its calibrations could have a built-in offset relative to the national laboratory.

Doing an adequate job of estimating the total systematic error in a complex measurement process requires some degree of technical sophistication. One must understand the measurement process and be sensitive to the possible sources of uncertainty.

To some extent upper bounds to systematic errors can be estimated a priori by the experienced metrologist, and represent an upper limit of error that might exist without being discovered by the metrologist. (Unfortunately, it is the systematic errors that we do not know about that cause the most trouble!) Another method is to utilize calibrations by a standardizing laboratory to assess a systematic error component. If one measures a standard and gets the same answer as the standardizing laboratory within acceptable limits, and repeats this process several times with consistent results, then one can assume that there are negligible offsets present. If one does not get the same answer but observes a consistent offset, then one has identified a bias with known magnitude and sign. Such a bias can then be eliminated by adjustments or a correction and will subsequently no longer contribute to the uncertainty except for that part of the offset that still remains in doubt. (Note however, that a calibration of a laboratory's standard by a standardizing laboratory, by itself, cannot shed any light on the imprecision of the measuring laboratory's process, nor does it necessarily identify all sources of systematic error.)

Another type of systematic error encountered in measurement processes stems from the preceding calibration of the laboratory's reference standard. Suppose the imprecision of a standards laboratory's process is much larger than its systematic error, so that its systematic error can be ignored. Then the uncertainty in the calibration of a reference standard for the measurement laboratory derives entirely from random error. For the measurement laboratory using the reference standard, however, that uncertainty causes a possible bias or offset, so that the uncertainty is considered by some as a systematic error. But because the resulting uncertainty can be estimated using statistical methods, it may have to be classified as a Type A uncertainty in another method of uncertainty evaluation.

It is recommended that metrologists estimate a ''standard deviation equivalent'' for non-data-based errors which can then be combined with the standard deviation from the random error of the process, using the root-sum-square (rss) method, and multiplied by an appropriate factor to arrive at the total uncertainty

(3). In recommending this method, a BIPM working group aimed to give a uniform rule, by agreement, for all metrologists.

There is wide agreement among metrologists that it is desirable to quote limits of uncertainty due to systematic error and those due to random error separately if possible. If this information is provided, the user of the uncertainty statement can combine the components in whatever fashion is deemed appropriate for the application.

3. *Case Study: Uncertainty Estimate for an Electrical Current Measurement*

In a large-scale, thermal-hydraulic test program, [4] electrical current measurement was made for the simulated nuclear reactor heat source using the circuit shown schematically in Figure 2. Voltage drop across the shunt was input to the data acquisition system and used to calculate the current supplied to the core using the following equation:

$$I = \frac{E - E_0}{R}$$

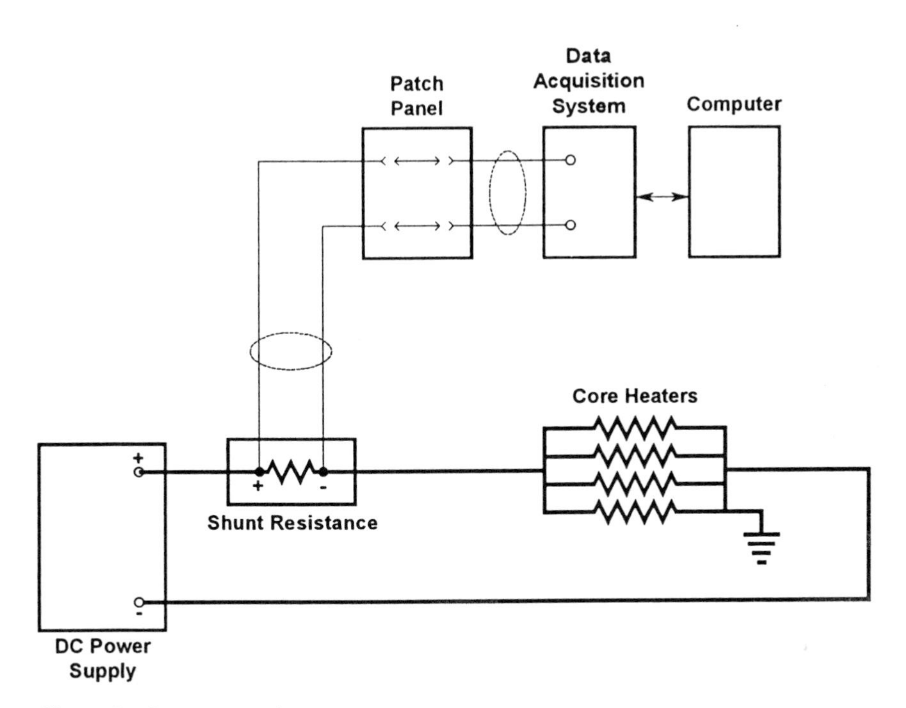

Figure 2 Reactor vessel core current measurement system.

where:

I = current to the core, amperes
E = voltage drop across the current shunt, volts
E_0 = voltage drop across the current shunt with no current through the circuit (zero value), volts
R = shunt resistance, ohms

The uncertainty in this current measurement included the errors associated with the following parameters:

- Uncertainty of the calibrated shunt resistance
- Error of the data acquisition system in measuring E and E_0.

The random component of the total uncertainty, S_I, *was calculated using Taylor's series expansion by the equation:*

$$S^I = \left[\left(\frac{\partial I}{\partial\,(E - E_0)}\, S_{(E-E_0)}\right)^2 + \left(\frac{\partial I}{\partial R}\, S_R\right)\right]^{1/2}$$

where:

$S_{(E-E_0)}$ = random error in the shunt voltage drop difference, volts
S_R = random error in the shunt resistance, ohms
$\partial I/\partial R$ = partial derivative of I with respect to R

Evaluation of the partial derivatives yields the following expression:

$$S_I = \left[2\left(\frac{S_{(E-E_0)}}{R}\right) + \left(\frac{-(E - E_0)}{R^2} S_R\right)2\right]^{1/2}$$

Similarly, the bias error in the current calculation, B^I, was propagated from the bias error of the measured variables. It was introduced as an expansion by using Taylor's series:

$$B_I = \left[\left(\frac{\partial I}{\partial (E - E_0)}\, B_{(E-E_0)}\right)^2 + \left(\frac{\partial I}{\partial R}\, B_R\right)^2\right]^{1/2}$$

where:

$B_{(E-E0)}$ = bias error in the shunt voltage drop difference, volts
B_R = bias error in the shunt resistance, ohms
= partial derivative of I with respect to R

During the testing program, it was determined that a bias in shunt resistance uncertainty, B^R, was present, while no bias was observed for the shunt voltage drop difference. The bias resulted from a connection difference of the power supply cables to the shunt during calibration and during test performance. Thus, the bias error in the current reduced to

$$B_I = \frac{E - E_0}{R^2} B_R$$

The governing equation for the core current measurement uncertainty, wI (amp), was defined by

$$W_I = B_I \pm S_I$$

or

$$wI = B_I \pm \left[2 \left(\frac{w(E - E_0)}{R} R \right)^2 + \left(\frac{w(E - E_0) wR}{R^2} \right)^2 \right]^{1/2}$$

where:

wR = uncertainty in the shunt resistance (random portion), ohms
$w(E - E_0)$ = uncertainty in the shunt voltage drop difference (random portion), volts

The nominal uncertainty in the shunt resistance was ±0.016 micro-ohms. The bias in shunt resistance was estimated to be −0.064 micro-ohms, equivalent to an error in current of −249 ($E - E^0$) amperes. Based on the analog-to-digital equipment, the accuracy (in both E and E^0) was ± 0.040 millivolts. The resulting measurement uncertainty for the core current measurement, shown in Figure 3, was obtained using these values in the above equations.

B. Measurement Process Control

Control charts and check standards are used for the most exacting measurement process controls. Their discussion here should not be interpreted as implying that they should be used to control most measurement processes. Their application normally requires additional resources to obtain information regarding the measurement processes which are not otherwise available. The expenditure of such resources should remain in a reasonable relationship to the value of this additional information. As the top of the metrological control pyramid, they provide, however, a valuable basis for deciding which tools for process controls should be used and which data should be obtained through the controls versus which data may be estimated or obtained from other sources.

1. Standards

The standards used in measurement process control are discussed in this section. These include check standards and reference standards.

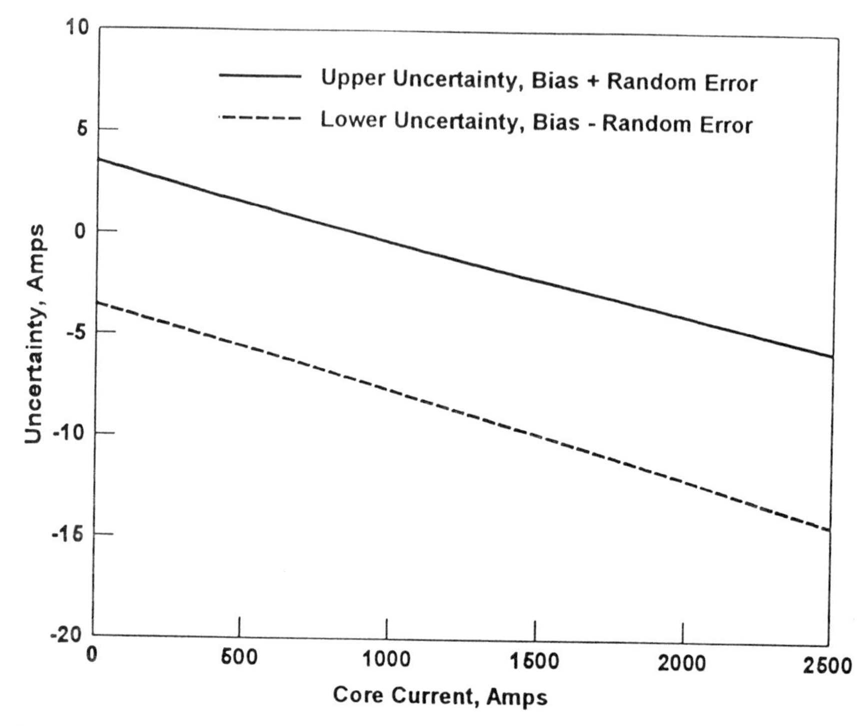

Figure 3 Measurement uncertainty for the core current.

a. Check Standards. A check standard is a device that is similar in kind to the items being measured by the measurement or calibration process that is to be controlled. By making regular measurements on a check standard that is expected to be stable and reproducible, and plotting the data on a control chart, one can observe graphically the measurement random error contribution to the uncertainty. When the average of the measurements of a check standard differs from the previously established value of that standard, one must suspect that an apparent bias has crept into the measurement process. This apparent bias must be tested to determine whether it is significant. Rigorous statistical tests are available for making such tests. A significant bias should be removed or compensated for, as appropriate.

A check standard need not be a singular object. Sometimes, the difference between two standards is used as a check standard, for instance. Hence, a check standard can be defined as measured equipment serving to collect a data base for the control of a measurement process.

For maximum utility, a check standard should be measured or calibrated by a process which is independent of the process used for control and which preferably is also more accurate than the process it controls. But even without such an independent measurement or calibration, which is not always possible, the check standard can nevertheless serve a useful function to control the process. However, an independent measurement or calibration traceable to a national standard is necessary if the check standard is to be used to assess the uncertainty due to the systematic error of the measurement process.

b. Reference Standards. Reference standards are calibrated by standardizing laboratories, often the national standards laboratory. They may be subject, therefore, to a different evaluation of their uncertainty contribution than other measuring instruments measured within a supplier's system. Normally, the values obtained during calibrations by the outside source and the limits of uncertainty bracketing that value are reported. These values are subject to change from one calibration to the next because of variable uncertainty elements experienced by the calibration source and short-term and long-term changes in the standards themselves. In assigning a value and its uncertainty limits to a reference standard, all these variations and uncertainty contributions must be considered. The calibration uncertainty limits reported by the calibration source should not be considered as the uncertainty limit of the value of the standard unless it can be shown that all other uncertainty contributions and variations are negligible with respect to the uncertainty limits reported by the source. Uncertainty limits thus established for the value of a standard may be considered as a systematic error contribution by the standard to calibrations performed against it or its Type A and Type B components may be considered separately.

A standard should normally not be used as a reference standard, if possible, before a history has been established from which its uncertainty limits can be determined. If it is necessary to use a standard without known history, uncertainty limits sufficiently wide should be assigned to its values to allow for all normally expected variations.

2. *Control Charts*

A control chart is a graphical technique for displaying data from some process that is to be controlled (5–7). As a variation from control charts used to control the quality of manufactured products, for instance, the data for charts to control measurement processes are obtained from measuring a check standard repeatedly. A control chart can be used as a basis for decision making concerning the process being controlled. Control charts were developed originally for controlling the quality of industrial production processes [2], but can be applied equally well to other types of processes, and in particular, can be effectively used in monitoring a measurement or calibration processes.

A control chart can identify unexpected temporary biases or process variabilities as well as long-term trends. If the measured value of the check standard changes with time, then one must conclude that either the check standard is changing or that the measurement process is changing, or both. In any case, this alerts the metrologist, who should halt the measurement process until the reason is found for the variation and statistical control of the process is restored. (Note that one may achieve statistical control of a time-dependent measurement process if the reasons for the time dependence are well understood and if the time-dependent behavior is fully predictable. If the measurement on the check standard indicates a possible out-of-control condition, then the measurement is rejected and must be repeated. A confirmed out-of-control condition should be identified and corrected.)

"A process is in statistical control when the variability results only from random causes. Once this acceptable level of variation is determined, any deviation from that level is assumed to be the result of assignable causes which should be identified and eliminated or reduced" (6).

Many measurement processes, however, are not sufficiently critical to warrant check standard measurements and the maintenance of control charts.

a. Types of Charts. For maximum utility in controlling measurement processes, a plotted measured value should be the average of a few observations. Three to five observations are commonly used. Because the plotted value is an average, it is customarily designated as $\bar{X}$ (X-bar), and the control chart is called an X-bar chart.

In the usual type of X-bar chart, one plots the actual measured value of the parameter of interest sequentially, but not necessarily as a function of time. Thus, in monitoring a measurement process for standard resistors, one might plot a point corresponding to the measured resistance of a particular standard resistor each time a measurement is made. When a measurement falls outside the control limits, one is alerted that the process may not be in a state of statistical control, and that corrective action may be needed. Similarly, other patterns of distribution of measured values can serve to detect existing or developing loss of statistical control or the existence of "assignable causes" (Shewhart) or "special causes." ("Common causes" are the sources of process variability which are usually present in the process under control.)

There are many different types of control charts for variables that can be applied to controlling measurement processes. Control charts for averages and ranges (Figure 4) are particularly convenient to use and are one of the most powerful tools for measurement control. In addition to plotting, on the X-bar chart, the measured value of the check standard as the average of a small number of observations, say three to five, one also plots simultaneously the range R of the observations in such a chart. (The range is the difference between the largest and the smallest observation.) In so doing, one obtains an R-chart, an indispensable tool for the control of the variability of the measurement process. The output of a process is fixed only when its mean (X-bar chart) and its variability (R-chart) are fixed. A change in either means a changed process.

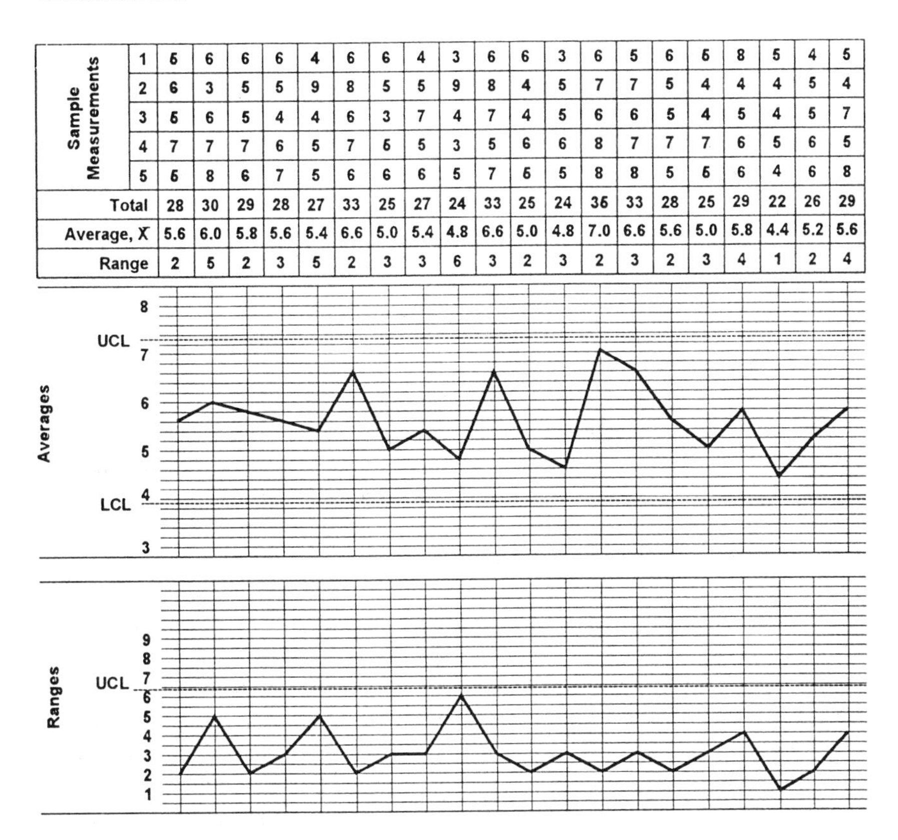

Sample Measurements																				
1	5	6	6	6	4	6	6	4	3	6	6	3	6	5	6	5	8	5	4	5
2	6	3	5	5	9	8	5	5	9	8	4	5	7	7	5	4	4	4	5	4
3	5	6	5	4	4	6	3	7	4	7	4	5	6	6	5	4	5	4	5	7
4	7	7	7	6	5	7	5	5	3	5	6	6	8	7	7	7	6	5	6	5
5	5	8	6	7	5	6	6	6	5	7	5	5	8	8	5	5	6	4	6	8
Total	28	30	29	28	27	33	25	27	24	33	25	24	35	33	28	25	29	22	26	29
Average, $\bar{X}$	5.6	6.0	5.8	5.6	5.4	6.6	5.0	5.4	4.8	6.6	5.0	4.8	7.0	6.6	5.6	5.0	5.8	4.4	5.2	5.6
Range	2	5	2	3	5	2	3	3	6	3	2	3	2	3	2	3	4	1	2	4

Figure 4 X-bar and R-chart examples.

Sometimes, sample standard deviations are used instead of ranges, resulting in s-charts. Ranges, however, are the preferred measures of process variability. They are more efficient than standard deviations as estimators of process variability when the samples are small. Standard deviations begin to be more efficient when more than about twelve observations are available. Usually, however, the number of observations has to be kept small because they must often be obtained over an appreciable period of time. Several hours to several days are frequently required to obtain the necessary observations for one measurement. Shorter time periods often cause the emergence of significant "between-group" variability which prevents the control of the variability of the process.

b. Frequency of Measurements. In the most rigorous approach to controlling a measurement process, one makes a measurement on a check standard each time a batch of like items are measured. If the measured value and range or standard deviation of the check standard measurement, when plotted on the control charts, fall within the previously established pattern, then the process is shown to be in a state of statistical control. For less rigorous control, check standard measurements are less often required.

If check standards and unknowns are measured by computing the average of a number of observations, and the range of the observations of the check standard is plotted on a control chart for ranges, a point on the chart for ranges falling within the control limits would indicate a random error typical for the process. The point on the control chart for averages falling within the control limits would indicate the probable absence of an atypical systematic error. Its relation to the preceding points can indicate the development of a systematic error long before it would otherwise become apparent.

The frequency at which check standard measurements are made depends on the amount of control and measurement assurance desired as well as on the degree of criticality of the measurement uncertainty. Normally one would remeasure the check standard rather frequently for a new process or check standard. Once experience is gained and the process is shown to be stable and remains in statistical control, the intervals between check standard measurements can often be lengthened.

Check standard measurements and their frequency must, of course, be specified by procedure, like all other important requirements of the measurement process. Well-established measurement processes with a history of few assignable causes may be deemed to require check standard measurements at some predetermined intervals which may be longer where the controls are less critical than for other measurement processes. It may be sufficient to make a measurement on a check standard at randomly selected times during a working day or week. If check standard measurements are not made with each measurement of unknown items, one should assure that such measurements are made at random times. A measurement that is always made at the same time may mask between-group random errors. The same may be true for any other condition that is held constant only for the check standard measurement.

One can usually begin to control a measurement process with twenty-five to thirty data points on a set of control charts (e.g., X-bar and R-charts) after some experiments have been made to determine the details of the process. Afterwards, some adjustments and changes may still have to be made until the process consistently and reliably remains within the desired state. Then, the experienced grand average, the center line of the X-bar chart, should remain unchanged until it is evident that is represents a bias or that the process is shifting or drifting. In other words, the initial grand average should be used as the standard against which the performance of the process is measured. Subsequent changes should be rare, well justified, and documented. The control charts should be visually analyzed for suspect patterns. The metrologist is encouraged to become familiar with various patterns on control charts which signal existing or potentially out-of-control conditions.

In monitoring calibration accuracies, it is important that all data be considered. If control charts are used, all points must be entered in the charts, including obvious outliers. The frequency with which such outliers are obtained is an indication of the degree of justification for maintaining previous uncertainty limits assigned to the calibration process. With control limits set at +/−2 times the standard deviation, usually called "warning limits," one does expect to observe outliers in the range of two to three standard deviations from time to time, and the occurrence of such a point does not necessarily mean that the process is out of control. The normal procedure would be to repeat the check standard measurement when an outlier is observed. If subsequent points fall within the control limits, the process is still in control. Repeated data falling outside the control limits are evidence of a loss of control. One must then find and eliminate the cause of the variability or shift. If this cannot be done, one must broaden the control limits commensurate with the new check standard data and increase the stated uncertainty accordingly. Ordinarily, one would not report data to customers until sufficient experience has been gained with the process operating within the new limits to ensure that the new limits are realistic.

3. *Case Study: SPC Applied to a Test Program*

Statistical process control (SPC) techniques have been applied to multiple-instrument experimental systems to provide data quality control and instrument performance monitoring [8]. This technique was applied to evaluate its usefulness as an alternative to the typical critical instrument "certification" process. The typical certification is provided by pre-test, post-test, and set-interval calibrations to ensure the quality of data during a test program.

In this case, the SPC technique used instrument redundancy designed into the experimental test arrangement to track the performance of an instrument relative to other instruments in a defined instrument group. This procedure made use of SPC techniques to track an instrument's or group of instrument's historical performance and set up the confidence intervals for individual and group performance.

Groups consisted of interchecking instruments whose responses, under certain operating conditions of the experimental apparatus, should be equal to one another, or should equal some obtainable value within a determined confidence interval. For example,

- A "summing instrument group" is characterized by the responses of several instruments being added to provide an expected response for one other instrument. An example is a series combination of differential pressure transmitters which should agree with an overall differential pressure transmitter (Figure 5a).
- A "symmetric instrument group" is applicable to those experimental systems that exhibit symmetry in parallel systems and can be operated so that the conditions in each symmetric portion are identical. An example of this

group is the symmetric positioning of temperature transmitters in a two-loop flow system (Figure 5b). The responses of symmetrically located temperature transmitters are expected to be equal under certain operating conditions.

- A "redundant instrument group" is one where more than one instrument responds directly to exactly the same input from the experimental appara-

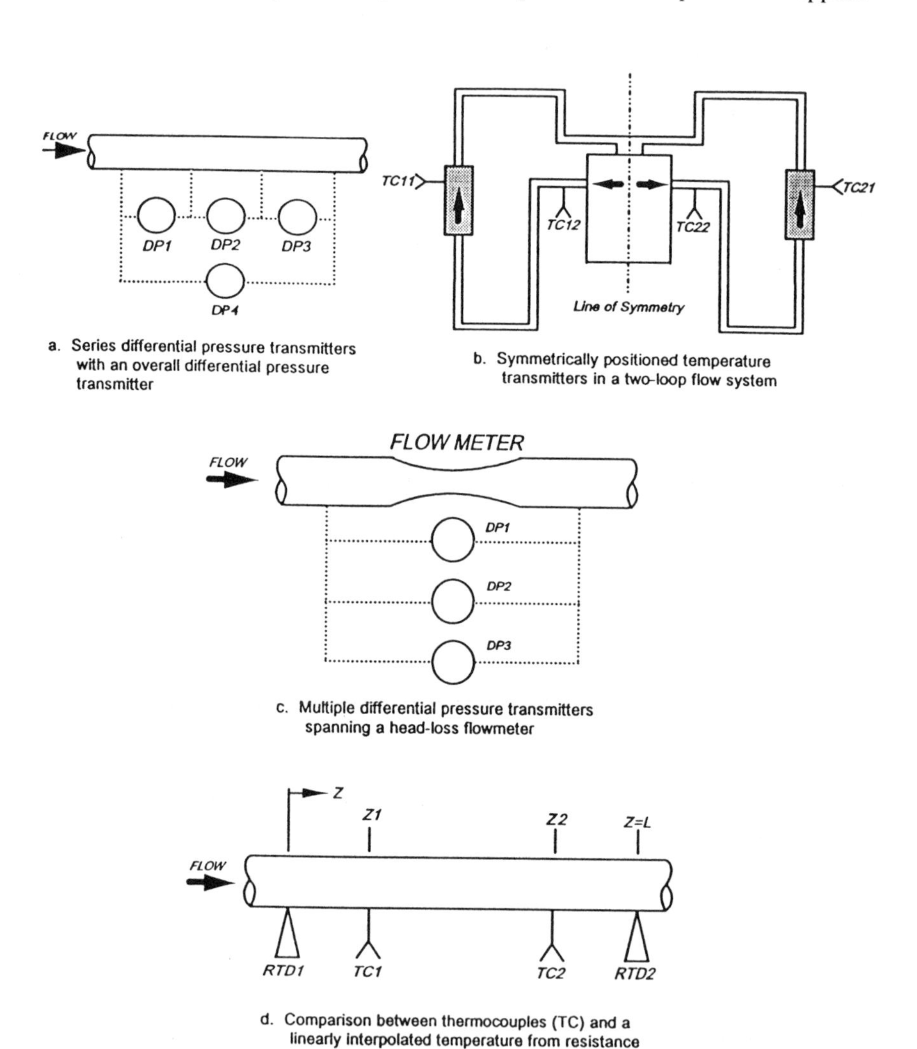

Figure 5 Instrument groupings to facilitate on-line performance tracking using SPC methods.

tus. An example of this type of grouping is where multiple differential pressure transmitters are installed on a head-loss-type flowmeter to minimize measurement uncertainty over a wide range of flow (Figure 5c). For certain operating conditions, the pressure drop across the flowmeter will be within the range of several of the transmitters, and the transmitter response would be expected to be equal.

- A "calculated comparative group" uses a comparison of an instrument response to a calculated expected value or a comparison among several calculated expected values. An example is a comparison of individual thermocouple-based temperatures with a temperature interpolated from bounding, higher-accuracy resistance temperature detectors (Figure 5d).

SPC techniques were used to compare the instruments in a group using their historical performance to the stated experimental uncertainty level and to the expected deviations if only random chance were causing variations in the group. The SPC techniques considered in this work include the X-bar (also "stability") chart and the R-chart (also "range" – chart). An example of the control charts for the case where the sum of three narrow-range differential pressure measurements is compared with an overall measurement (see Figure 5a) is shown in Figure 6. The ideal condition was for both charts to be in control; that is, all points should fall within the control limits and are randomly distributed about the mean. This condition would indicate the instrument/group was operating stably and accurately. If the compared responses did not fall within the predefined confidence interval, the certification group was flagged as requiring further attention. This procedure could be used to track the instrument's performance throughout the test program and provide evidence of the instrument's reliability and stability.

The stability chart of Figure 6a showed control problems, but the range chart of 6b indicated that the instruments were performing within the stated tolerances. Samples 4 and 5 in the stability chart show out-of-control situations for two tests. It was found from the program test log that this could be attributed to instruments being inadvertently out-of-service during tests 4 and 5 of the seventeen-test series.

Redundant information coupled with SPC techniques can provide a useful tool for tracking historical performance of a group of instruments. In this particular case study, the SPC analysis was performed post-test, and thus provided limited benefit. Use of the SPC techniques on line may have prevented the collection of data with important instrumentation unavailable, and resulted in improved data quality for the test program.

III. QUALIFICATION OF DATA

Data qualification is defined as the set of activities performed to verify compliance with the data quality objectives. These activities are performed to assure the test

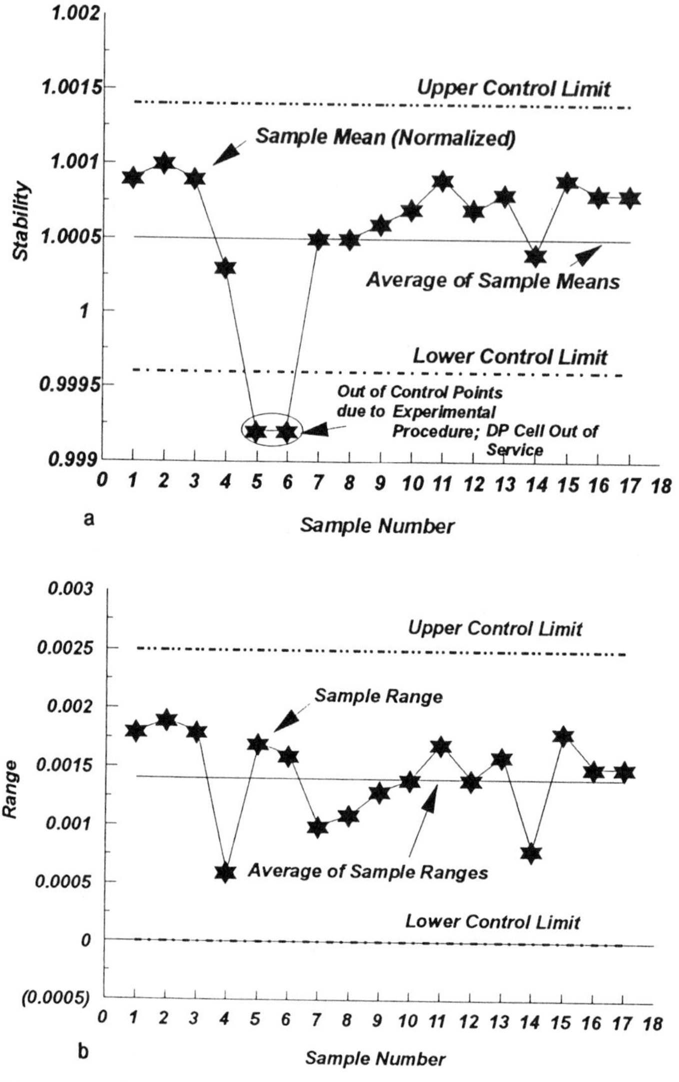

Figure 6 (a) X-bar or stability chart for the differential pressure transmitter grouping. (b) R-chart or range chart for the differential pressure transmitter grouping.

data acquired for a program has been reviewed and qualified. The product of these activities is documentation that provides subsequent users of the data with an understanding of those measures that have been applied to the data during its acquisition and review processes to arrive at its qualification status. The qualification status of individual entries in a test data set refers to the level of restriction that accompanies the subsequent use of each entry and is one of three levels:

1. Fully qualified with no restrictions after applying the data qualification processes identified for the test
2. Qualified with defined restrictions that limit the use of the entry in the data set
3. Data rejection based upon failure of one or more qualification steps

The methodology associated with data qualification involves a series of steps that begin with initial characterization of the measurement system and ends with review (manual and computer assisted) of the acquired test data set. Each step in this qualification process is considered of equal importance. As a result, each step must be defined so that at the end of the process a qualification statement can be assigned to each entry of the data set that assures consistency among all test data sets. Since the qualification process involves a sequence of steps, it is convenient to discuss this process in well-defined groupings that are consistent with the particular test program. In particular, these groups may consist of:

1. Measurement system characterization
2. Facility characterization checks
3. Pretest checks
4. Real-time checks during test performance
5. Post-test data qualification activities
6. Overall program qualification activities

Each of these groups is discussed in the following sections.

A. Measurement System Characterization

The measurement system may consist of the instrument, connecting cabling, signal conditioning electronics, the patch panel, and data acquisition system that are used to measure and acquire the physical quantity of interest. Characterization of the measurement system provides the foundation for subsequent qualification steps. As shown in Figure 7, the characterization of the measurement system involves steps that begin with the initial definition of the measurement requirements and continue through instrument procurement, installation at the test site, and connection to the data acquisition system. In parallel with the instrumentation, similar activities are shown for the data acquisition system that include definition of the data acquisition system requirements, procurement, and installation into the test site. Procurement of instrumentation and associated instrument calibrations are closely monitored and guided by quality assurance procedures. These procedures establish the basis for selection, evaluation, and verification of suppliers.

Once the data acquisition system and instrumentation are connected, a continuity check of the measurement system may be performed. This check verifies continuity from each instrument, through the connecting cabling to the data acquisition system, intermediate signal conditioning electronics, and data acquisi-

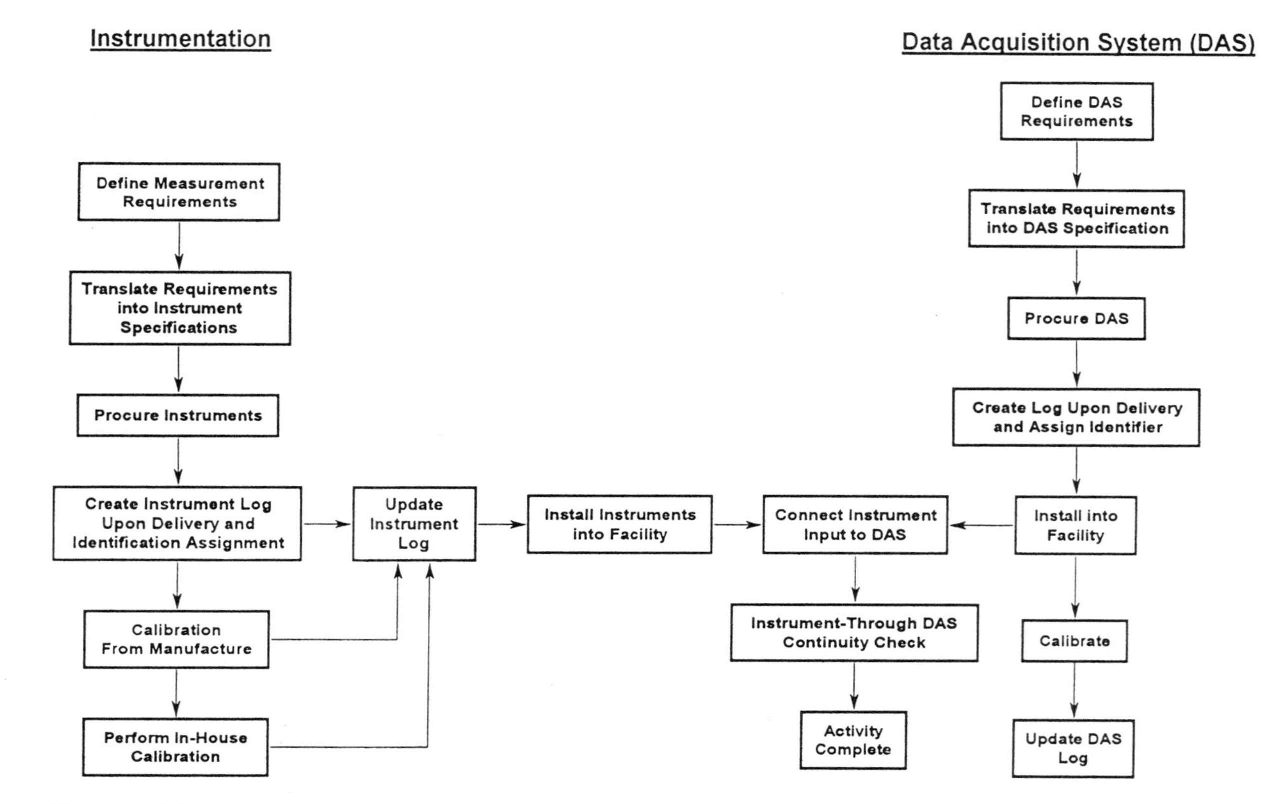

Figure 7 Measurement system characterization.

tion software. The continuity check may consist of disconnecting the cabling at the instrument and either:

- Monitoring the change in signal through the data acquisition system using the data acquisition software
- Inputting a known source at the disconnected instrument cable and verifying the presence of that value through the data acquisition software

B. Facility Characterization Checks

At the completion of instrument and data acquisition system installation and verification of instrument continuity through the data acquisition system software, a number of qualification steps may be performed to further ensure the integrity of the measurement and data acquisition system. In particular, this sequence of characterization measures may involve:

- "In-place" calibration of selected instruments
- Initial comparison of neighbor and redundant instruments for consistency
- Checks of the derived quantity or computed parameters present in the data base.

The experimenter may find based on experience that, whenever practical and possible, some instruments, such as thermocouples, should be calibrated in-place after installation. During installation, thermocouples often are brazed into position and sheaths are bent to attain proper junction position. These operations may impart a bias to the thermocouple reading that can be calibrated out after installation is complete. In addition, the thermocouple lead wire and reference junction associated with each of the thermocouples make up a part of the measurement system and are included as a part of the calibration.

During the time of facility characterization and after calibrations are completed, tests can be performed that will lend themselves to data acquisition and comparison of instruments that would be expected to read the same. Or, with simple and well-defined corrections, differences could be accounted for that would allow for direct comparisons. The purpose of such comparisons before the test program is initiated is to begin collecting and correlating data that will allow valid conclusions to be made when similar comparisons are made later during the test program. For example, during steady-state test apparatus operation with fluid temperatures in excess of ambient, characteristic temperature gradients exist around the facility that lend to comparison of temperature measurements. By trending all the temperature measurements with respect to their physical position to each other, it is then possible to flag those measurements that lie outside of expected deviations as determined by some criteria (possibly measurement uncertainty estimates) and prior experience. Other examples of redundancy checks that may be possible during the time of facility characterization include:

- Comparison of pressure transmitter readings with static pressure corrected to a common elevation
- Comparison of differential pressure readings to the difference in pressure readings (if both types of instruments are being used)
- Comparison of total fluid flow with the sum of the branch flows

Other comparisons may become apparent for a particular test design.

Derived quantities or computed parameters are those values in the data base that use one or more instrument readings together with physical relationships and property routines to compute engineering quantities of interest to the data base user. A derived quantity may require input from other derived quantities to perform the computation. For example, derived quantities include items like the saturation temperature computed from a pressure measurement, void fraction or collapsed water level from differential pressure measurements, and component mass and energy. During the period of facility characterization, the calculations associated with these quantities can be checked for accuracy as compared with the expected uncertainty.

Examination of the derived quantities during the time of facility characterization prior to the start of the test program allows an independent check of a number of instruments through a single calculated parameter. For example, calculations that involve an energy balance on either the whole facility or specific components, such as a heat exchanger, require input from several derived quantities. As such, these calculations tie a number of instruments and derived quantities together. A successful calculation therefore gives the data base user more confidence in the measurement system and physical relationships that are used to compute the derived quantities of interest.

Statistical process control (SPC) techniques described earlier can be applied to experimental systems to provide data quality control and instrument performance monitoring [8,9,10]. These techniques provide the assurances that one obtains with the above-described consistency checks. However, the SPC techniques have an accepted statistical basis, and can be used as a predictive tool for identifying future instrument failures. The SPC techniques are used to compare the instruments in a group using their historical performance to the stated experimental uncertainty level and to the expected deviations if only random chance were causing variations in the group. Control charts are maintained on an instrument group to track instrument performance. The ideal condition is for all charts to be in control; that is, all points should fall within the control limits and are randomly distributed about the mean. This condition would indicate that the instrument/group is operating stably and accurately. The information provided by these techniques could be extended to provide sufficient justification to omit post-test calibrations, provided the instrument has operated stably and accurately.

C. Pretest Checks

The discussions thus far have covered those activities that transpire during instrument calibration and measurement system continuity checks, and the period of facility characterization. During facility characterization, the data acquisition system is used in full capacity to record instrument readings that are subsequently used to establish the data base for proposed redundant instrument checks. When the test program begins, the information acquired during the period of measurement system characterization and facility characterization may be used to establish guidelines whereby instrument failure or out-of-specification conditions can be identified. Prior to test performance, a series of pretest checks can be performed. These checks serve to establish or check the current calibration of the measurement system, and provide manual and computer-aided review of steady-state operating data to flag erroneous instrument measurements. The following case study provides an example of how this concept may be applied to a test program.

1. Case Study: Pretest Checks in a Large-Scale Test Program

In a large-scale thermal-hydraulic test program [4] conducted at Babcock & Wilcox Company's Alliance Research Center, on-line checks were performed before each test to verify performance of the analog-to-digital conversion equipment, to examine transmitter "zero" values, to verify instrument raw data was within expected limits, and to examine consistency among instruments. This test program employed over 800 instruments (e.g., thermocouples, pressure transmitters) and an additional 300 engineering quantities derived from the instrument set (e.g., mass flow rate from a venturi flowmeter using input from differential pressure, pressure, and temperature measurements). Computer automation was required to provide the needed quick assurances of acceptable instrument performance prior to data collection.

The software structure for the pretest checks for this large program is shown in Figure 8. The data acquisition system was centered about the "Global Common," which contained instrument information needed for acquiring the raw data and reducing it to engineering data. The information was shared by all routines (shown here are routines SCANIT, ZEROS, RANGE, and CONSIS.) New raw data from each instrument was digitized by the analog-to-digital conversion equipment and loaded into the common area at user-specified time intervals.

To aid in making pretest checks, measurement uncertainty models were developed for all instruments and many of the derived engineering quantities in the test data set. These models permitted effective monitoring of instrumentation and control system performance during the multi-year test program. A number of applications were developed and were the basis for both on-line and off-line instrumentation monitoring.

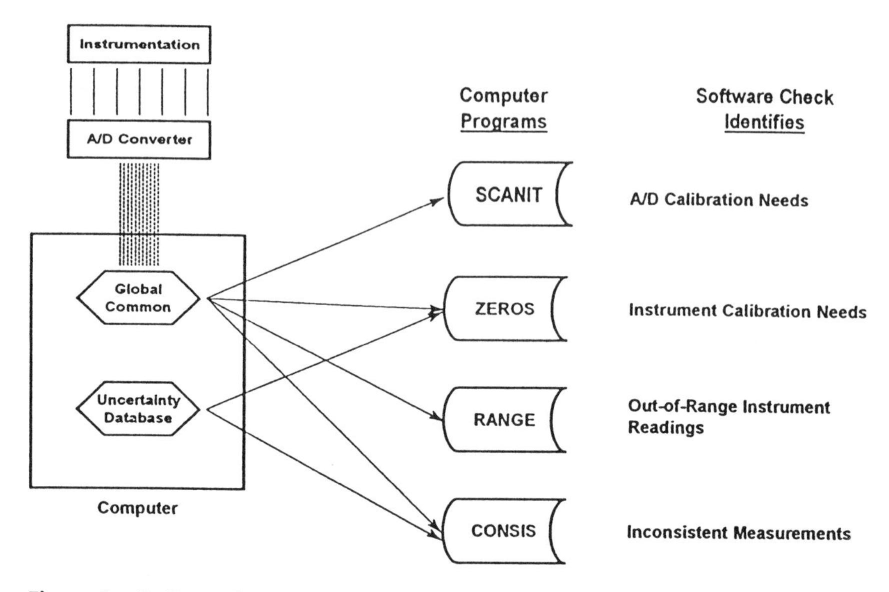

Figure 8 On-line software structure for measurement system qualification.

Also shown in Figure 8 is the uncertainty data base. The uncertainty data base included all additional information needed to compute uncertainty estimates for each instrument. For example, for a differential pressure transmitter, this data base included the transmitter-calibrated span, the manufacturer's specifications for accuracy, ambient temperature effect on span and zero value, static pressure effect on span and zero value, and long-term stability. The uncertainty data base information was updated as required when measurement system changes occurred, such as replacing or recalibrating a component in the measurement system.

The simplest check for the instrument set was performed by the SCANIT program. This routine used the known uncertainty of the data acquisition system analog-to-digital conversion equipment to perform pretest (also, post-test) calibration checks. SCANIT identified analog-to-digital conversion equipment channels that were approaching or had exceeded the specification based on the measurement uncertainty band. Figure 9 is an example of the type of information displayed when the SCANIT program was executed. The sample output shows that for a negative voltage input of -0.078 volts, diagnostics are present. These diagnostics indicate the calibration has drifted for some channels to a value that is greater than the acceptable calibration shift (± 0.056 millivolts). Depending on the degree of importance of the out-of-specification channels, actions by the test operator (including possible recalibration of the channel) may be prescribed in the test procedure.

SCANIT - Statistical Data Acquisition

Date: 30-AUGUST	13:52:38.40		
Chassis:	0	First Channel:	1
A/D Type:	7	Last Channel:	64
Gain:	1	Number of Samples:	100

CALIBRATION MODE IS: NEGATIVE
INPUT VOLTAGE IS: -0.078 Volts
ACCEPTABLE DIFFERENCE: 0.000056 Volts

CHASSIS	CHANNEL	VTAB	READING AVERAGE (volts)	DIFFERENCE (volts)	DEVIATION (volts)
0	3	SPARE	-0.077932	-0.000068	0.000012
0	4	IFMT07	-0.077941	-0.000059	0.000003
0	6	PSTC02	-0.077942	-0.000058	0.000002
0	21	IFMT04	-0.077939	-0.000061	0.000005
0	26	IFMT09	-0.077943	-0.000057	0.000001
0	42	OFMT02	-0.077944	-0.000056	-0.000000
0	43	OFMT03	-0.077941	-0.000059	0.000003
0	57	OFMT17	-0.077941	-0.000059	0.000003
0	59	PSAM01	-0.077937	-0.000063	0.000007

Figure 9 Output from Program SCANIT.

"Zero" checks of all transmitters (pressure, differential pressure, load cells, and others having a zero offset) were performed by executing the ZEROS program. This routine assessed transmitter performance by examining changes in the zero value over time. An example of a ZEROS program output is shown in Figure 10. For each instrument (or "VTAB"), the zero recorded when the unit was calibrated provides the baseline value for comparison for all future checks. The uncertainty in the zero reading was used as the acceptance criteria. The uncertainty database contained information needed to compute the acceptable zero band (e.g., manufacturers' specifications, calibration temperature, and zero value.) The program would identify whether or not the new zero value was acceptable, and would insert the new value into the global common data base if the test operator desired. The ZEROS check was responsible for early detection of a generic hardware problem in a differential pressure transmitter model.

Another simple check was performed by the RANGE program. The current raw signal reading was supplied to the routine for each instrument. The program compared the raw signal with the expected minimum and maximum readings. An off-line version of the program, OFFRANGE, was used to identify out-of-

ZEROS RUN JOURNAL FILE
30-AUG- 14:04:32.

LOOP PRESSURE 14.09 psia (IPAP01)
AMBIENT TEMPERATURE 87.59 deg F

VTAB NAME	CALIBRATION ZERO	NEW ZERO VALUE	+ / - UNCERTAINTY	ERROR FLAG	UPDATE FLAG
BFAP01	0.073015	0.073297	0.000628	OK	Y
C2AP01	0.073121	0.073526	0.000630	OK	Y
C2AP02	0.073778	0.073708	0.000629	OK	Y
C2AP03	0.073054	0.073166	0.000630	OK	Y
C2AP04	0.073052	0.073276	0.000630	OK	Y
C2AP05	0.073245	0.073567	0.000629	OK	Y
C2AP06	0.073245	0.073528	0.000629	OK	Y
C2AP07	0.073866	0.073763	0.000629	OK	Y
C2AP08	0.073584	0.073498	0.000629	OK	Y
C2AP09	0.073280	0.073555	0.000627	OK	Y
C2AP10	0.073307	0.073635	0.000627	OK	Y
C3AP01	0.073745	0.073625	0.000629	OK	Y
C3AP02	0.073020	0.073262	0.000629	OK	Y
C3AP03	0.073261	0.073407	0.000627	OK	Y
C4AP01	0.074099	0.073975	0.000629	OK	Y
C4AP02	0.073322	0.073339	0.000627	OK	Y
C4AP03	0.073290	0.073403	0.000627	OK	Y
IPAP01	0.073073	0.073393	0.000627	OK	Y
OPAP01	0.073233	0.073663	0.000629	OK	Y
TFAP01	0.073881	0.073765	0.000629	OK	Y
WSAP01	0.072937	0.073173	0.000634	OK	Y
MSTC02	0.000000	0.001851	0.000000	F	N

Figure 10 Output from Program ZEROS.

range conditions for each instrument, for all raw data in the test data file. An example of the OFFRANGE program output is provided in Figure 11.

This check was successful in identifying malfunctioning measurement systems for thermocouples, resistance temperature detectors, conductivity probes, and gamma densitometers. The range check identified time intervals when two-phase flow conditions were present at venturi and orifice flowmeters which were intended for single-phase flow conditions. These two-phase flow conditions resulted in rapidly varying flowmeter differential pressure, with many readings outside the expected limit. Finally, the range check successfully identified anomalies in the automatic control system performance. For example, a problem in the automatic auxiliary feedwater flow controller was flagged as a result of an over-ranged flow rate measurement. The range check provided timely feedback on instrument problems and allowed corrective action prior to further testing.

```
                                                              Page No.   1

                   RANGE CHECK FOR TEST 043_14

DEGCLB LOCA SERIES 3 TEST 1; 6.2 MW. 150 PSI, 278 GPM, 80 F
RAW DATA FILENAME: 043_14.R3A
TOTAL NUMBER OF SCANS:  7833
TEST TIME INTERVAL: 30-AUG-     17:33:11.05 - 30-AUG-1991 17:42:12.94
TIME OF RANGE CHECK EXECUTION: 31-AUG-91 07:28:31

   NAME        ALLOWABLE         NO.  DIR   START OF          DEVIATION
                 RANGE          SCANS        INTERVAL     AVERAGE    MAXIMUM
  ------   ------------------   -----  -   ----------    --------  ---------

  BFTC01   0.9240E-3,0.4041E-2             ALL RDGS IN RANGE
  IFMT07   0.9240E-3,0.1375E-1   7833 L     17:33:11.05   0.327E-1  0.458E-1
  C4TC04   0.9240E-3,0.1110E-1             ALL RDGS IN RANGE
  C2TC01   0.9240E-3,0.1050E-1             ALL RDGS IN RANGE
  C2TC02   0.9240E-3,0.1050E-1             ALL RDGS IN RANGE
  C2TC03   0.9240E-3,0.1050E-1             ALL RDGS IN RANGE
  C3TC01   0.9240E-3,0.1050E-1             ALL RDGS IN RANGE
  C3TC02   0.9240E-3,0.1050E-1             ALL RDGS IN RANGE
  C3TC03   0.9240E-3,0.1050E-1             ALL RDGS IN RANGE
  C4TC01   0.9240E-3,0.1050E-1             ALL RDGS IN RANGE
  C4TC02   0.9240E-3,0.1050E-1      1 L     17:34:00.95   0.960E-5  0.960E-5
                                    2 L     17:34:01.15   0.316E-4  0.340E-4
                                  182 L     17:34:01.45   0.264E-3  0.539E-3
                                    1 L     17:34:20.25   0.169E-4  0.169E-4
                                   21 L     17:34:20.55   0.839E-4  0.149E-3
                                   12 L     17:34:22.85   0.847E-4  0.185E-3
                                    5 L     17:34:24.15   0.491E-4  0.117E-3
                                    5 L     17:34:24.75   0.443E-4  0.780E-4
                                    2 L     17:36:15.55   0.401E-4  0.657E-4
                                   24 L     17:36:15.95   0.448E-4  0.828E-4
                                 4669 L     17:36:18.65   0.155E-2  0.286E-2
                                   29 H     17:41:59.60   0.131E-2  0.174E-2
  C4TC03   0.9240E-3,0.1050E-1             ALL RDGS IN RANGE
  IFMT01   0.9240E-3,0.1375E-1             ALL RDGS IN RANGE
  IFMT02   0.9240E-3,0.1375E-1             ALL RDGS IN RANGE
  IFMT03   0.9240E-3,0.1375E-1             ALL RDGS IN RANGE
  IFMT04   0.9240E-3,0.1375E-1             ALL RDGS IN RANGE
```

Figure 11 Output from Program OFFRANGE.

Measurement consistency checks were applied to neighbor and/or redundant measurements to assess instrument performance. The CONSIS program performed a number of these checks during the steady-state, pretransient conditions. Instrument readings, the uncertainty data base, and assumptions regarding measurement relationships were required for this check. An example of the output from the CONSIS program is provided in Figure 12. The check performed in this example compared a differential pressure transmitter indication with the difference in two pressure measurements. Because the instruments have common taps into the test section, agreement to within measurement uncertainties was expected.

Differential pressure measurements were tested for measurement consistency in a number of ways. During the design of the test facility, instrument locations were selected when practical to facilitate instrument checks. As in the earlier example, some differential pressure measurements shared vessel taps so that tap-to-tap continuity was achieved for the flow loop. Because of this, the measurements could be combined to verify that the algebraic summation of the loop pressure drop equaled zero, within the expected uncertainty. Also, orifice and venturi flowmeters were generally spanned by multiple differential pressure transmitters to provide better flow rate accuracy over the entire flow range. The multiple transmitters were tested for consistency during operations in which they were in service and within their calibrated range. These consistency checks have been valuable in identifying measurement problems. Investigation of the exception reports over the course of the test program revealed isolated occurrences of improperly filled process lines, incorrect calibration coefficients, inadvertent process line isolation, and malfunctioning transmitter manifolds. The conditions were identified and rectified before test performance.

One of the consistency checks compared temperature readings from thermocouples with those of more accurate resistance temperature detectors (RTDs). RTDs were installed at the inlet and outlet of most loop components, and many thermocouples were used between the RTDs. For most piping components, the temperature gradient from inlet to outlet was small (and linear) during steady-state, pretest conditions. For this condition, the thermocouple readings were compared to interpolated RTD indications. Figure 13 illustrates the comparison for a piping component with twenty-eight thermocouples bounded by two RTDs. If the difference between a thermocouple-based temperature and the interpolated RTD temperature exceeded the combined uncertainty for the thermocouple and the RTD temperatures, then an exception report was created for the thermocouple. In one instance, this check was successful in identifying a faulty current supply in the RTD circuit. This particular hardware problem caused a negative bias for RTD temperatures that resulted in many exception reports from this consistency check.

```
DEGCLB LOCA SERIES 3 TEST 1; 6.2 MW, 150 PSI, 278 GPM, 80 F

        Execution Time = 31-AUG-    07:27:30

        Data Scan Time = 30-AUG-      17:42:31.49

             Test Time =   0 00:00:33.24 after zero

        Header File DAS$TEST:[LOG]043_14.HDA;1

        Raw Data File DAS$TEST:[RAWLO]043_14.ROA;1


Chk# Incl.                        Description
---- -----                        -----------

  1    T    Check fuel assembly power (& check resistances).
  2    T    Check inlet flowrate (turbine against venturis).
  3    T    Pressure agreement at same elevation (IF no flow, no power).
  4    T    Venturi & pitot zero check (IF no flow, no power).
  5    T    Venturi temperatures (IF flow).
  6    T    Fuel assembly temperatures (IF flow, no power).
  7    T    Redundant pressure differences (dP against Ps).
  8    T    Heat balance on Fuel Assembly.

     Check Number   7 :

  Redundant pressure differences (dP against Ps).

                    VTAB       Pressure      (unc.)              (flag)
                   ------    ------------  -----------           ------

     Group   1 :
                   OPAP01      51.5366      +/-.4355      psia
                   IPAP01      78.9284      +/-.4198      psia
                   IPDP03      27.4334      +/-.2148      psid
            (discrepancy) 0.416565E-01      +/-.6419      psid   OK

     Group   2 :
                   BFAP01      73.2089      +/-.4153      psia
                   IPAP01      78.9284      +/-.4198      psia
                   BFDP01      6.02579      +/-.1363      psid
            (discrepancy) 0.306267          +/-.6061      psid   OK

     Group   3 :
                   TFAP01      57.9081      +/-.4671      psia
                   BFAP01      73.2089      +/-.4153      psia
                   BFDP02      14.8475      +/-.1360      psid
            (discrepancy) -.453354          +/-.6397      psid   OK

     Group   4 :
                   C2AP09      59.1917      +/-.4121      psia
                   C2AP07      71.0889      +/-.4199      psia
                   C2DP01      11.5118      +/-.1369      psid
            (discrepancy) -.385431          +/-.6040      psid   OK
```

Figure 12 Output from Program CONSIS.

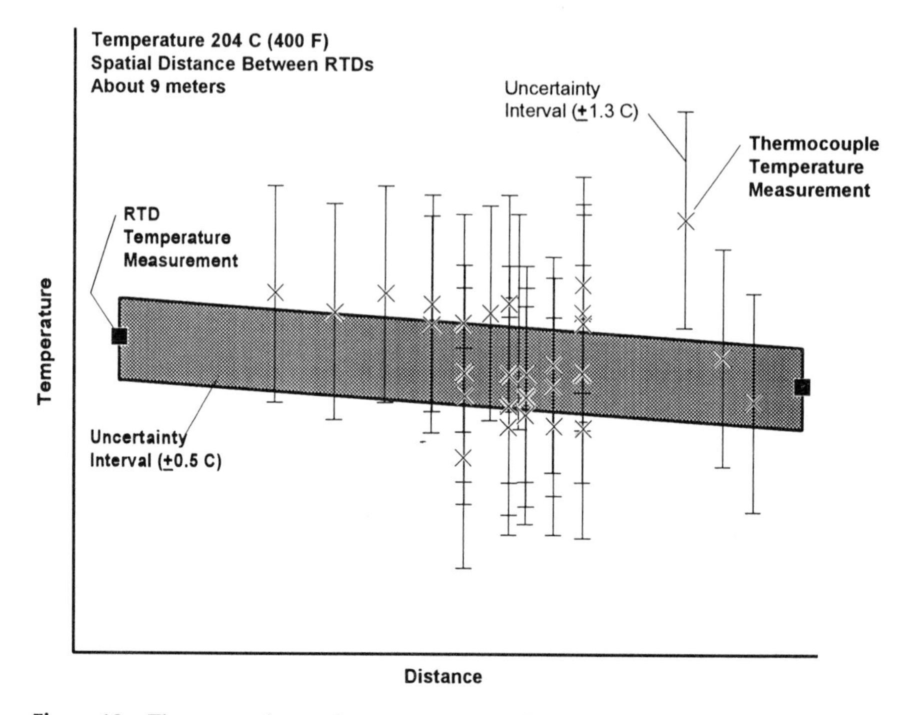

Figure 13 Thermocouple to resistance temperature detector (RTD) temperature measurement consistency.

Additional consistency checks and utilities that used uncertainty models to assess instrument and control performance were developed for the test program. The consistency checks proved to be a valuable tool for identifying measurement problems before running tests, thus minimizing costly test reruns. As part of the test design, consideration was given for instrument selection and placement so that instrument consistency checks could be performed. Uncertainty estimates are an appropriate means of quantifying acceptable agreement. For large data bases, a computer-based approach such as that described above, permits thorough, yet timely, measurement qualification.

D. Real-Time Checks During Test Performance

With the pretest instrumentation and data base checks complete, test performance and real-time data acquisition commence. The data qualification process can be augmented during testing if the data acquisition system features an "alarm" monitor, and through the use of a control room data display. The "alarm" monitor can be used to identify when key measurements are outside expected values. If the checked raw data value is outside of the expected range, a message should be displayed to the test personnel and a corresponding entry made to a disk file for post-test review. If available, a schematic of the test apparatus with key measurements displayed can be a useful tool for detecting data anomalies. The display can be used to enhance the testing personnel's information required for test control, as well as to understand the physical processes that are occurring. During the viewing process, if questionable or erroneous data is displayed, then the operator should log such observations for review at the end of testing. The logged entry should note the time of the observed anomaly and a brief description that will later aid reviewers in understanding the observation.

E. Post-Test Data Qualification Activities

Following the completion of a test and the acquisition of a raw data file for the set of test instrumentation, the process of data review and qualification continues, culminating in a report that documents guidelines for subsequent users of the data base. This review process extends from the measurement system through the converted engineering data base, noting any deviations from the pretest computed specifications of uncertainty. This review should also assign a qualification statement to each data base variable, for both instruments and derived quantities. To achieve this end product, a review of the measurement system checks and the test data set is required.

The foundation for all subsequent data review and qualification statements relies on a good understanding of the calibration status for each instrument and data acquisition channel of the measurement system. This infers that a review of all pretest and post-test calibration data be examined. It also means that post-test calibration data (''zero'' readings) be acquired for the critical instruments, repeating many of the pretest instrument checks.

An early step in the post-test data processing activity can be a comparison for each acquired raw data value, for each instrument, with its expected range of raw data values. Previous range comparisons during real-time data acquisition may have considered ''critical'' instruments only. This check is most beneficial if completed prior to starting the next test.

The review of the engineering unit data base can require substantial computer resources and manpower. This data base contains not only the instrument readings but also the derived quantities that will be of significant use to subsequent users of the test data for test event analyses. Several review steps, in addition to those described earlier for the raw data file, can be performed to lead to qualification of the engineering data base. These review steps include:

- Review of critical instruments and derived quantity trend plots
- Review of neighbor and redundant instrument checks
- Review of analysis plots

Trend plots (data base variable versus time) should be prepared at the completion of a test. These plots, which may be specified in the formal test procedure governing the test, should be reviewed to identify anomalous instrument performance, and to assess whether the major test objectives have been achieved prior to performing the next test. This review process provides input to later qualification statements regarding the examined data base variables. The test section boundary conditions, such as pressure, temperature, mass flow rate, and power, could be reviewed for adequacy using the trend plots. The neighbor and redundant instrument checks, or consistency checks, performed prior to the test should be repeated using the data in the acquired data set. These checks should be performed using data from steady-state or quasi-steady-state portions of the acquired data set. If data from a nonsteady portion of the test is used, the consistency check may generate a considerable number of useless exception reports. This might occur due to differences between instruments exceeding the uncertainty-based tolerances, which may not consider transient effects. If the consistency checks can be made at the start and end of the test, often it can be assumed (with high confidence) that such checks provide closure on the checked instruments during the entire test.

Additional trend plots may be generated that present the test data in a concise format to facilitate test review, test analysis, and test data qualification. These plots can contrast measured and calculated quantities such as mass balance and energy balance for test apparatus components as well as the overall system.

These post-test reviews should be thorough, providing the basis for test analysis and presentation in initial reports, and instrument uncertainty analyses. It is likely that anomalies present in the engineering data base will be found during the review. A record of these anomalies should be logged in an open items list. The open items list provides the basis for future data base and/or software revisions. Each of the open items should be closed out prior to considering the test data to be "final."

F. Overall Program Qualification Activities

In conjunction with those activities described earlier that support the tests and data qualification in an on-going manner, a number of reporting and data base management activities may be performed to disseminate the results of the data qualification process to users of the test data and to maintain the test data files in a retrievable and current format. These activities include:

- Uncertainty analysis report
- Initial and final reports
- Test data base management

An uncertainty analysis should be performed during the planning phase of the experiment to identify uncertainty estimates (typically at 95 percent confidence) for each of the instruments. The uncertainty analysis report should be revised at the completion of the test program to define the nominal uncertainty estimates for each of the derived quantities, as well as to identify instrument uncertainties for those measurements that exceeded the pretest estimates.

The data qualification process can be supported by the production of the "initial" and "final" reports. These documents can be the culmination of an extensive examination and evaluation of the test data. The initial report is started within days of test completion and includes trend plots of key measured and calculated variables. These plots can facilitate the rapid review of instrument indications, test procedures, and test apparatus behavior. These reports should consist of a thorough discussion of test conduct, observations, test results, and conclusions. During this process of data review, failed or "off-nominal" deviations to the pretest uncertainty estimates should be reported. In the event instrument anomalies are found following the issuance of the initial report, the revised uncertainty estimate should be included in the uncertainty analysis report. The uncertainty analysis report may be considered to be a portion of the final report, accompanying the release of the final certified engineering data to the end user.

Data base management activities are required to support the overall qualification program. In particular, the following activities should be performed:

1. The raw data file for each test should be backed up on a suitable storage media, thereby providing multiple copies of the raw data to ensure its long-term integrity.
2. Maintenance of directories that provide access to the test data files archived, and a distribution list of those users to whom data tapes have been distributed, and what test data files were distributed
3. Maintenance of an open items list that describes those anomalies present in the data base that have been identified during the data review.
4. As directed by the documented data base anomalies that appear in the open items list, data base reruns of the affected data sets should be performed to correct those anomalies where possible. The new engineering data file should be distributed to the users at the completion of the rerun.

One can rest assured that users of the test data will appreciate the organized information made available through these data base management activities.

REFERENCES

1. Coleman, H. W., and Steele Jr., W. G., *Experimentation and Uncertainty Analysis for Engineers*, John Wiley & Sons, New York, 1989.
2. Shewhart, W. A., *Economic Control of Quality of Manufactured Product*, D. Van Nostrand Company, New Jersey, 1931.
3. Kaarls, R., rapporteur. "Report of the BIPM working on the statement of uncertainties to the Committee International des Poids et Mesures," *CIPM Process Verbaux*

des Seances, Vol. 49, 70th Session, 1981, p. Al. (Reprinted in *NCSL Newsletter* 23, No. 3, Sept. 1983, pp. 35A1, with a statement that the recommendations were adopted by CIPM in Oct. 1981.)
4. G. C. Rush and M. T. Childerson, ''Application of Measurement Uncertainty to Data Qualification for Large Test Programs,'' Instrument Society of America 35th International Instrumentation Symposium, May, 1989.
5. ISO 7870.2, *Control Charts: General guide and introduction.*
6. ISO 8258, *Shewhart Control Charts.*
7. *ASTM Manual on Presentation of Data and Control Chart Analysis*, 6th ed., ASTM Manual Series: MNL 7. Revision of Special Technical Publication (STP) 15D. Philadelphia: American Society for Testing and Materials, 1990.
8. Habib, T. F, and Moskal, T. E., ''Application of Statistical Process Control Techniques for Instrument Evaluation,'' presented at the Eastern Energy Quality Assurance Conference, Hollywood, Florida, April 8, 1991.
9. Montgomery, D. C., *Introduction to Statistical Quality Control*, John Wiley & Sons, New York, 1990.
10. Wadsworth, H. R., Stephens, K. S., and Godfrey, A. B., *Modern Methods for Quality Control and Improvement*, John Wiley & Sons, New York, 1986.

9

SOFTWARE QUALITY ASSURANCE

Taz Daughtrey
Babcock & Wilcox
Lynchburg, Virginia

I. INTRODUCTION

Quality assurance of software is a specialized discipline where quality principles traditionally applied to hardware environments have to be modified or redirected [1].

Of the many definitions of quality, "fitness for use" is the most succinct. A three-word definition that contains but two troublesome words. Certainly software must be "fit for use," but how do you intend to use it? Furthermore, what will be the basis for determining fitness?

Who will be using the software: Novices? Experienced users? What will they be expecting the software to do and to be? How far will they push it beyond its original intent?

The key to defining use is to state, quite explicitly, a specification of requirements. That specification needs to be reviewed and agreed to by all parties involved. They must also decide on the subset of requirements that are most important and which other requirements are to be given lower priority in development and evaluation. Whatever assumptions are made should also be stated explicitly.

How will the fitness of the software be determined? All these specifications of requirements are themselves rather useless if they are not expressed in terms of acceptance criteria. How do you know when you've done enough evaluation unless you have established specific, preferably quantitative, goals for the evaluations?

Here's where software differs so much from more tangible products. No amount of inspection can make visible the quality of the finished product. No amount of product testing can hope to exercise all the possible combinations of inputs and conditions that the software can experience in operation. No one evaluation technique can provide enough confidence, so a range of inspections, walkthroughs, reviews, tests, and other analyses need to be applied.

II. THE SOFTWARE QUALITY CHALLENGE

It is too easy to write software, and too hard to write good software . . . or to know when you have it. This is certainly a troubling observation, given the great range of computerization in the R & D setting and the ever-increasing reliance placed on the correct operation of that software. Poor quality in software has been implicated recently in nationwide telephone service disruptions, multimillion-dollar financial losses, and injuries and deaths in computer-controlled equipment from factory floors to medical clinics to aircraft. [2,3]. How much risk is the R & D effort willing to bear in its use of software? What are the practical alternatives?

Software is typically written to perform data- and computation-intensive, often time-critical, tasks. The inherent complexity of such applications, however, means there is great uncertainty that a given software product will satisfy its requirements while also not introducing significant new failure modes. The task of quality assurance is to decrease that uncertainty to an acceptable level.

The complex nature of the software and the means by which it is produced and controlled preclude a simple or unitary means of ensuring that any given software item is sufficiently fit for use. Instead, a set of complementary activities have to be undertaken, encompassing prevention, appraisal, and consequence mitigation. These approaches can be thought of as addressing the products, processes, and personnel involved in software development.

Many efforts focus on the software *products* themselves. Are all the required documents present and in suitable form? Can the static and dynamic properties of the software be assessed as acceptable? Appraisals are typically based on inspections, walkthroughs, and a variety of testing techniques.

Other activities center on software development *processes*. Were the actions well planned and executed? Did forethought and discipline characterize the work? Auditing is the primary tool for this class of appraisal, along with capabilities assessments.

Finally, the *personnel* working on the software are the subject of certain investigations. What selection criteria were used in assigning individuals to specific projects? Were these individuals appropriately trained and supported?

Qualifications, including both educational achievements and professional experience, are most typically controlled through internal procedures or contractual arrangements.

A. Products

Products can be inspected for possession of specified characteristics. The development of a computer-based system involves the generation of incremental software deliverables (requirements specifications, design descriptions, source code listings) as well as the plans for controlling and assessing those deliverables (configuration management, quality assurance, verification and validation). Standards may be applied to evaluate:

- The existence and format of documentation items
- Compliance with computer-language syntax and style
- Satisfaction of acceptance criteria for performance

Figure 1 indicates some representative product evaluation methods, along with typical categories for acceptance criteria.

A substantial body of accepted product standards has emerged over the past decade, sponsored by a variety of professional organizations such as the IEEE Computer Society [4]. In fact, most consensus software standards have dealt with products, perhaps by analogy to well-established quality control techniques. Increasingly, these standards are addressing quantitative elements, or software quality metrics. Such metrics may define the structural complexity of a software

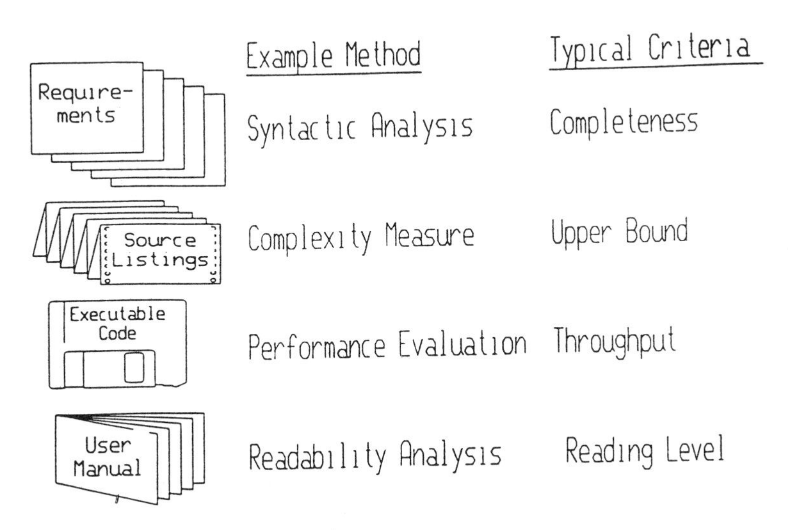

Figure 1 Software product evaluations.

product, its projected reliability, or the degree to which it has been exercised during testing [5].

B. Process

Process concerns are of more recent origin. One organizing principle has been the notion that software development practices can be thought of as maturing through a series of stages or maturity levels. (This model has been popularized most successfully by the Software Engineering Institute [6].) Evaluation of an organization and its practices can be used to assign that organization a development maturity level and to prescribe remedial or improvement actions. Figure 2 shows one model for evolution through the various development process maturity levels.

Internationally accepted quality management systems standards, known as the ISO 9000 series [7], not only address the general issue of managing processes but also contain a software-specific guidance document. Quality systems are audited for compliance to the requirements of the appropriate ISO 9000 standard and can achieve certification by a third-party assessor. On the other hand, the process assessment methodology developed by the Software Engineering Institute is based on use of a questionnaire and in-depth technical interviews. The ideal in either case is a software-development process under statistical control, where it can generate products within anticipated limits of cost, schedule, and confidence in quality [8].

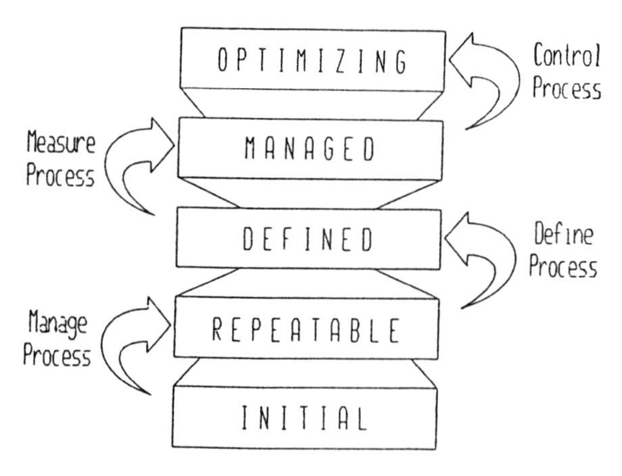

Figure 2 Evolution through Software Engineering Institute process capability maturity model.

C. Personnel

Personnel qualifications could be established at a project, corporate, or industry level. One guideline might be the emerging certification being sponsored by the Software Division of the American Society for Quality Control (ASQC), which is illustrated in Table 1 [9]. Several other organizations also now grant individuals in specialized technical fields a qualification certificate, based on a mixture of formal training, job experience, examination results, and personal recommendations.

III. FORMALIZING SOFTWARE QUALITY

Wernher von Braun is supposed to have said, "Research is what I'm doing when I don't know what I'm doing." In a research setting it is often claimed that there are no formal requirements for beginning much of the work, nor clear-cut

Table 1 Proposed Body of Knowledge for ASQC Certified Software Quality Engineers

I. Software Quality Management
- Total Quality Management
- Software Quality Planning
- The Software Quality Function
- Quality Information Systems
- Quality Management Tools
- Quality Education and Training
- Professionalism

II. Software Engineering
- Basic Concepts
- Software Eng. Techniques
- Software Eng. Life Cycle

III. Project Management
- Planning and Control
- Managing People

IV. Appraisal
- Software Inspections
- Testing
- Verification and Validation
- Assessments, Audits, and Reviews

V. Issues
- Controls
- Data Integrity
- Disaster Planning
- Liability
- Maintainability
- Reliability
- Risk Management
- Safety
- Security

VI. Analytical Methods
- Metrics and Measurement
- Probability and Statistics
- Statistical Process Control

VII. Quality Systems
- Software Corrective Action
- Configuration Management
- Standards and Procedures
- Improvement and Innovation
- Software Quality Function Deployment
- Procurement

acceptance criteria at the end. However, that does not excuse the acquisition of software—any more than any other activity—being done in a haphazard or undocumented manner.

Software development activities need to be formalized at different organizational levels, ranging from the facility-wide to project specific. At the highest level there is an institutional quality program, often described in a document or set of documents called a quality manual. This program is implemented through a full set of procedures available for use as specified within each project. A progression down this organizational ladder might contain elements as shown below.

Procedures: Facility-wide
Design standard, coding standard, testing standard

Instructions: Specific to each organizational unit
Yourdon methodology, C language conventions, statistical validation

Plans: Project-specific
Project design plan, project coding plan, project test plan

A. Procedures

The standards are called out by facility-wide procedures such as:

System development
Software development
System evaluation and certification
System documentation
Configuration management

Each procedure contains high-level requirements, mapped to various industry-accepted standards, which could be satisfied in several different ways.

B. Instructions

The instructions, on the other hand, are written as the default mode of performing tasks in a given organization. If, for example, the organization typically used a particular design methodology or programming language, those details would be captured as operating instructions.

C. Plans

Finally, any given project will need plans to describe activities unique to that specific effort. The plans are written from boilerplate fill-in-the-blanks provided in the corresponding instruction, which in turn satisfied the higher-level procedure and thus the industry standards ultimately levied by the customer. Table 2 indi-

Table 2 Candidate Industry-Consensus Standards to Be Satisfied in Procedures

Software Quality Assurance
ANSI/IEEE Std 730-1989, Standard for Software Quality Assurance Plans.

Software Configuration Management
ANSI/IEEE Std 828-1990, Software Configuration Management Plans.

Software Evaluation
ANSI/IEEE Std 829-1983, Software Test Documentation
ANSI/IEEE Std 1012-1986, Software Verification and Validation Plans
ANSI/IEEE Std 1028-1988, Software Reviews and Audits

Software Documentation
ANSI/IEEE Std 830-1984, Software Requirements Specifications.
ANSI/IEEE Std 1016-1987, Software Design Descriptions.
ANSI/IEEE 1063-1987, Software User Documentation

Software Development Process
Software Engineering Institute: Capability Maturity Model

cates typical industry consensus standards that may be imposed, which software development projects could reasonably be expected to satisfy.

IV. TAILORING FOR RISK

Ideally, each project will involve an appropriate matching of assurance efforts with risks. Different types and degrees of assurance would surely be sought for different software that ranged from the routine to the life-critical. Defining a given application at a specific risk level will guide quality efforts in ways that might avoid both wasteful overkill and tragic undershoot [10].

A. Risk Level Assignment

Risk level assignments may be based on an evaluation of the size of the software and its impact on the system in which it is to function. Additionally, any system whose failure has the potential for threatening human health or safety must receive more stringent attention. There is a trade-off between the simplicity of fewer levels and the discrimination possible with more levels. Three to five levels is probably most manageable.

In the example contained in Table 3, all nonsafety systems are assigned to Class D, C, or B. Safety systems receive an assignment one class more stringent than an equivalent nonsafety system. (That is, if size and impact would otherwise indicate assignment to Class C, a safety system would be assigned to Class B.)

Table 3 Risk Level Assignment

Class	Criteria
Class D:	fewer than *n* function-points of delivered functionality AND less than *x* work-months of development (including documentation) AND failure of software will not degrade system performance AND non-safety system
Class C:	safety system otherwise Class D OR (non-safety system AND fewer than *5n* function-points of delivered functionality AND less than *3x* work-months of development AND failure of software does not preclude work-around to accomplish system mission)
Class B:	safety system otherwise Class C OR (non-safety system AND (more than *5n* function-points of delivered functionality OR more than *3x* work-months of development OR failure of software precludes system accomplishing its mission))
Class A:	safety system otherwise Class B

B. Corresponding Tasks

Assignment to a particular risk level is then associated with a specific set of development and assurance tasks mapped across the software development life cycle. Typical sets of tasks that could be identified for each risk level are illustrated in the following matrix (Table 4).

V. A CONCEPTUAL MODEL FOR SOFTWARE QUALITY

Consider any project that involves software development as taking place in a multi-dimensional reference frame. Various aspects of the project (budget, tech-

Table 4 Task/Risk Level Matrix

RISK LEVEL:	A	B	C	D
TASK				
		REQUIREMENTS PHASE		
Software Requirement Specification				
functional requirements	X	X	X	X
performance requirements	X	X	X	
interface requirements	X	X	X	
design constraints	X	X		
quality attributes	X			
Software Requirements Review				
peer review			X	
independent review		X		
safety review board	X			
System Hazard Analysis	X			
		DESIGN PHASE		
Software Design Description				
control flow	X	X	X	
pseudo code	X			
Software Design Review				
peer review			X	
independent review		X		
safety review board	X			
Test Plans and Report [for test(s) shown below]				
independent review	X	X		
		IMPLEMENTATION PHASE		
Source Listings				
convention, style guides	X	X	X	X
choice of language	X	X		
metric standards	X			
Source Code Review				
peer review			X	
independent review		X		
safety review board	X			
Perform Unit Testing	X			
		TEST PHASE		
Perform Integration Testing	X	X		
Perform System Testing	X	X	X	
		INSTALLATION PHASE		
Functional Audit	X	X		
Physical Audit	X	X		
Perform Installation Testing	X	X	X	X

nical progress, and so on) can be thought of as axes along which measurements can be made. Often, two characteristics are related by drawing a two-dimension graph of, say, budget expenditure versus time, or work units completed versus budget expenditures. Three such axes, and their interrelations, are particularly helpful in looking at software development. They represent time, abstraction, and management visibility.

A. Time Axis

Obviously, projects will be tracked against calendar time. Deadlines for progress reports, holdpoints, and deliverables will be estimated, negotiated, slipped, and revised. Time is the dimension across which various efforts will be integrated: person-months of design, machine-hours of testing, or the like.

The standard models of software (and indeed system) development agree in representing time-phased activities. That is, a certain sequence of tasks is established. Classic representations divide the development effort into a series of phases, one of which predominates at any given time. Thus, we talk about the Requirements Phase or the Implementation Phase of a software project. Not that other types of activities aren't also going on simultaneously, but the focus at a certain time is on one dominant undertaking.

Control over the time dimension is usually accomplished by demarcating each of these phases with some sort of checkpoint or holdpoint. At the boundary between the Implementation Phase and the Test Phase, for instance, we may place a Test Readiness Review. At the boundary between the Installation Phase and the Operation Phase we may place an Operational Readiness Audit. To the extent that the application is inherently critical or risky there may be more phases, and hence more checkpoints, designed into the project.

Often, different specialized individuals or groups will perform the tasks associated with each of the development phases. The phase boundaries then also become the transitions where information is passed over from a design analyst to a programmer or from a coding team to a testing team, much as the baton is passed from one runner to another in a relay race.

Figure 3 shows a typical sequencing of software development activities, divided into some five phases.

B. Abstraction Axis

The various intermediate products during software development range all the way from documents written in natural language (English narratives) to computer coding specific to a given programming language, operating system, and hardware configuration. The trend is from the abstract to the specific. Highly abstract means lacking in implementation details: many different specific programs could be written to satisfy any one given set of high-level requirements. This progres-

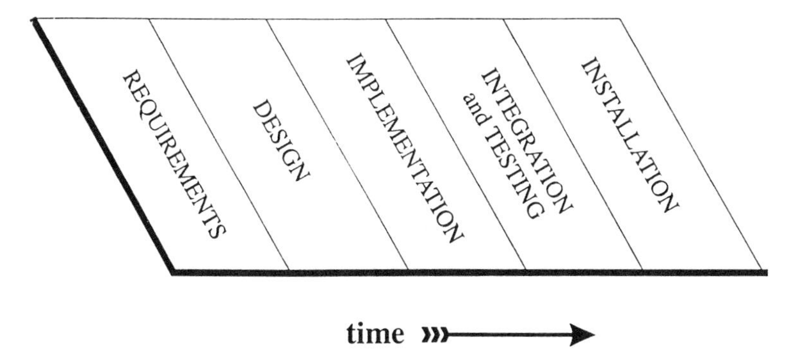

Figure 3 Time-phased activities.

sion is also from the relatively accessible (even if all English documents—such as legal contracts—are not that directly comprehensible to the layman) to the relatively inaccessible, requiring specialized knowledge and training both to create and to interpret.

Figure 4 indicates some of the levels of abstraction in the representation of what will eventually be a software-based system.

Combining these first two axes gives a plot of typical products throughout the development life cycle, shown in Figure 5. Early in the project relatively abstract items, such as requirements specifications, are generated. Later the deliverables become more specific and inaccessible (being represented, for instance, in some particular symbology). At the latest times one finds the least abstract products, which are also the least accessible.

C. Management Visibility Axis

A third axis of consideration is that of management visibility. Some activities and products are scrutinized at the highest levels of management and by the ultimate customers or users. Other items are not reviewed except at very low levels, as by peers or first-level managers. The choice of how much visibility to accord any activity may depend on perceived risk, organizational custom, per-

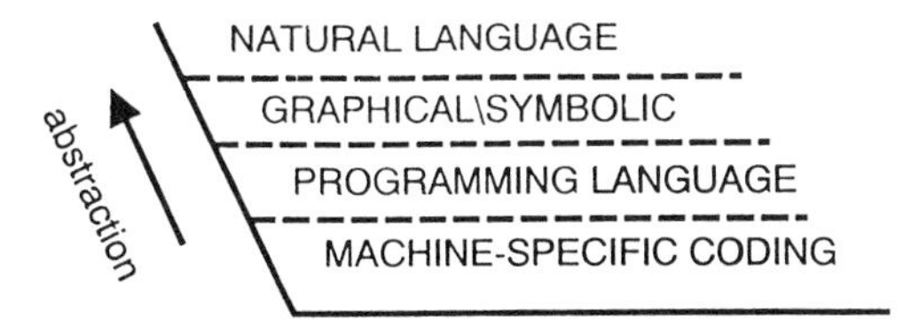

Figure 4 Abstraction of representations.

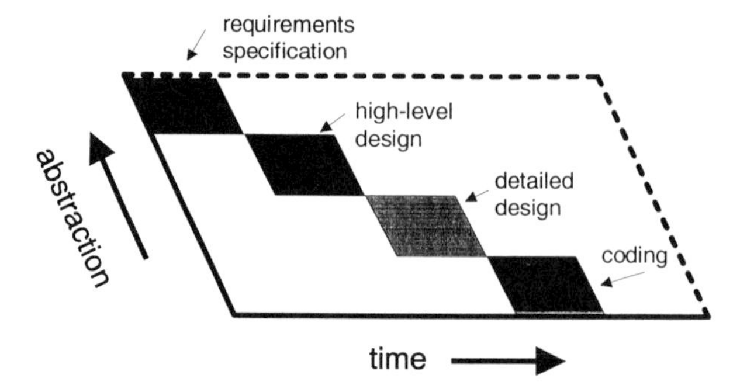

Figure 5 Life-cycle products.

sonal interest, or any of several other factors. The challenge is in matching each item to the appropriate level of attention, so there is neither too little oversight nor too much ''second guessing.''

Figure 6 superimposes this additional axis onto the other two already discussed. Typical review activities occurring throughout a software development project are illustrated in Figure 7.

On such a diagram the end points of any development effort—project initialization and project completion—are positioned as shown in Figure 8. The drama lies entirely in getting from one to the other. Look at Figures 9 through 12 to see both unacceptable (''pathological'') and acceptable trajectories through time-abstraction-visibility space. Look for the twin evils of too little oversight (Figure 9) and too much ''micromanaging'' (Figure 10).

The elements of software total quality that need to be managed may be projected onto our three axes: configuration management along the time axis,

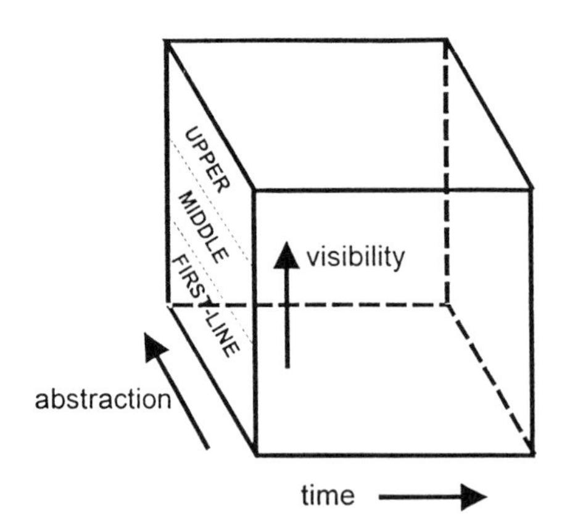

Figure 6 Levels of management visibility.

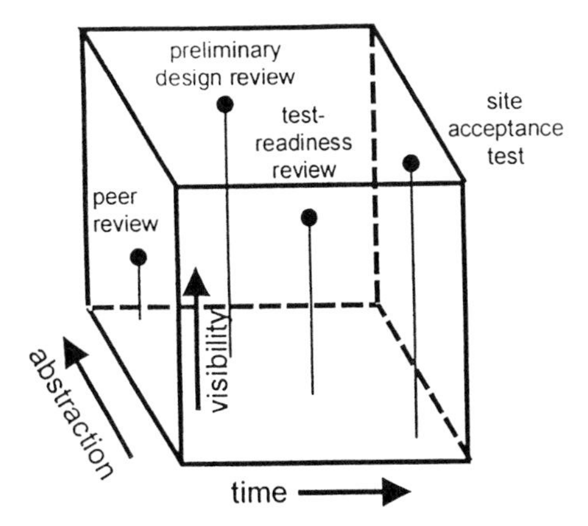

Figure 7 Typical reviews.

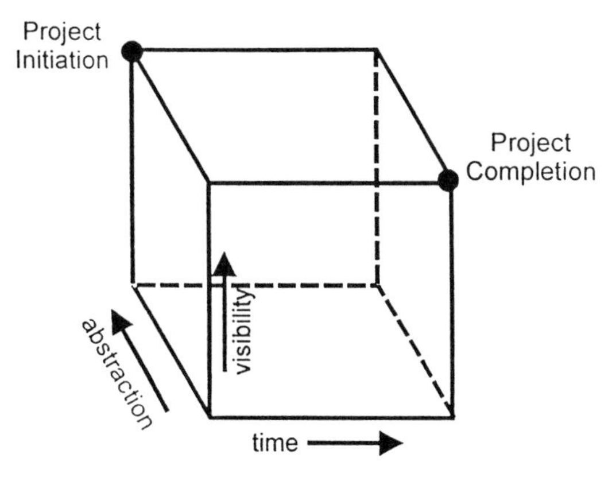

Figure 8 Project endpoints.

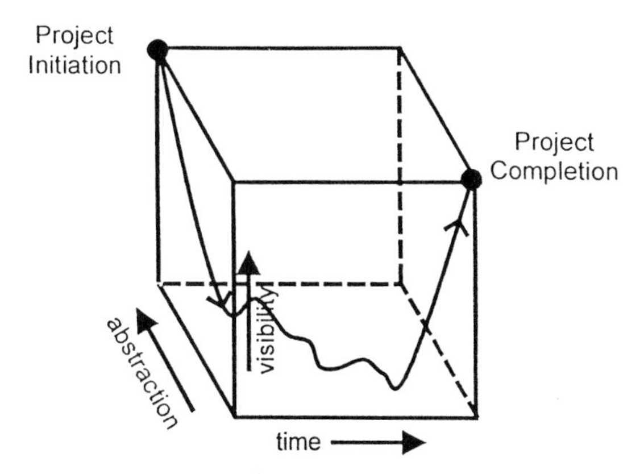

Figure 9 "No visibility" pathological trajectory.

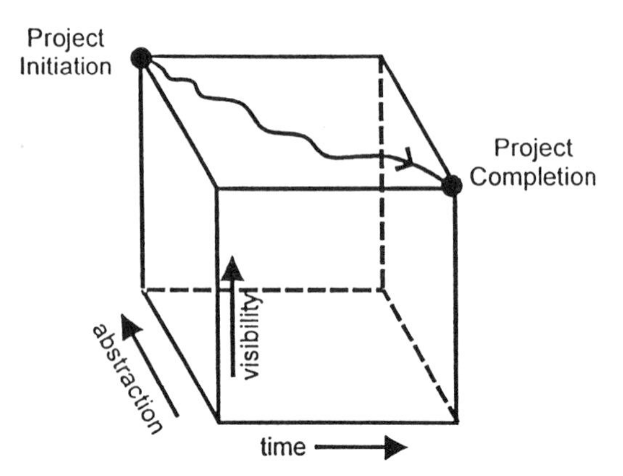

Figure 10 "Second-guessing" pathological trajectory.

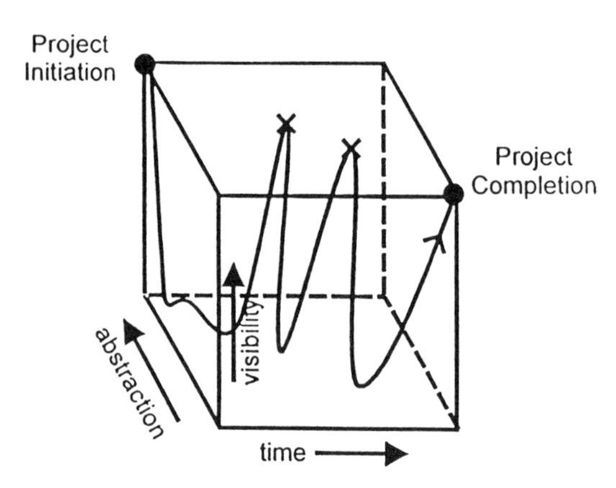

Figure 11 Improved trajectory.

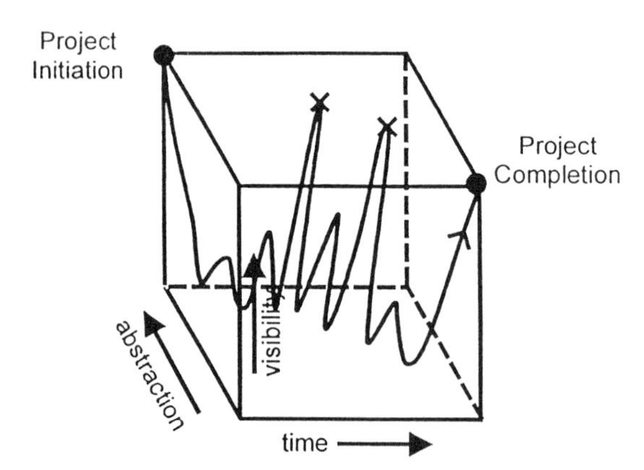

Figure 12 Preferred trajectory.

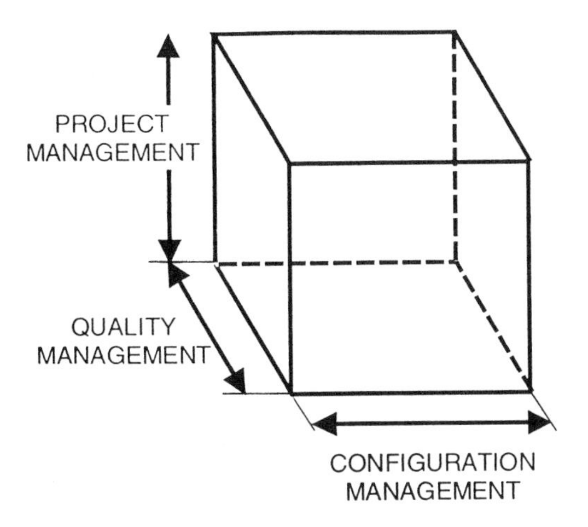

Figure 13 Elements of software total quality.

quality management along the abstraction axis, and project management spanning the management visibility axis (Figure 13). This mapping is just a convenient way of addressing the different management issues involved in developing software. Typical concerns of each of these management elements are shown in Table 5.

VI. ENGINEERING OF SOFTWARE FOR ACCEPTABILITY

Current models of quality management emphasize that any business process should view itself as both a customer (of its suppliers "upstream") and a producer (with its own customers "downstream"). In fact, many processes have multiple customers, with potentially conflicting expectations and demands. Software-based products in an R & D setting illustrate such a range of customers and expectations.

Table 5 Concerns of Software Total Quality

Project management	Configuration management	Quality management
Proposal preparation	Identification scheme	Quality engineering
Estimating	Document preparation	Development standards
Budgeting	Review and release	Technical reviews
Risk management	Change control	Vendor audits
Technique selection	Status accounting	Testing
Managerial reviews		Error reporting
Status tracking		Corrective action

Any software, if it is to be considered a quality product, must be acceptable to its several audiences. Those who must accept a software product include the sponsors who pay for the work, the users who exercise the finished software, and, in certain critical applications, the regulators who approve the use or results of the product [11].

Each of these customers characterize acceptability in a different way. The sponsor seeks confidence that the software does indeed satisfy specified requirements. The user values the ability of the software to perform consistently and correctly within the established work environment. The regulator is charged to see that software does not diminish the safety of the system where it is applied.

How is software made acceptable to these different customers? The sponsor needs visibility of the development process and self-descriptiveness within the product. The user wants ease of learning and use, as well as reliability in operation. The regulator requires reviewability and, to the extent it is present in other system components, quantifiability.

The term *software engineering* was coined to advance the claim that principles and practices already successfully applied in the engineering of a wide range of material artifacts could also address the challenge of developing software and software-based systems [12]. Basic good engineering practice can certainly be applied to addressing the fundamental questions for any software-based system. What is the software needed for? How must it interact with other systems components (hardware and software), as well as human users? How well must it perform? What are the requirements for documentation of the product and of the process by which it was produced? What provision must be made for test data? What are the required comparisons to other software programs, plus lab and field data?

The engineering of software relies on several complementary approaches to providing acceptable products:

- Full and formal communication of everyone's expectations
- Trained and well-equipped project personnel
- A disciplined, phased development process, and diverse product evaluation methods.

Each of these considerations must be incorporated into software development and evaluation procedures. The engineering approach for software can be shown to enhance shared perceptions, analysis of hazards, effective translation into operational form, and visibility of both process and product.

Software must be both verified and validated. Though often used loosely and even interchangeably, these words have precise and distinct meanings within the software quality discipline [13]. To verify software is to confirm that it satisfies its specified requirements. In other words, is it doing the job right? On the other hand, to validate software or a software-based system is to map its performance against an objective external reference. This reality check asks, is it doing the right job? Perfectly self-consistent and sufficiently verified software may still be invalid if the requirements don't match reality.

A number of evaluation approaches, including various analyses, inspections, and test methods, need to be organized within a comprehensive framework for characterizing software. Well-designed and well-conducted quality evaluations can then establish an adequate level of confidence in the fitness of the software for its intended use.

REFERENCES

1. American Society for Quality Control, Special issue on Software Quality Assurance. *Quality Progress*, XXI (November 1988):29–70.
2. Neumann, Peter G., compiler, "Illustrative Risks to the Public in the Use of Computer Systems and Related Technology," *Software Engineering Notes*, 16 (January 1991):2–9.
3. Leveson, Nancy G., and Clark S. Turner, "An Investigation of the Therac-25 Accidents," *Computer*, 26 (July 1993):18–41.
4. Institute of Electrical and Electronics Engineers, *Software Engineering Standards*, Spring 1993 Edition, New York, 1993. (These standards are included: ANSI/IEEE Std 610.12-1990, Standard Glossary of Software Engineering Terminology; ANSI/IEEE Std 730–1989, Standard for Software Quality Assurance Plans; ANSI/IEEE Std 828–1990, Standard for Software Configuration Management Plans; ANSI/IEEE Std 829-1983, Standard for Software Test Documentation; ANSI/IEEE Std 830-1984, Guide to Software Requirements Specifications; ANSI/IEEE Std 1012–1986, Standard for Software Verification and Validation Plans; ANSI/IEEE Std 1016–1987, Recommended Practice for Software Design Descriptions; ANSI/IEEE Std 1028–1988, Standard for Software Reviews and Audits; ANSI/IEEE Std 1042–1987, Guide to Software Configuration Management.)
5. Musa, J. A., A. Iannino, and K. Okumoto, *Software Reliability: Measurement, Prediction, Application*, Professional Edition, McGraw-Hill, New York, 1990.
6. Humphrey, Watts S., "Characterizing the Software Process: A Maturity Framework," *IEEE Software*, 5 (March 1988):73–78.
7. International Organization for Standardization, ISO 9000, Quality Management and Quality Assurance Standards—Guidelines for Selection and Use, Geneva, 1987.
   International Organization for Standardization, ISO 9000-3, Quality Management and Quality Assurance Standards - Part 3: Guidelines for the application of ISO 9001 to the development, supply, and maintenance of software, Geneva, 1991.
   International Organization for Standardization, ISO 9001, Quality Systems—Model for Quality Assurance in design/development, production, installation, and servicing, Geneva, 1987.
8. *IEEE Software* (special theme articles on software process), 10 (July 1993):14–64.
9. Daughtrey, Taz, "Certifying Software Products, Processes, Personnel," *Transactions of the 46th Annual Quality Congress*, American Society for Quality Control, Milwaukee, 1992.

10. Levinson, Stanley, and Taz Daughtrey, ''Risk Analysis of Software-Dependent Systems,'' paper presented at American Nuclear Society Probabilistic Safety Assessment International Topical Meeting, Clearwater, Florida, January 1993.
11. Miller, Douglas R., ''The Role of Statistical Modeling and Inference in Software Quality Assurance,'' Centre for Software Reliability Workshop on Software Certification, Gatwick, England, September 1988.
12. Shaw, Mary (interview with), ''Seeking a foundation for software engineering,'' *IEEE Software*, 7 (March 1990): 102–103.
13. American Nuclear Society, ANSI/ANS-10.4-1987, Guidelines for the Verification and Validation of Scientific and Engineering Computer Programs for the Nuclear Industry, LaGrange Park, Illinois, 1987.

10

VERIFICATION

George W. Roberts
Babcock & Wilcox
Alliance, Ohio

As in controlling project execution, various tools are available to verify technical aspects of the work. Some of these, such as design reviews and inspections, seem more appropriate to the construction of the test facility or test specimen. Others, such as peer reviews, seem more effective when applied to the technical results. Technical reviews may be used in both situations and audits are useful to evaluate the full scope of research activities, at the project and at the institutional level.

I. FACILITY VERIFICATIONS

A. Design Review

Historically, the initial purpose of the design review was the early detection and remedy of design deficiencies which could jeopardize the successful performance of a product or process being issued, the cost to build or fabricate the product, or the ease of maintaining the product (1). To be effective, a design review should be conducted before the commitment of major resources.

Design reviews were initially established on a cost–benefit basis. With the advent of the military and aerospace programs, however, design reviews have become more of a mandatory process and are dictated by either customer or government requirements. Specifications with design review requirements [for example, Army AMC 703-2 (2) and NASA's NHB 5300.4 (3)] are normally associated with the reliability programs. Design review is permitted by 10 CFR 50, Appendix B (4) as one alternative form of checking the design; other choices include independently verifying calculations or prototype/qualification testing.

In the past, courts have been reluctant to impose liability in design, particularly for products made by established manufacturers. They did not want a jury to have to evaluate a product prepared by experts in their field. The courts also realized that once a judgment was made against a contractor or manufacturer, it could open the door for additional claims against that manufacturer by other people. However, liability is being imposed today in an increasing number of cases for negligence of design (5). Design negligence is predicated on one of three theories (6):

1. Failure to use material of adequate strength or quality
2. The incorporation of a concealed hazard
3. The failure to employ needed or reasonably required safety devices.

This requires the manufacturer to know about the latest developments in the field, including safety precautions. It is interesting to note that: "Noncompliance with a standard established by statutes, ordinance, or administrative regulation or order furnishes evidence of deficient design, and may, under certain circumstances, constitute negligence per se" (6).

The reverse is not necessarily true. Mere compliance with codes and standards may not prove that the manufacturer has exercised sufficient reasonable care in designing a safe product. Therefore, compliance with codes should not be taken as a substitute for a design review.

Although the initial objectives of a design review program may be to minimize the risk and cost associated with new products or processes developed by the R & D organization, the use of the design reviews can also be extended to new or redesigned experimental facilities within the division. Identification of the need for this type of design review would be determined by the appropriate laboratory manager, who would make the decision either in accordance with division policy or on a case-by-case basis. Either way, a design review is usually called when the facility is expensive, is complex, or incorporates new technologies. While the provisions for formal design review are offered as guidelines, the actual review of experimental facilities could be conducted at varying levels of formality or management attention, appropriate for the particular situation.

The formal design review process generally includes at least two reviews. The first review, often referred to as a conceptual or requirements review, is usually conducted during the initial stages of planning and focuses on the specifications or functional requirements of the design. The second review, referred to as a final review, is performed at the completion of the design, before vendors of major components are released to begin work, and prior to construction of the facility. If the facility is complicated to operate or if the consequences of failure are significant, a third review is conducted. This review is referred to as a readiness review, the purpose of which is to review all of the preparations for starting and operating the facility. Additional review meetings may be scheduled as necessary if there are modifications to the specifications or to the design.

Each of these reviews is conducted by a board of technically experienced and skilled personnel. They can be drawn from either inside or outside of the organization, but are not directly associated with the development of the design being reviewed. A typical review board consists of five to eight people including a chairman. There is often a tendency to staff a board with too many people. A large board places heavy demands on the organizational and leadership skills of the chairman. Conversely, a small board has difficulty covering the range of technical content inherent in large projects and sophisticated facilities. The laboratory manager contributes considerable value to the review by carefully matching the technical talents and number of board members with the complexity of the project and leadership skills of the chairman.

At the conclusion of each design review, the findings and recommendations of the review board are documented and signed by all participants. The signed document denotes an understanding of the recommendations and includes the opinions of participants who disagree with the recommendations.

Each recommendation of the design review board should be addressed by the design team. If the design team commits to adopting a recommendation, they should explain how they intend to meet that commitment. If they decide not to adopt a recommendation, they should explain the reasoning for that decision. The organization should have a clearly identified process for monitoring and assuring completion of these commitments. There is often a tendency for management to charge the design review board or its chairman with the responsibility to monitor the design team and to assure that its commitments are completed. This rarely works. The typical board members are from other parts of the organization, they do not have routine access to information on status, and they do not control the resources necessary to meet the commitments. This responsibility clearly belongs with management and must remain there.

Design review meetings, particularly those which have as their purpose the final approval of the design, rely on a number of a characteristics to ensure their success. Primary among these is the element of formality. Design reviews should be planned and scheduled like any other legitimate activity. The meeting should be structured around an agenda, which typically includes discussion of the following:

1. *Requirements:* These include codes, standards, regulations, quality assurance, handling, storage, cleaning, shipping, and identification.
2. *Design methods:* These include inputs, outputs, analyses, computer programs, and standards.

3. *Adequacy of the design:* This includes an evaluation of the ability of the design to meet each of the requirements. Also considered are tests, test results, inspections, acceptance criteria, pre-operational checks, experience with similar designs, failure modes, and documentation.
4. *Design interfaces:* These include manufacturability, constructability, vendor equipment, environments, transportation, guarantees, existing systems, maintenance, inspection, and operator training.
5. *Safety:* This includes compliance with safety regulations in addition to anticipated failure modes.

Design review checklists are often used to develop the agenda and the design information package. The checklist will vary depending on the industry, product, or service involved, but the list above is illustrative of the scope of the review. A copy of the agenda should be sent out to the design review board members prior to the meeting, along with sufficient additional information enabling them to become familiar with the design characteristics and to develop pertinent questions to be discussed during the design review meeting. The design information package should at least include the agenda, alternatives to the design, results of any special tests performed to develop or verify the design, the design drawings, the recommendations developed during the specification review, a comparison of the design to the specifications, costs to manufacture to the recommended design, and costs for alternative designs which have been considered.

The design review closes with a report. The design review report should include a discussion of the objectives of the design, specifications, design drawings, backup information, recommendations of the design review board, minority opinions, and responses to each of the recommendations of the board. Once the design review report has been generated, it should be signed by all members of the design review board signifying their agreement with the contents of the report. A file should be maintained for all information pertinent to the design process, including the information package that has been distributed to participants and the documentation of the review meeting. The design review report and backup file should then be retained for as long as the item is being marketed or used.

B. Technical Review

If *all* designs were to be verified or challenged, the use of formal design reviews would be very costly. An alternative that is less expensive and less rigorous is a system whereby the reviewing of technical data is done on a day-to-day basis involving one or two people. In a research project, a competent person who hasn't worked on the data can accomplish the task quite easily and efficiently. A project may involve in-depth design analyses for a field of interest which may be less broad than the total design of a given product, for example, vibration analysis of a tube bundle assembly within a large steam generator. The independent technical review is oriented toward that tube bundle analysis only. Later, the results of that analysis and several additional analyses of different components of that steam generator may be taken into account during a major design review activity for the overall steam generator. Taking this reduced scope into consideration, the requirements for independent technical review for a component can be modified from the requirements for design review at the system level.

If there is a design of a test facility, the design calculations and drawings should be submitted for an independent review. The test facility design should be evaluated for:

1. Conformance with project specifications
2. Material compatibility with the environment, such as pressure temperature, corrosion fluids, water chemistry, etc.
3. Design interfaces, as related to the fitup of parts
4. Dimensional stability, including the coefficient of expansion, creep, fatigue, loading, etc.
5. Conformance with safety requirements
6. The correctness of information used for the calculations
7. Adequacy and correctness of any computer programs involved
8. Appropriateness of specified quality standards

C. Inspection

Once a purchased item is received at the research center, it should be immediately inspected to verify that the material does in fact meet the purchase order requirements. This inspection may be by a receiving inspector if such an inspection is warranted, but at the very minimum, it should be done by the project leader or one of the technicians. The receiving inspection should be documented and should require immediate corrective action if there are any discrepancies. This is to permit the research center to collect for any adjustments due to defective material, and to prevent schedule and construction delays later on down the line if the material is pulled out of stock and found to be unserviceable. The purchase order should not be closed out nor should the supplier be paid until the receiving inspection is performed and signed off. Appropriate tags should identify the material being held for inspection, that which has been inspected and accepted, and that which is discrepant.

Inspections during test article or facility fabrication are usually performed for those physical characteristics identified by the project leader as being critical to the test results. The need for inspection can be identified on receiving copies of purchase orders, route sheets, and technical procedures or inspection checklists. If desired, a special identification system can be use to denote critical dimensions on the drawings. Dimensions so identified on the drawing would then have a corresponding checklist or route sheet to direct the inspection of that critical dimension. In this manner, inspection efforts will be confined to those attributes that are truly important to the test results. If there are certain construction codes involving test apparatus, the inspection of the attributes that are required by the code can be handled similarly.

Inspections can be performed by any person familiar with the operation who is competent to do the inspection, but some regulatory requirements insist that the inspector should not have been responsible for performing the operation which is being inspected. If compliance is required to an inspection standard or to a construction code or if the project is a fairly large one with construction over a period of years, then it would probably be prudent to use professional inspectors reporting to a separate QA organization. An alternative would be to have an inspection verification by QA. That is, critical inspections may be witnessed by the QA organization even though those inspections may actually be performed by project personnel, machine shop personnel, or someone otherwise technically competent to handle the operation. Appropriate controls over this type of activity could include a program of certification of designated inspectors. A program of training and evaluation could be required along with periodic recertification. This gives the QA organization a more direct control over the inspection function.

Occasionally, the customer will want to watch inspection or testing operations, and those "witness" points or "hold" points can be included in route sheets or technical procedures. It is then up to the project leader to ensure that work does not proceed past any inspection points without appropriate documented authorization from the customer. Documentation associated with inspections will vary to serve the needs of the inspector or technical section and may include specially designed checklists, technical procedures, route sheets, laboratory notebooks, and computer files. Regardless of format, inspection results should include the identification of the inspector, the type of inspection and observation performed (such as visual inspection, dimensional inspection, or the type of nondestructive examination used), the results of the inspection, and information regarding the action taken in connection with any deficiencies.

II. SCIENTIFIC VERIFICATIONS

A. Peer Review

Peer review is a venerable institution that apparently had its beginning in the mid-1600s when the Royal Society of England sought advice from its members about what papers to publish in its journal (7). While it is noteworthy for serving many scientific institutions, the practice has received the most attention for its effect on the approval of research grants, not surprising since the issue boils down to money and personal prestige. Both the National Science Foundation

(NSF) and the National Institutes of Health (NIH) use the peer review process to select worthy research projects for funding.

The second major use for peer review is for judgment of acceptability prior to publishing in a technical journal. It is a form of final inspection before officially declaring the results of a research effort.

Peer review may be used as a personal assessment of a professional's conduct. One member of the Physician Insurers Association of America invokes a peer review upon receipt of a single claim that is questioned by the insurance underwriter. The case is examined by the peer review committee and a vote is taken to determine whether the physician performed at an acceptable level (8).

Sometimes peer review may be used as a broader operational evaluation of an entire research facility. This application approaches that of an operational audit and emphasizes the importance of technical credibility when performing any evaluation.

The use of peer review has not been without substantial problems. Representative John B. Conlan (R-Arizona) has charged that NSF's management practices were closed and unaccountable to the scientific community (9). During congressional hearings in 1975, he charged that "the NSF staff appears to have purposely misrepresented reviewers' comments to the programs committee of the National Science Board in order to get approval of $2.2 million in further funding." He cited an instance where an NSF program manager and his superiors misrepresented peer review comments. The program manager excerpted the peer reviewer's remarks in such a way as to indicate the reviewer's unqualified support for a report that the reviewer had, in fact, criticized severely. Others defended the process with equal vigor. With such acrimony, a series of congressional hearings lasted over nine years. Although battered about, the process has survived intact and is a cornerstone of the scientific verification methodology.

There is substantial evidence that a wide variation exists in the application of peer review and the concomitant results. The National Academy of Sciences' Committee on Science and Public Policy (COSPUP) conducted a five-year study of NSF's peer review. (10) The first part of the study, reported in 1978, established that there was essentially no systemic bias, although there was a high correlation between reviewer's ratings and grants made. Of 1200 proposals, there was not a high correlation between grants awarded and measures of the previous scientific performance of the applicants. Nor was there evidence that reviewers at major institutions treated proposals from applicants at major institutions more favorably. However, there were low to moderate correlations between reviewer ratings and the rank and prestige of the applicant.

The second phase of the COSPUP experiment did establish that there was a greater degree of variability injected into the process by the reviewer than that injected by the proposal itself. Using a second set of reviewers for proposals previously accepted, there was a 25 percent incidence of reversal in recommendations to fund the work. The COSPUP report was careful to stipulate that the results applied only to basic science programs at NSF and that the NIH study section form of peer review differs significantly.

Variation occurs between disciplines as well. Hargens (11) examined referee's evaluations of submissions to the *Astrophysical Journal*, *Physiological Zoology*, and the *American Sociological Review* and found striking differences in outright rejection rates as well as in the average number of revisions of eventually accepted papers and in the average time lags between submission and final editorial decisions. He concluded that the data supported claims that "disciplinary differences in consensus on research priorities and procedures contribute to variation in typical journal peer review systems."

Ceci and Peters (12) seem to take issue with the COSPUP experiment in one major area: The authors resubmitted ten previously approved articles to the same journals. They retyped the articles but only changed some minor wording in the abstracts. They replaced the author's names with names that had no meaning in the field of psychology and replaced the names of the prestigious institutions with those having no status in the field. Other than that, there were virtually no changes to the texts. The results were remarkable—nine out of ten of the manuscripts were recommended for rejection without a hint that they might be acceptable with revision or rewriting.

Much criticism and subsequent analysis has been focused on NSF while holding that the NIH system is superior (13). But McCutchen (14) spares neither, stating that "a lynch mob is also a peer panel," without the rules and procedures required for a jury trial. He believes that "specialist peer review is fraught with biasing influences." Having worked within the systems of both the NSF and the NIH, he accuses NIH of cronyism among the reviewers (who are more independent than at NSF) and complains of a lack of accountability. He goes on to say that, "Unfortunately, the power of referees, usually anonymous, permits self-interest, jealousy, revenge, and other unworthy motives to influence decisions."

There are some inherent weaknesses in the peer review concept. Crawford and Stucki (15) describe the research process as having the following series of tasks:

Inception of idea
Formulate problem statement or hypothesis
Develop methodology
Develop proposal, obtain funds, identify team of co-workers
Set up laboratory/field group
Test hypothesis—perform the research, collect data
Process and analyze data, make inferences
Report results which are peer-reviewed

They maintain that peer review will not detect errors or falsification in collecting or analyzing data. This is an area that can only be verified by interim checks that are more appropriately done during the conduct of the experiment (with an in-house review process such as technical reviews). It is their opinion that it is during the final three steps that misrepresentation is most likely to occur. Since peer reviewers are reading about the work as it is reported in a manuscript, they do not have access to procedures during the data collection and analysis stages. As Arnold Relman, editor in chief at the *New England Journal of Medicine* said, it's "impossible for journal editors to know who's cooking data" (7). It is this inability for the journal type of peer reviewer to get at the actual laboratory setups or to see the raw data that makes the process an "either–or" situation. Another editor, Daniel Koshland, Jr., of *Science*, said "Peer review identifies the very good and the very bad papers. It's the middle ones that are difficult to judge. That's what editors are for" (7).

As an aside, while doing the research for this book, a number of papers and texts were reviewed that purportedly would give some illumination or definition for the "scientific method." The model by Crawford and Stucki seems to be quite close to defining what it is supposed to be, but the National Academy of Science says there is no such thing (16). There is no scientific method, but there are a variety of methods used by scientists. While searching further to discern what, if any, guidance could be obtained as to improved methodology, some help was found: A. G. Webster, one of the founders of the American Physical Society seemed to sum it up quite well: "The proper order of procedure may be stated: 'Think, evaluate, plan, experiment, think,—and first, last, and all the time think.' " His observation of some actual practice was less encouraging. "The method often pursued is: 'Wonder, guess, putter, guess again, theorize, and above all avoid calculation' " (17).

Back to the question at hand. With all the problems, there are still significant advantages to performing peer review, particularly in the absence of any other form of verification save replication of the experiment. In many instances, the cost of that option would be prohibitive—how many supercolliders can anyone afford? One particular application is prior to publishing. Even McCutcheon observes that when peers referee journal articles, they may find mistakes and sometimes fraud.

And there is the increasingly disturbing problem of professional liability. There was an increase of almost 300 percent in the frequency and severity of liability claims against lawyers between 1980 and 1989 (8). Similarly, suits against architects and engineers rose steadily from the 1960s through the mid-1980s before starting to abate somewhat. At one time, an estimated 750 claims were filed annually against this group of professionals in California, alone (18). Insurers of professional liability coverage and professional associations took note of how medical insurers use peer review when underwriters raise questions about claims. And they took a lesson from the medical profession and began using peer review to reduce malpractice claims (8). Now, Design Professionals Insurance Co. of Monterey, California, will fully reimburse its policy holders for the cost of peer review (up to a limit). The program is administered by the American Consulting Engineers Council (19).

But, given that you are going to conduct a peer review, the first step is to identify someone to do the review; you need a peer. Bodnarczuk maintains that the prerequisite for being a peer is first having authority in the field, i.e., having a proven record as a doer before being a checker (20). The second requirement is that to be a peer, one must also be an "active competitor who pursues the same type of research." By that narrow definition, Bodnarczuk states that "Einstein maintained much of his authority, but ceased to be a peer in regard to the Copenhagen interpretation of quantum mechanics." Bodnarczuk does shed some light on the success of peer review over the ages by stating that it is this element of active competition required of a peer that maintains the sharpness and integrity of the peer review process.

Others don't feel the criteria should be that restrictive. Consulting Engineer Don Kline from Kimley Horne Associates in Raleigh, North Carolina, says, "You don't need to pick someone who's an everyday competitor. For instance, the reviewer might be from a different geographical area or from a firm whose size precludes it from competing with the designer. You could choose a reviewer who has retired but is still considered an authority" (19). Certainly, Kline agrees on the requirement for having authority.

Once you've identified the peers, how should they go about the review? One of Ralph Nader's lawyers has suggested five broad areas for improvement of the peer review process (21):

- Make records accessible.
- Keep complete records.
- Keep minutes of panel meetings.
- Notify applicants of derogatory comments.
- Guard against conflict of interest.
- Make it easier to appeal rejections.

The U.S. Army Inventory Research Office has established some guidelines as to when to do peer review for projects: (22)

- Every project is subjected to frequent peer review.
- Project analysts set the date and time for reviews.
- Attendance at reviews is voluntary, but the proper mix is ensured.
- Reviews are conducted by the project analyst.
- Decisions are made during the review about project direction, rework, etc.

Dominenic Cicchetti, a psychologist with the Veteran's Administration Medical Center in West Haven, Connecticut, evaluated peer review over a twenty-year period and came up with his own suggestions (23):

- Send manuscripts to at least three independent reviewers
- Allow authors to request anonymity from reviewers.
- Encourage referees to take more responsibility by signing their comments.
- Solicit author reviews of referees for periodic evaluation.
- Reward referees who provide consistent high-quality evaluations.
- Develop systems for peer-review appeals.
- Train reviewers by means including distributing guidelines for what makes a good review.

The Office of Naval Research (ONR) has developed review criteria and rating factors to be applied to proposed research projects (24). Reviews are by panels arranged to address areas of similar science and consisting of from ten to twelve reviewers on a panel. An example rating sheet and definitions of the criteria are shown in Figures 1 and 2. ONR projects are normally funded over a five-year period and are subjected to a second peer review midway through the work. A slightly modified rating sheet and criteria are shown in Figures 3 and 4.

For both types of reviews, ONR follows the practice of "priming" reviewers and presenters with as much advance information as possible including copies of the evaluation forms so that both parties understand the process about to be followed. In addition, reviewers are given copies of the narratives and viewgraphs so they may be better prepared to ask questions.

However flawed and despite constant revelations of the "warts" in the system, peer review is still the method commonly accepted as the primary form of quality verification. Its use over the centuries has caused it to become ingrained into the culture of the scientific community. It is becoming even more widely accepted within the engineering profession as more and more of their colleagues come to accept peer review as "the highest order of quality management you can have" (19).

B. Technical Review

The project technical plan (see Chapter 5) identifies the technical requirements and objectives of a research project. The plan includes the philosophy and basic principles involved in the choice of the test or experimental design concept, and any inherent limitations of that design concept. Equipment designs or calculations that provide a basis upon which the experiment is to be conducted, the data is to be evaluated, or the final conclusions are to be drawn may be documented and subjected to an independent review by a someone other than the one who performed the original design or calculation. The overall scope of an independent review should include any calculations made based upon the customer's inputs, those required to reduce experimental test results, and calculations required to complete the final report. The entire project may need to be reviewed to ensure the appropriateness of the experimental design, proper selection and use of measuring apparatus, and the proper reduction and interpretation of results.

RESEARCH OPTION EVALUATION FORM

Title of Research Option ______________________________

Reviewer Name ______________________________

1A. RESEARCH MERIT (Circle one number or –)
1 - 2 - 3 - 4 - 5 - 6 - 7 - 8 - 9 - 10
I--- LOW ---I I------ FAIR -------I I------ AVERAGE ------I I----- GOOD -----I I---- HIGH ---I

1B. RESEARCH APPROACH/PLAN/FOCUS/COORDINATION (Circle one number or –)
1 - 2 - 3 - 4 - 5 - 6 - 7 - 8 - 9 - 10
I--- LOW ---I I------ FAIR -------I I------ AVERAGE ------I I----- GOOD -----I I---- HIGH ---I

1C. MATCH BETWEEN RESOURCES AND OBJECTIVES (Circle one number or –)
1 - 2 - 3 - 4 - 5 - 6 - 7 - 8 - 9 - 10
I--- LOW ---I I------ FAIR -------I I------ AVERAGE ------I I----- GOOD -----I I---- HIGH ---I

1D. BALANCE BETWEEN EXPERIMENT AND THEORY (Circle one number or –)
1 - 2 - 3 - 4 - 5 - 6 - 7 - 8 - 9 - 10
I--- LOW ---I I------ FAIR -------I I------ AVERAGE ------I I----- GOOD -----I I---- HIGH ---I

1E. PROBABILITY OF ACHIEVING RESEARCH OBJECTIVES (Circle one number or –)
1 - 2 - 3 - 4 - 5 - 6 - 7 - 8 - 9 - 10
I--- LOW ---I I------ FAIR -------I I------ AVERAGE ------I I----- GOOD -----I I---- HIGH ---I

2A. NAVAL NEED (Problem or need which this research addresses)

2B. POTENTIAL IMPACT ON NAVAL NEEDS (Research/Technology/Operations) (Circle one number or –)
1 - 2 - 3 - 4 - 5 - 6 - 7 - 8 - 9 - 10
I--- LOW ---I I------ FAIR -------I I------ AVERAGE ------I I----- GOOD -----I I---- HIGH ---I

2C. PROBABILITY OF ACHIEVING POTENTIAL IMPACT ON NAVAL NEEDS (Circle one number or –)
1 - 2 - 3 - 4 - 5 - 6 - 7 - 8 - 9 - 10
I--- LOW ---I I------ FAIR -------I I------ AVERAGE ------I I----- GOOD -----I I---- HIGH ---I

2D. POTENTIAL FOR TRANSITION OR UTILITY (Circle one number or –)
1 - 2 - 3 - 4 - 5 - 6 - 7 - 8 - 9 - 10
I--- LOW ---I I------ FAIR -------I I------ AVERAGE ------I I----- GOOD -----I I---- HIGH ---I

2E. PHASE OF RESEARCH & DEVELOPMENT (Circle one number)
6.1 - - - - - - - - - - 6.2 - - - - - - - - - - 6.3
I--- BASIC RES. ---I I----- APPLIED RES. ----I I----EXPLORATORY DEV. ----I ADVANCED DEV.

3. REVIEWER'S EXPERTISE IN RESEARCH AREA COVERED BY THIS PROGRAM (Circle one number or –)
1 - 2 - 3 - 4 - 5 - 6 - 7 - 8 - 9 - 10
I--- LOW ---I I------ FAIR -------I I------ AVERAGE ------I I----- GOOD -----I I---- HIGH ---I

4. OVERALL PROGRAM EVALUATION (Circle one number or –)
1 - 2 - 3 - 4 - 5 - 6 - 7 - 8 - 9 - 10
I--- LOW ---I I------ FAIR -------I I------ AVERAGE ------I I----- GOOD -----I I---- HIGH ---I

5. COMMENTS (Use reverse side as necessary)

Figure 1 Science and mission impact questionnaire (proposed programs). (Courtesy of U.S. Office of Naval Research.)

RESEARCH OPTION SCORING CRITERIA

The research option evaluation form contains factors generally related to research and naval relevance issues. The scoring bands for all criteria except 2A, 2E, and 5 are identical: 1 - 2 (LOW), 2.5 - 4 (FAIR), 4.5 - 6.5 (AVERAGE), 7 - 8.5 (GOOD), and 9 - 10 (HIGH). Criterion 2A has no scoring range; Criterion 2E has its own scoring range defined; Criterion 5 is for comments.

DEFINITIONS OF CRITERIA ON RESEARCH OPTION EVALUATION FORM

1A. RESEARCH MERIT — Importance to the advancement of science of the question or problem addressed by the program. Consider the technical objectives, potential advancement of state-of-the-art, and uniqueness of contribution.

1B. RESEARCH APPROACH/PLAN/FOCUS/COORDINATION — Quality of process employed to solve the research problem, including the quality and focus of the research plan, definition of research milestones, degree of innovation, understanding of field, and coordination (or cognizance of) other related program to minimize duplication or gaps.

1C. MATCH BETWEEN RESOURCES AND OBJECTIVES — Relationship between scientific objectives proposed and total resources requested.

1D. BALANCE BETWEEN EXPERIMENT AND THEORY — Balance between experiment and theory proposed relative to optimum required to achieve performance targets.

1E. PROBABILITY OF ACHIEVING RESEARCH OBJECTIVES — Probability that the program's research objectives will be achieved.

2A. NAVAL NEED — Identify the naval need or problem (operational, technological, research) to which this research relates.

2B. POTENTIAL IMPACT ON NAVAL NEEDS — Potential impact of this program on naval research/technology/operational needs if successful.

2C. PROBABILITY OF ACHIEVING POTENTIAL IMPACT ON NAVAL NEEDS — Probability that the program will achieve its potential naval impact assuming that its research objectives have been met.

2D. POTENTIAL FOR TRANSITION OR UTILITY — Probability that results from this program will be transitioned to or utilized by naval technical community assuming that its research objectives have been met.

2E. PHASE OF RESEARCH & DEVELOPMENT — Level of program development. Scale ranges from basic research (6.1) through exploratory developement (6.2) to advanced development (6.3).

3. REVIEWER'S EXPERTISE IN RESEARCH AREA COVERED BY THIS PROGRAM — Self-explanatory.

4. OVERALL PROGRAM EVALUATION — Single number description of overall program quality based on all relevant criteria. Provide detailed narrative of pros and cons and any recommendations under COMMENTS.

5. COMMENTS — Self-explanatory.

Figure 2 Scoring criteria (proposed programs). (Courtesy of U.S. Office of Naval Research.)

PROJECT EVALUATION FORM

Title of Project ______________________________

Reviewer Name ______________________________

1A. RESEARCH MERIT (Circle one number or –)
1 - 2 - 3 - 4 - 5 - 6 - 7 - 8 - 9 - 10
I--- LOW ---I I------ FAIR -------I I------ AVERAGE ------I I----- GOOD -----I I---- HIGH ---I

1B. RESEARCH APPROACH/PLAN/FOCUS/COORDINATION (Circle one number or –)
1 - 2 - 3 - 4 - 5 - 6 - 7 - 8 - 9 - 10
I--- LOW ---I I------ FAIR -------I I------ AVERAGE ------I I----- GOOD -----I I---- HIGH ---I

1C. MATCH BETWEEN RESOURCES AND OBJECTIVES (Circle one number or –)
1 - 2 - 3 - 4 - 5 - 6 - 7 - 8 - 9 - 10
I--- LOW ---I I------ FAIR -------I I------ AVERAGE ------I I----- GOOD -----I I---- HIGH ---I

1D. QUALITY OF RESEARCH PERFORMERS (Circle one number or –)
1 - 2 - 3 - 4 - 5 - 6 - 7 - 8 - 9 - 10
I--- LOW ---I I------ FAIR -------I I------ AVERAGE ------I I----- GOOD -----I I---- HIGH ---I

1E. PROBABILITY OF ACHIEVING RESEARCH OBJECTIVES (Circle one number or –)
1 - 2 - 3 - 4 - 5 - 6 - 7 - 8 - 9 - 10
I--- LOW ---I I------ FAIR -------I I------ AVERAGE ------I I----- GOOD -----I I---- HIGH ---I

1F. PROJECT PRODUCTIVITY (Circle one number or –)
1 - 2 - 3 - 4 - 5 - 6 - 7 - 8 - 9 - 10
I--- LOW ---I I------ FAIR -------I I------ AVERAGE ------I I----- GOOD -----I I---- HIGH ---I

2A. POTENTIAL IMPACT ON NAVAL NEEDS (Research/Technology/Operations) (Circle one number or –)
1 - 2 - 3 - 4 - 5 - 6 - 7 - 8 - 9 - 10
I--- LOW ---I I------ FAIR -------I I------ AVERAGE ------I I----- GOOD -----I I---- HIGH ---I

2B. PROBABILITY OF ACHIEVING POTENTIAL IMPACT ON NAVAL NEEDS (Circle one number or –)
1 - 2 - 3 - 4 - 5 - 6 - 7 - 8 - 9 - 10
I--- LOW ---I I------ FAIR -------I I------ AVERAGE ------I I----- GOOD -----I I---- HIGH ---I

2C. POTENTIAL FOR TRANSITION OR UTILITY (Circle one number or –)
1 - 2 - 3 - 4 - 5 - 6 - 7 - 8 - 9 - 10
I--- LOW ---I I------ FAIR -------I I------ AVERAGE ------I I----- GOOD -----I I---- HIGH ---I

2D. PHASE OF RESEARCH & DEVELOPMENT (Circle one number)
6.1 - - - - - - - - - 6.2 - - - - - - - - - 6.3
I--- BASIC RES. ---I I----- APPLIED RES. ----I I----EXPLORATORY DEV. ----I ADVANCED DEV.

3. REVIEWER'S EXPERTISE IN RESEARCH AREA COVERED BY THIS PROJECT (Circle one number or –)
1 - 2 - 3 - 4 - 5 - 6 - 7 - 8 - 9 - 10
I--- LOW ---I I------ FAIR -------I I------ AVERAGE ------I I----- GOOD -----I I---- HIGH ---I

4. OVERALL PROJECT EVALUATION (Circle one number or –)
1 - 2 - 3 - 4 - 5 - 6 - 7 - 8 - 9 - 10
I--- LOW ---I I------ FAIR -------I I------ AVERAGE ------I I----- GOOD -----I I---- HIGH ---I

5. COMMENTS (Use reverse side as necessary)

Figure 3 Science and mission impact questionnaire (existing programs). (Courtesy of U.S. Office of Naval Research.)

PROJECT EVALUATION SCORING CRITERIA

The project evaluation form contains factors generally related to research and naval relevance issues. The scoring bands for all criteria except 2D and 5 are identical: 1 - 2 (LOW), 2.5 - 4 (FAIR), 4.5 - 6.5 (AVERAGE), 7 - 8.5 (GOOD), and 9 - 10 (HIGH). Criterion 2D has its own scoring range defined; Criterion 5 is for comments.

DEFINITIONS OF CRITERIA ON PROJECT EVALUATION FORM

1A. RESEARCH MERIT — Importance to the advancement of science of the question or problem addressed by the program. Consider the technical objectives, potential advancement of state-of-the-art, and uniqueness of contribution.

1B. RESEARCH APPROACH/PLAN/FOCUS/COORDINATION — Quality of process employed to solve the research problem, including the quality and focus of the research plan, definition of research milestones, degree of innovation, understanding of field, and coordination (or cognizance of) other related program to minimize duplication or gaps.

1C. MATCH BETWEEN RESOURCES AND OBJECTIVES — Relationship between scientific objectives proposed and total resources requested.

1D. QUALITY OF RESEARCH PERFORMERS — Consider publications, honors and awards, relevant experience, and other less tangible factors which contribute to team quality.

1E. PROBABILITY OF ACHIEVING RESEARCH OBJECTIVES — Probability that the program's research objectives will be achieved.

1F. PROJECT PRODUCTIVITY — Volume and quality of work produced and relationship of this output to the resources available, costs incurred, and time elapsed since project initiation.

2A. POTENTIAL IMPACT ON NAVAL NEEDS — Potential impact of this program on naval research/technology/operational needs if successful.

2B. PROBABILITY OF ACHIEVING POTENTIAL IMPACT ON NAVAL NEEDS — Probability that the project will achieve its potential naval impact assuming that its research objectives have been met.

2C. POTENTIAL FOR TRANSITION OR UTILITY — Probability that results from this project will be transitioned to or utilized by naval technical community assuming that its research objectives have been met.

2D. PHASE OF RESEARCH & DEVELOPMENT — Level of project development. Scale ranges from basic research (6.1) through exploratory developement (6.2) to advanced development (6.3).

3. REVIEWER'S EXPERTISE IN RESEARCH AREA COVERED BY THIS PROJECT — Self-explanatory.

4. OVERALL PROJECT EVALUATION — Single number description of overall project quality based on all relevant criteria. Provide detailed narrative of pros and cons and any recommendations under COMMENTS.

5. COMMENTS — Self-explanatory.

Figure 4 Scoring criteria (existing programs). (Courtesy of U.S. Office of Naval Research.)

III. PROJECT ASSESSMENTS

A. Assessments or Audits

There is a substantial body of knowledge accumulated on the subject of auditing quality systems. For years, government specifications have required formal audit programs and there is increasing emphasis toward using the audit process in preference to direct oversight of activities. Initially, audits were used as a means to check up on the implementation of the quality control programs dictated by customers or government fiat. It was a means of inspecting the inspectors. There was an emphasis on independence to make sure the auditors were not unduly influenced in their decision making by the need to maintain good career dynamics. There was also a concern that auditors have the proper qualifications. The nuclear industry established certification requirements for their lead auditors early in the 1970s. The American Society for Quality Control established a formal certification program for auditors in 1987 and the International Organization for Standardization has its own requirements for lead assessors for ISO 9000 certification. Audit teams selected projects themselves and took great pains to ensure randomness and appropriate sample sizes. Some audit standards stressed the need for ''random, unannounced'' audits, presumably to surprise the audited organization before it had time to hide the evidence.

The weakness in all of this was that, while the audits excelled in their ability to operate in an antagonistic environment, they were not respected as particularly useful for management except to scrub the operation from possible findings by like-minded audit teams from outside the organization. All too often, the audit teams, although qualified to auditing standards, were not technically competent in the area they were examining. Being uncomfortable in the technology, they could not make adjustments in their interpretations of the written word to suit the occasion. This led to frustration by the research organizations and a growing belief that all the auditors were interested in was counting ''beans,'' dotting i's and crossing t's. In response, research organizations hired their own professional QA staffs to defend themselves against the onslaught and to devise means of exempting certain operations from oversight. As further protection, organizations participated in attempts to revise the more onerous standards to make them more amenable to research and thereby provide protection from ''renegade auditors'' (25).

Contrary to intent, the inspect-everything mentality was shown, particularly in the nuclear industry, to perpetuate an obsession with keeping the paperwork straight but to the detriment of actually improving quality (26). In response, the nuclear industry and the U.S. Department of Energy began to emphasize ''assessments'' that focused more on assessing the effectiveness of the activities than on their strict adherence to requirements. A subset of that is the management assessment, whereby the organization's upper management performs or (often) obtains an assessment of its quality affecting activities by persons at a high level but not associated with the quality assurance function. All this is aimed at getting the line management more directly involved in the assessment activities and thereby increasing the effectiveness of the assessment as well as the activity being assessed.

The use of technical management in the process of assessing research projects has been shown to be very productive. Examples of good technical management practices are:

1. Standards may not be rigid. All researchers know that creativity is often employed in various ways during the project and not just in designing the experiment. Research management must thoroughly buy into what they are expecting of their technical staff. Management has to know where judgment can be permitted and where to draw the line. Recognizing that each laboratory director might have an individual interpretation of specifically how records are kept or data transmitted to customers, those interpretations should be acknowledged up front and expressed openly. Managers should not espouse one set of procedures and then act to subvert them in actual practice. Employees are smart enough to spot hypocrisy quickly. The laboratory directors should publish their guidelines for conduct of projects and use those guidelines as the basis for assessments of the work. The Babcock & Wilcox *Standard Practice Manual* (27) was developed by laboratory directors as a quick ''tour guide'' for their researchers. The directors recognized that researchers didn't always have time to read detailed procedures, particularly when a project was starting up, so the essence of the procedures were reduced to a manageable level, one or two key points. It was enough to alert the researcher to whether the detailed procedure should be consulted. Certain issues were mandated by procedure and emphasized in the manual. Most of the contents were in the form of recommendations for good engineering and scientific practice, but were ultimately left to the final judgment of the researcher. The specific interpretations were left to the laboratory director, not to QA.
2. Let managers select their own projects. If managers have set the guidelines, they are naturally curious about how their researchers are responding. They also know which projects may be in trouble or are of particular importance and will want to know how well the work is progressing. If some of the best were selected, fine. A lot can be learned by studying good projects as well.
3. Let managers identify the technical reviewers. Managers understand that any review is going to take time away from normal business. They want to get the most good from the effort. They can identify people they will respect for examining work in their laboratories. If those reviewers come from the same technical group as the one being audited, so be it. The reviewers will learn, particularly from the participation of the lab managers, and will return to their own work better prepared from the experience.

4. Ask managers to personally participate in the review process. Of all the concepts of assessment, this will reap the most value. When the laboratory director sits in on the evaluation discussion, there is a world of learning taking place. The project leader learns directly what the director's attitude and interpretations are; the director learns what is really happening in the lab. The assessments are not witchhunts. They are a valuable training tool for new and veteran researchers alike. They bring together experience and new ideas and enthusiasm.

At the conclusion of the assessment, the principal investigator has pretty much received direct feedback on how well the project is progressing. There has been an opportunity to discuss the project with possibly several levels of laboratory management and there has been ample time for personal recognition as well as constructive criticism. The assessors have had an opportunity to be "calibrated" as to management's interpretations of the procedures and guidelines in actual situations. Management feels more directly involved in the day-to-day work.

As the assessments move from project to project, the team leaders compile the results in a manner that can be sorted and analyzed in various ways. The laboratory directors may wish to review results pertaining only to their labs or may want them compared with results from the overall research center.

A presentation and discussion at the research center top management level (see Steering Committee in Chapter 3) is useful. The results can be used as an internal measure and, over a period of successive assessments, tracked for improvement.

REFERENCES

1. J. Juran, *Quality Control Handbook*, McGraw-Hill, New York, 1974
2. AMC Pamphlet No. 702–3, *Quality Assurance Reliability Handbook*, Headquarters, U.S. Army Material Command, Washington, D.C., 1968
3. National Aeronautics and Space Administration NHB 5300.4 (1a), *Reliability Program Provisions for Aeronautical and Space System Contractors* (formerly NPC 250–1), U.S. Government Printing Office, Washington, D.C., 1971
4. *Code of Federal Regulations 10 CFR 50, Appendix B*, Quality Assurance Criteria for Nuclear Power Plants and Fuel Reprocessing Plants, U.S. Government Printing Office, Washington, D.C., 1970
5. *Product Liability, The Present Attack*, American Management Association, 1970
6. Corpus Juris Secundum 72 CJS Supplement 21, Products Liability, West Publishing Co., St. Paul, 1975
7. M. Sun, "Peer Review Comes Under Peer Review," *Science*, Vol. 244, May 26, 1989, p. 910
8. W. L. Clapp, Jr., "Controlling the Professional Liability Risk," *Rough Notes*, June 1990, p. 40
9. J. Walsh, "NSF Peer Review Hearings: House Panel Starts with Critics," News and Comment, *Science*, Vol. 189, August 8, 1975, p. 435

10. S. Cole, J. R. Cole, and G. A. Simon, "Chance and Consensus in Peer Review," *Science*, Vol. 214, November 20, 1981, p. 881
11. L. L. Hargens, "Variation in Journal Peer Review Systems," JAMA, Vol. 263, No. 10, March 9, 1990, p. 1348
12. S. J. Ceci and D. P. Peters, "Peer Review, A Study of Reliability" *Change*, September, 1982, p. 44
13. T. Gustafson, "The Controversy over Peer Review," *Science*, Vol. 190, December, 1975, p. 1060
14. C. W. McCutchen, "Peer Review: Treacherous Servant, Disastrous Master," *Technology Review*, October, 1991, p. 29
15. S. Crawford and L. Stucki, "Peer Review and the Changing Research Record," *Journal of the American Society for Information Science*, April, 1990, p. 223
16. Committee on the Conduct of Science, National Academy of Sciences, "On Being a Scientist," National Academy Press, Washington DC, 1989
17. *Applied Optics*, Vol. 15, No. 7, July, 1976
18. S. Taravella, "California Architects Design Peer Review to Trim Losses," Business Insurance, August 25, 1986
19. D. Preziosi, "Reviewing Peer Review," Civil Engineering, November 1988, p. 46
20. M. Bodnarczuk, "Peer Review, Basic Research, and Engineering; Defining a Role for QA Professionals in Basic Research Environments," 16th Annual National ASQC Energy Division Conference, Ft. Lauderdale, FL, 1989
21. E. Marshall, "NSF Peer Review Under Fire from Nader Group," Science, Vol 245, p. 250
22. B. B. Rosenman, "Managing an MS/OR Activity by Peer Review," Interfaces 13:6, December, 1983, p. 110
23. B. Bower, "Peer Review Under Fire," Science News, Vol. 139, June, 1991, p. 394
24. R. N. Kostoff, "Evaluation of Proposed and Existing Accelerated Research Programs by the Office of Naval Research," IEEE Transactions on Engineering Management, Vol 35, November, 1988, p. 271
25. J. J. Dronkers, "Report on the Activities of the ASME NQA Committee Working Group on Quality Assurance Requirements for Research and Development," Eighteenth Annual National Energy Division Conference, American Society for Quality Control, Danvers, MA, 1991
26. W. Altman, T. Ankrum, and W. Brach, NUREG-1055, "Improving Quality and the Assurance of Quality in the Design and Construction of Nuclear Power Plants. A Report to Congress." U.S. Nuclear Regulatory Commission, Washington, D.C. (1984)
27. Babcock & Wilcox, *Standard Practice Manual*, Alliance, OH, 1991

11

THE IMPROVEMENT PROCESS

Nick G. Sandru
Babcock & Wilcox
Alliance, Ohio

David D. Boath and Richard P. Maddams
PA Consulting Group
Hightstown, New Jersey

William H. Hamilton
Consulting Engineer
Ligonier, Pennsylvania

I. CONTINUOUS IMPROVEMENT

"How does one go about eating an elephant?" Everyone knows the answer is "one bite at a time." The research and development milieu is like the elephant, too large to characterize as a whole; however, it is somewhat more easily understood by its component parts. Each element of the research and development function, whether it is the actual scientific effort or the support work, can be described in terms of some input to a transformation effort (R & D process) that results in an output or product to a customer. There may be some adjustment (feedback) based on the expected results, which, in turn, affects the input. This paradigm can be described in the classical process model shown in Figure 1.

Many R & D activities (both white- and blue-collar) can be considered processes and controlled much the same way as manufacturing processes are controlled. In fact, many white-collar processes are equally as complex as manufacturing processes—such activities as engineering, distribution, personnel, data processing, and others. In the past, most of the attention has been directed at process controls for the manufacturing areas only. Today, the real payoff comes from applying the proven manufacturing control and feedback techniques to all key activities in the business, treating the entire company as a complex process that contains many subprocesses. Included in these is the series of processes that produces research for the corporation (1).

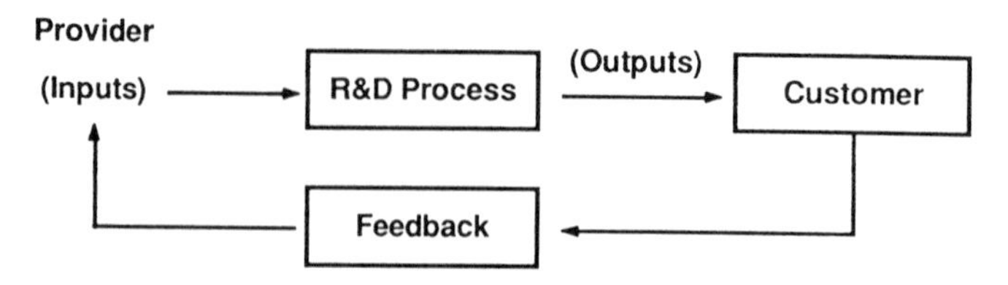

Figure 1 Classical process model.

The act of quality improvement (of research or any other process) must be viewed as an ongoing journey, not a program or a project. It has a beginning but no end. It encompasses all, is all-consuming. Why is it then that so few master the process, and those who do stand out above all others? What causes some organizations to achieve so much more than the ordinary? The answer lies in the concept that only those who totally embrace the customer and work to satisfy or exceed the customer's needs will become quality masters (2).

The application of an improvement process to research and development is, at best, looked upon skeptically because of the perceived nature of the R & D function. Research and development does not fit the mold! Researchers are different; therefore, it won't work for them. Anthony Montana (3) describes why R & D is slow to accept quality improvement. Quality improvement calls on experience-based thought versus empirical science. Quality is perceived as "perfection"—excellence in results rather than meeting customer expectations. Quality improvement measures improvement in project performance, not research results. Researchers have a limited understanding of the "soft" side. Once they can hurdle the barrier of accepting that this systematic process can do some good for them, the initial successful application of the process can take over and foster more use and more success. Keeping in mind that "doing it the right *way* the first time" is the best start of any work task, the researcher can recognize the contribution to be made from the process. An example of this is the preparation of an R & D proposal to an operating division or business unit customer. The process is the same regardless of whether the project involves wet chemistry or thermal hydraulics. Each step, from identification of the problem, to designing the experiment, to estimating the costs, to preparing the final report, is part of the same process and is always subject to improvement. This chapter describes an overall perspective of the improvement process with some very basic tools. Other texts are filled with much more detail on a wide variety of tools, with new ones being identified or developed each day. As competent as the scientists and engineers are at using sophisticated tools to solve technical problems, sometimes it only requires consistent execution of the basics to get at many of today's problems.

A. The High-Level Model

Figure 2 has modified the classical process model to show that processes are not only driven by customer requirements and dependent upon supplier inputs, but they also consist of value-added (and non-value-added) work which is done in context of the work area's mission. The Area Activity Analysis described in Chapter 5 addresses each element of this model in detail to aid researchers and support personnel to completely replan area functions.

1. Input Sources

The paradigm described in Figure 2 becomes dysfunctional if the feedback loop, customer to measurement system and back to supplier, becomes disconnected or fails due to unclear communication. Major deficiencies in information-sharing plague many companies since most of them operate on a need-to-know basis. Griffith (2) states that even though everyone knows the importance of communication, it cannot be left to a specific department alone to handle. "It must be recognized as a task for which everyone is responsible, not something to be delegated to another area . . . and then forgotten." Communication is the channel through which organizations can check themselves against their customers' requirements and expectations. Here, one must consider both internal (proximate) and external (ultimate customer). A dialogue must be initiated to provide regular (not just annual) identification of needs and measurement of accomplishment and satisfaction. The customer needs (and changes to those needs) assessment should never be considered only in the context of one's own experience or knowledge of customer needs. The inclination is to think one *knows* what the customer needs. There is but one way to really assess customer needs and that is to *ask* the customer.

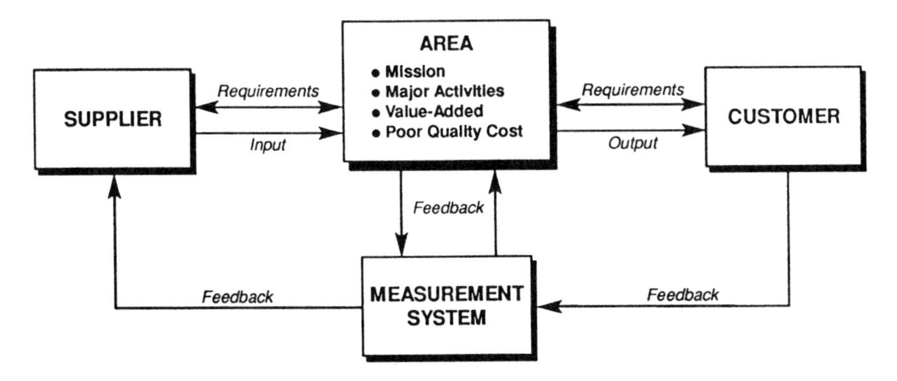

Figure 2 The stable process.

2. *Check Against Customer Requirements/Expectations*

Communication means sharing with the customer(s). The dialogue between internal departments, the meeting with an outside customer, or even conversations with the boss are examples of opportunities to more clearly understand what these customer expectations are. Teams help to improve interdepartmental communications by discussing deficiencies during cross-departmental information sharing (2). Instead of operating on a need-to-know basis, or having to deal with filtered information, departments should continually work toward open communications. Joint staff meetings, report sharing, and regular meetings between department management can all help to bring down the walls.

For external customers, regular face-to-face meetings, plant tours, project updates, and videotapes will open up the flow of information needed by *both* sides. Partnering is developed between supplier and customer only after lengthy and thorough exchanges of information, sharing of philosophy, and revealing of the strengths and weaknesses of both. Mutual trust and understanding can develop to the point where the customer will not consider another supplier for similar services. They will have one preferred source for R & D. The feeling of customer versus supplier will be replaced by a recognition that everyone is on the same team.

How often do people take the time to sit down with their boss to share information? The boss is a key customer for R & D or anywhere else. For the scientist in the laboratory developing a new additive for paint as well as the principal engineer replacing a water line in the boiler room, the boss needs to be kept informed. This customer needs to know on a regular basis the progress being made with the work, any operational improvements, the attitudes and expectations of the people in the organization, and the current problems being worked on.

3. *Feedback*

It is essential that an R & D organization make some overt moves to communicate with its customers to periodically assess how well it is doing. Several techniques have been used to elicit feedback such as specific requests by letter, tear-out sheets included in the research reports, and surveys.

Customer comments may be obtained by sending either short response forms or request letters to internal or external customers at the end of a certain phase of work, a milestone such as a shipping date, or the end of a particular test run. If comments are requested early on, they can provide some important feedback at a point in the project life where changes can be effectively made. Experience says that being asked for a response is a positive action which the customer appreciates. An example is included in Figure 3.

Tear-out sheets can be included in the final report to a customer. They help to obtain immediate comments on the success of the work from the direct recipient

September 11, 19__

Recently, ______________________ completed the contract, ________________ for the Research Center.

My colleagues and I have a commitment to continuous quality improvement in the work that we do. We recognize that our contract research customers and program sponsors are the ultimate judge of our success. I am writing to ask you to help us in this effort by commenting on the quality and usefulness of the work we performed in connection with the contract. Did we perform, through the lifetime of the project, in accord with our commitments and your expectations? Was the research carried out competitively and were the results communicated effectively? Did we perform in a timely and cost effective way?

We would be grateful for your comments. I hope that you will include among them any suggestions which you feel would enable us to be more effective in the future. I want to assure you of our appreciation for the opportunity to perform this work and to thank you in advance for your comments on our quality. I look forward to hearing from you.

Sincerely yours,

Vice President

kt

Figure 3 Customer appreciation letter.

of the report. The response to this method usually is less than with an overt solicitation by memo because it requires an unprompted action on the customer's part to fill it out. Results can be codified for trend analysis or used to develop a specific response to adverse feedback if it occurs. Figure 4 shows an example of a customer tear-out form included in final R & D reports.

Surveys tend to be longer and more demanding of the customer's time. However, the amount of information that can be gathered is quite good. A well-prepared survey not only gives a wide range of feedback, but it identifies similar comments from various customers that can highlight issues or problems which have arisen. When reissued periodically, surveys are excellent for developing trend information over a longer period of time. Figures 5, 6, and 7, are from an extensive customer perception survey.

CUSTOMER FEEDBACK QUESTIONNAIRE

Please help us do a better job for you.

Report Title: EasyFIV -- Version 3.0, Programmer's Manual

Report No.: RDD:94:62316-006-001:01

Customer feedback on our work is important to us. We would like to have your comments concerning this report. Your cooperation in completing this form and returning it to the address shown will assist us in improving our service to you.

Do the contents of this report meet your needs? ☑ Yes Not Entirely, because ____________

Did you receive this report when we promised it? ☑ Yes ☐ No

Comment: We had (mutually) changed the initial deadline

How could we have made this report more useful to you? We have gone through a useful draft review exercise in which all of my comments were satisfactorily addressed – no further comment.

Thank you.

Name: ____________ Date: Dec. 1/

Organization: ____________

Please fold and return this form to address shown on the reverse side.

Figure 4 Customer feedback questionnaire.

Research and Development Division

CUSTOMER SURVEY

FOR OPERATING DIVISIONS

Here's your chance to tell us how we're doing by giving us some feedback. Please respond to the following and return this survey to:

R&D Division TQM Survey Team

Target Return Date: June 21, 19__

Date: ______________

Name ______________ Title ______________

Division ______________ Location ______________

Job Function: ❑ General Management ❑ Project Management ❑ Technical ❑ Manufacturing
❑ General Engineering ❑ Marketing / Sales

What R&DD sections / technologies do you mainly deal / interface with (please check all appropriate boxes)?

❑ Quality Assurance
❑ Technology Development and Evaluation
❑ Chemical Engineering
❑ Combustion and Advanced Energy Systems
❑ Chemistry
❑ Nondestructive Methods and Diagnostics
❑ Ceramics and Metals
❑ Materials Performance
❑ Metallurgy and Manufacturing Technology
❑ Nuclear Steam Systems and Components
❑ Heat Transfer and Fluid Mechanics
❑ Structural Mechanics
❑ Experimental Apparatus
❑ Accounting
❑ Others (please indicate) ______________

Your level of involvement ($ and/or contacts) with R&DD: ❑ High ❑ Medium ❑ Low ❑ None

(Even if your level of involvement is none, please provide us with your perceptions to the following items.)

I. **PERFORMANCE. Please grade us on the following items A through K (Circle: A = Excellent, B = Good, C = Average, D = Fair, F = Poor, N = No Opinion):**
Also, identify the three (3) performance items of highest importance to you by placing a number (1 being highest) in the box next to the following letters **A** through **J**.

A. ❑ Project Execution / Management:

Planning and estimating	A	B	C	D	F	N
Adherence to workscope	A	B	C	D	F	N
Adherence to schedule	A	B	C	D	F	N
Adherence to budget	A	B	C	D	F	N
Workscope defined to meet customer needs	A	B	C	D	F	N
Flexibility to change workscope to meet new needs / opportunities	A	B	C	D	F	N
Quality of technical work	A	B	C	D	F	N

B. ❑ Reports:

Technical content	A	B	C	D	F	N
Quality	A	B	C	D	F	N
Timeliness	A	B	C	D	F	N

Next ➡

Figure 5 Customer survey, page 1.

I. **PERFORMANCE (Continued)**

C. ❑ **Innovation:**							
Source for new ideas / patents / projects	A	B	C	D	F	N
Creativity / resourcefulness for problem solving	A	B	C	D	F	N
Source for new or advanced technology	A	B	C	D	F	N
Identifies / explores new opportunities and ideas during project	A	B	C	D	F	N
Ability to define alternate approaches for projects	A	B	C	D	F	N
D. ❑ **Cost Competitiveness/Value:**						
For long-range R&D projects	A	B	C	D	F	N
For unplanned, short-term "firefighting"	A	B	C	D	F	N
E. ❑ **Problem Solving:**						
Responsiveness when called on short notice	A	B	C	D	F	N
Ability to find a solution and get the job done	A	B	C	D	F	N
F. ❑ R&DD personnel understanding of your business and products	A	B	C	D	F	N
G. ❑ R&DD personnel help in identifying technical risks for new designs / products / processes	A	B	C	D	F	N
H. ❑ Adequacy of laboratory equipment and other facilities	A	B	C	D	F	N
I. ❑ Adequacy of R&DD technical personnel	A	B	C	D	F	N
J. ❑ Working relationship with R&DD personnel	A	B	C	D	F	N
K. Overall opinion of how R&DD performs for you	A	B	C	D	F	N

II. **Please rate us on these *Potential Strengths* of R&DD (Circle: 5 = Strongly Agree; 1 = Strongly Disagree; N = No Opinion)**

A. R&DD responds quickly when you call for help on important problems	5	4	3	2	1	N
B. R&DD properly plans long-range projects and executes them well technically	5	4	3	2	1	N
C. R&DD adequately anticipates your technical needs for new / advanced products, processes, and technologies, and conceives project ideas to support them	5	4	3	2	1	N
D. R&DD can focus on your longer range problems and projects without being side-tracked due to day-to-day operational problems	5	4	3	2	1	N
E. R&DD can integrate several technical disciplines to attack a problem	5	4	3	2	1	N
F. R&DD prepares detailed workscopes and proposals very well	5	4	3	2	1	N
G. R&DD has specialized equipment which you need to solve problems or gather data	5	4	3	2	1	N
H. R&DD supports marketing efforts effectively	5	4	3	2	1	N

III. **Please rate us on these *Potential Criticisms* of R&DD (Circle: 5 = Strongly Agree; 1 = Strongly Disagree; N = No Opinion)**

A. Too expensive	5	4	3	2	1	N
B. Inadequate facilities	5	4	3	2	1	N
C. Quicker and easier to do internally	5	4	3	2	1	N
D. Too slow to respond	5	4	3	2	1	N

Next ➡

Figure 6 Customer survey, page 2.

III. *Potential Criticisms* of R&DD (Continued)

E. Lacking technical expertise	5	4	3	2	1	N
F. Doesn't understand your products / businesses / customers	5	4	3	2	1	N
G. Tries to add undesired, not needed workscope	5	4	3	2	1	N
H. Too much paperwork	5	4	3	2	1	N
I. Runs off on tangents	5	4	3	2	1	N
J. Provides higher quality than needed	5	4	3	2	1	N
K. Not local; travel expenses, etc.	5	4	3	2	1	N
L. Overruns budget	5	4	3	2	1	N
M. Slow-pace workers	5	4	3	2	1	N
N. Afraid to take risks and make hard decisions	5	4	3	2	1	N
O. Operating division loses control	5	4	3	2	1	N
P. Insensitive to profit picture	5	4	3	2	1	N
Q. R&DD has too high regard for itself; "ivory tower" attitude	5	4	3	2	1	N
R. Focuses on long-term too much	5	4	3	2	1	N
S. R&DD's skills / capabilities are a big unknown	5	4	3	2	1	N
T. Contracts get higher priority than your projects	5	4	3	2	1	N

IV. COMMENTS (Please attach a separate sheet if additional space is needed.)

A. What R&DD ***strengths*** are important to you? ______________________

B. What are the R&DD's ***weaknesses*** as you see them? ______________________

C. Would you prefer to conduct R&D projects within your own division using your own people? ❑ Yes ❑ No

Why? ______________________

D. What would the R&DD need to change in order to participate in more projects with you and your division?

____

E. How can the R&DD better satisfy your needs? ______________________

F. Other comments: ______________________

THANK YOU FOR YOUR TIME AND OPINIONS!

Figure 7 Customer survey, page 3.

This survey was designed as a primary feedback tool to elicit information from operating divisions for use by the central R & D function. It was designed as a universal form to be applied to various levels of management and technical section people in the operating division. It can be coded to separate the feedback by respondent level, division, and location. In general, it was intended to convey a feeling that, regardless of degree of interaction between the customer and the R & D Division, the response was valued and important to the relationship.

A scale of response levels, A through F, or 5 through 1, allows for a graded response. The value "N" is included for "no opinion," if it is appropriate. The scoring is not affected, however, by the "N" option as it is not entered into the calculations of average or standard deviation.

The first portion requests respondent information by division, job function and location. More specifically, it requests the customer to identify the R & D section with whom the respondent deals most often. In some cases, a particular respondent might interface with several R & D sections. Selectively, those respondents could be asked to fill out a separate survey for each.

In Figures 5 and 6, the questions in Section I, "Performance," take the respondent through two elements of performance. Items A and B elicit feedback on execution, management and reports, intending to get the customer's feelings on R & D's ability to do successful projects for them. Items C through J ask for customer opinions on more basic issues of R & D contribution to the customer's needs. How well does the R & D Division perform as a source of new ideas and help bring them into practice? How well does the R & D Division understand the business and the products or services of the operating division? These types of questions give the R & D Division meaningful information on the perceived value of the work the operating division has received from the R & D Division in the past.

The second portion of the survey, in Figure 6, deals with potential strengths. This element was designed expressly to obtain feedback on the perception of future needs which might arise. Will the operating division be comfortable with coming to the R & D Division for solutions to its problems and needs? Will the work be only as needed to get out of a last-minute problem or will there be some longer-range planning for future needs? These kinds of questions help the R & D Division assess the perceived value of the division in the longer time frame.

The third portion allows the respondent to provide criticism to the R & D Division without it being masked in any way. It provides some clear, direct areas for improvement.

The fourth portion, Figure 7, provides space for narrative responses. In many cases, the respondents want to put into their own words the feelings they have toward the R & D Division. These comments, when included in the survey results, add a degree of completeness that a numerical scale cannot provide. (It was also used by one respondent to heap praise on the R & D Division just for asking his opinion.) Half the battle of good relations between divisions is the ability to ask "How are we doing?" "Can we be doing something better?"

Figure 8 shows the results of an initial (baseline) perception survey. Results of surveys need to be shared with employees as well as the customer. Assistance in preparing a meaningful survey is essential since asking the right questions is at best a difficult task. Being careful to not change the survey questions is another caveat, in this case to avoid a problem in comparing baseline with subsequent results.

4. *Measurement*

Managers and employees alike often rely on "gut feel" to solve problems. At times, this method works, but more often it fails. It is essential to have an objective understanding of the problem before trying to solve it. Thus, measurement is absolutely essential to progress. Quantitative measurements are better than qualitative ones, but sometimes error-related measurements are not available. Most importantly, one has to rely on the customer's perception. There is nothing wrong with asking someone what they think about the people, the operation, or the output (1). Customers must be included in the selection of measurement criteria to make the process improvement effort meaningful. The cycle time of a component failure analysis by the R & D Division when a utility boiler is disabled (and costs are mounting by the tens of thousands of dollars each day) has far more meaning to the Service Department than the time to get the written report on the same failure analysis. The customer (the Service Department) can exceed its customer's (the Utility's) expectations if the information regarding the cause of the failure is known quickly. Repairs can be started early. Water chemistry operating procedures can be modified quickly, saving the customer millions of dollars. How quickly the final report is delivered, while important, is not as meaningful to the customer's success as is the immediate failure mode of the component. Customer-driven and meaningful—two essential criteria for measurement success.

Measurement points need to be placed close to the activity being performed. Ideally they should be part of the activity. Reducing the feedback time saves money in two ways. First, the employee does not continue to make errors, and second, additional resources are not added to an already defective item (1). Not finding that a data point is in error because an instrument's is out of calibration will cost the project thousands of dollars in retesting. IBM studies indicate there is a 50 to 1 leverage in finding defective components in their subassembly test areas over finding the same defect in the customer's plant. The leverage for software errors is 80 to 1 (1).

An essential part of any measurement and feedback system is an independent audit system that will ensure compliance to the procedures. It is nonsense to

Research and Development Division

CUSTOMER SURVEY

FOR OPERATING DIVISIONS

RESPONSE SUMMARY

Group: POWER GENERATION

Divisions: ____________________

No. of Surveys Issued: ________ No. of Responses: ________ Return Rate Percentage: ________ %

CAT Identifiers (No. of Responses): Division Heads _____; Controllers _____; Sales/Marketing _____; Technical/General Management _____; Project Management _____; Operating Division Project Leaders _____

Other _____ R&DD Vision Composite Score (1 – 5): _____

(*NOTE*: Bar shown [bar] in responses to Parts I, II, and III indicates the mean plus two standard deviations.)

Job Function: 30 General Management 23 Project Management 21 Technical 3 Manufacturing 14 General Engineering 33 Marketing / Sales

What R&DD sections / technologies do you mainly deal / interface with (please check all appropriate boxes)?

3 Quality Assurance
46 Technology Development and Evaluation
27 Chemical Engineering
47 Combustion and Advanced Energy Systems
26 Chemistry
17 Nondestructive Methods and Diagnostics
5 Ceramics and Metals
9 Others (please indicate) ____________________
20 Materials Performance
32 Metallurgy and Manufacturing Technology
5 Nuclear Steam Systems and Components
34 Heat Transfer and Fluid Mechanics
11 Structural Mechanics
2 Experimental Apparatus
6 Accounting

Your level of involvement ($ and/or contacts) with R&DD: 19 High 32 Medium 52 Low 15 None

(Even if your level of involvement is none, please provide us with your perceptions to the following items.)

I. **PERFORMANCE. Please grade us on the following items A through K (Circle: A = Excellent, B = Good, C = Average, D = Fair, F = Poor, N = No Opinion):**
Also, identify the three (3) performance items of highest importance to you by placing a number (1 being highest) in the box next to the following letters **A** through **J**.

A. ☐ **Project Execution / Management:**

Item	Grade	No opinion (N)
Planning and estimating	A B C D F N	29
Adherence to workscope	A B C D F N	31
Adherence to schedule	A B C D F N	30
Adherence to budget	A B C D F N	29
Workscope defined to meet customer needs	A B C D F N	34
Flexibility to change workscope to meet new needs / opportunities	A B C D F N	28
Quality of technical work	A B C D F N	35

B. ☐ **Reports:**

Item	Grade	No opinion (N)
Technical content	A B C D F N	20
Quality	A B C D F N	20
Timeliness	A B C D F N	17

Next ➡

Figure 8 Customer survey, response summary.

blindly accept data that are generated without proper checks and balances, not because people will falsify data—most people won't—but because most employees have a strong desire to satisfy management and tell them what they want to hear.

The process improvement effort should also ensure that the product specifications are documented and truly reflect customer expectations. Customer feedback loops should be in place to ensure that the degree of customer satisfaction is being measured and changes in expectation levels can be recognized. Adapting to changes in customer expectations is a necessary skill of the organization of the future.

B. The Process

A technique which provides the organization a systematic approach to problem identification and solution is the key to success. Otherwise, the method for achieving the intended goal is, at best, a guess.

1. Why a Problem-Solving Process?

A systematic problem-solving process:

- Provides a common set of skills
- Increases the effectiveness of feedback
- Builds organizational cohesiveness
- Helps solve large and small problems
- Enhances customer satisfaction
- Provides consistent terminology

Without a methodology practiced by everyone in the organization, chaos will set in as rival approaches clash and employees get the feeling that anything goes. Figure 9 describes a simple, yet effective, seven-step process for problem solving used widely by the Ernst & Young organization.

The importance of adopting a systematic improvement process cannot be overstated. Whether the approach is by team problem solving or individual involvement, the premise is the same. To avoid the possibility of developing a suboptimal solution to the problem at hand, a systematic process needs to be followed.

One of the pressures an organization encounters is to circumvent the problem-solving process because the solution is "obvious." The fear of wasting people's time on trivial or simple matters is very real, especially in knowledge-worker organizations such as research and development. Upper management may be even more prone to be convinced by their past experience—or current position—that they know what is best. When this happens, they fail their organization in at least two ways: They have not permitted proper analysis of the issue, and they have not sought the ideas of others or listened to those offered. Obviously,

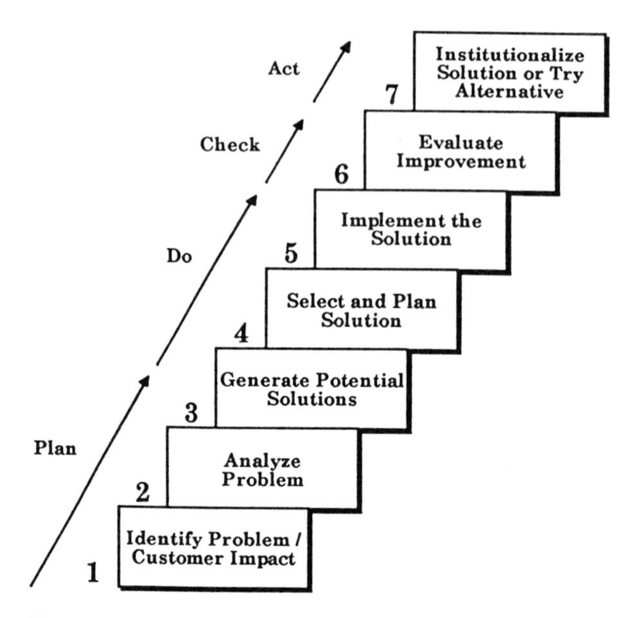

Figure 9 Seven-step problem-solving process.

such failure seriously affects the attitudes of those around the issue as well as limiting the power of combining more than one idea with a thorough analysis of options (2). Management must be mindful of the potential to stray from the process and teach and coach employees in the importance of "sticking to the script."

2. *Establish the Boundaries—Define the Problem*

Even before the first step to solving the problem is taken, the team or individual must recognize the need to put some boundaries around the issue at hand. This can take the form of a simple charter or position description which clearly states the goal of the exercise. The charter is a concise statement of the condition which exists: "It is taking too long to obtain failure analysis results and report them to our customer." Previous chapters have emphasized that teams should not be asked to "solve world hunger." Even if the problem charter does not seem to get at all of the problem, it is better to keep the scope to a smaller, more manageable size that can be handled within a reasonable time, say six to twelve months. But, to paraphrase Churchill, "so many problems, so little time." Where to begin.

3. *Pareto Analysis*

Juran (4) describes the use of the Pareto Analysis for determining which projects will provide the most benefit to the organization. This tool, which Juran mistakenly (5) attributed to Vilfredo Paredo, an Italian economist, diagrams the "80–20" principle, which says that 20 percent of the problem areas will give 80 percent of the total grief to any organization. Often, financial data bases are analyzed with this tool to determine potential areas for significant improvement, but other data will work just as well, with some created just for that purpose. Although the Pareto Analysis appears to be a simplistic approach, it can be cleverly employed in conjunction with other tools to arrive at fairly significant and sophisticated decisions. For example, a weld process improvement study was performed over the course of three shifts, using three processes and two types of materials. A check sheet was developed (Figure 10) to document the types of defects being encountered. Of course there was the usual assumption by each shift that the other shifts were less capable. The data did not bear out this prejudice. There was no significant difference between shifts, but there was a significant pattern to the defect types. When plotted on a typical Pareto curve (Figure 11), it appeared that defects associated with "pre/post weld heat treating," "weld shrinkage," and "cavities" accounted for 75 % of all defects and should be immediately targeted for improvement projects. But the analyst did not stop there. Using additional data provided by weld research engineers a priority index was created (Figure 12) which took into account the losses due to each problem, the years to recover given current knowledge of the technology, the anticipated investment to correct the problem, and the probability of success in resolving the problem. The priority index was used to calculate a priority ranking that altered the selection of improvement projects. The two biggest problems, "pre/post weld heat treatment" and "weld shrinkage" were still Number 1 and 2, but the third project to be selected was "low crown" because it had a loss value almost as high as "cavities," but with less investment and greater probability of success.

4. *The Seven-Step Process*

Step 1 requires that the problem be stated correctly. It will be suitable not to try to solve the problem at this stage, but to identify it and agree upon the validity of the problem. Customers, whether proximate or ultimate, need to be identified. The customer's requirements, both explicitly stated and implicitly expected, need to be identified. It is important during this step not to overlook a requirement either by oversight, ignorance, or assumption that it is known. Verification of requirements is a must.

Data measures must be agreed upon. Measures must be meaningful to the process element being studied. "It is taking six hours to enter the failure analysis request to the Analytical Metallurgy Section; four hours to run the analysis on the Auger analyzer; three hours to write the report; and five days to get approval on the report before results can be reported to the customer." If certain metrics are unknown, baseline measurements need to be made by making observations, tracking input/output cycles, or reasonable estimates.

Shift	1						2						3						
Weld Process	A		B		C		A		B		C		A		B		C		
Material Code	I	II	I	II	I	II	I	II	I	II	I	II	I	II	I	II	I	II	
Defect Type																			Frequency
A. Dimensional Var.	I	I	I	II			II	II				I	I	I	I		I		14
B. Low Crown		I			I		II	I		I			I	I		II	I	I	12
C. Slag			I					I								I	I		4
D. Set-Up					I									I					2
E. Pre/Post Heat T.	III	II	II 卌	II	I 卌	III	I 卌	卌	IIII	III	I 卌	II	II	III		卌	III	II	64
F. Equipment Malfunc.				I						I						I	I		4
G. Cavities	I	II		III	I	I	III	II		II	II	II	I	I	II		II	III	28
H. Weld Shrinkage	卌	I	III	I 卌	III	IIII		I 卌	III	II	III	I 卌	I	II	III	II	II	I 卌	58
I. Fill Material		I	I	II			II		I		I		I			II		I	12
J. Miscellaneous					I									I					2
Shift totals	68						72						60						

Figure 10 Check sheet of weld defects.

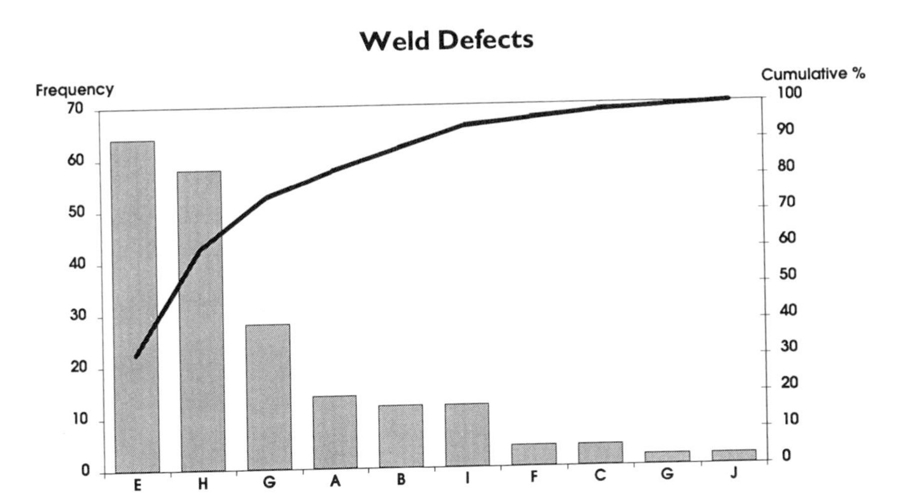

Figure 11 Pareto curve.

Step 2 requires that data be collected for the process being studied. Such things as how the work order gets entered, who enters it, how long it takes to enter it, where it goes, and so on, need to be recorded. The process flow chart illustrated in Figure 13 describes a typical request format that could be used by a research project engineer (internal customer) for design work performed by a mechanical engineering group (supplier) for support of a research project which is funded by an operating division (external customer). It is indicative of a multistep process which can be analyzed in its entirety or broken down into components for microanalysis. In either case, the boundary limits of the input and output points must be clearly defined to bracket the work elements to be reviewed. If one was to isolate the decision point surrounding the ''equipment availability'' issue, (see Figure 14) the process can be analyzed in greater detail. The research project engineer has the responsibility to initiate an equipment availability review if, in the first analysis, equipment does not appear to be available. He/she would step through Tab #7, prepared to help the support group mechanical engineer with the decisions. Tab #7 could include such items as:

Correct equipment not available in the lab:

- request capital to procure
  - rent or lease
  - borrow from another lab
  - etc.

Element	E	H	G	A	B	I	F	C	G	J
Value	64	58	28	14	12	12	4	4	2	2
Cumulative Value	64	122	150	164	176	188	192	196	198	200
Percent	32	29	14	7	6	6	2	2	1	1
Cumulative Percent	32	61	75	82	88	94	96	98	99	100
Annual Dollar Value	35	45	300	54	240	20	130	20	10	10
Years to recover	0.8	1.0	5.0	1.5	2.0	1.5	3.5	2.5	1.0	2.0
Investment	5	10	150	15	50	10	80	8	6	4
Probability of Success	0.70	0.85	0.65	0.50	0.80	0.70	0.60	0.95	0.80	0.40
Priority Index	6.13	3.83	0.26	1.00	2.00	0.93	0.28	0.45	1.33	0.50
Priority Rank	1	2			3				4	

$$\text{Priority Index} = \frac{\text{Annual Value x Probability of Success}}{\text{Years to Recover x Investment}} \qquad \frac{35 \times 0.70}{0.8 \times 5} = 6.125$$

Figure 12 Pareto worksheet.

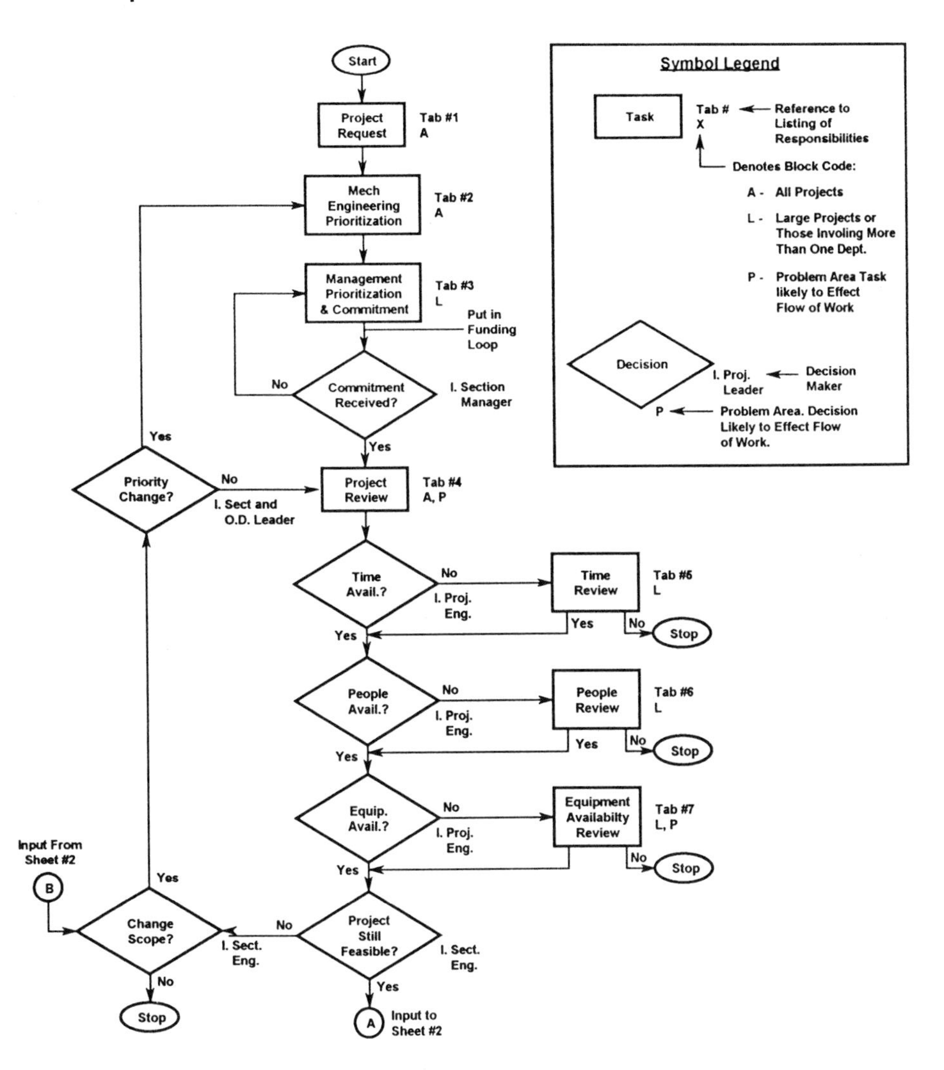

Figure 13 Process flow diagram.

Correct equipment available in the lab; however, not at this time:

- obtain approval to improve priority of this project
  - get calibration services to calibrate equipment
  - get maintenance services to correct deficiencies
  - and so on

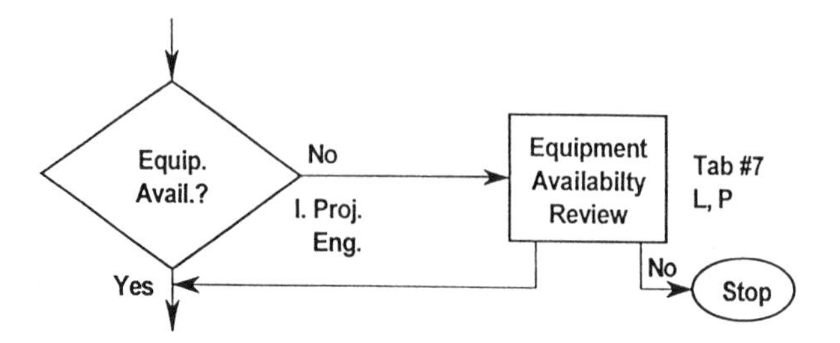

Figure 14 Process section detail.

If the review is successful (''yes'' answer), the project is still feasible. Otherwise, the process stops.

Process flow diagramming must be done with sufficient accuracy to reflect how the process takes place *now*. The next step is the interface with the ''customer'' and the ''supplier'' to establish the areas of measurement (and improvement) expected by the customer. As in this example, the customer is the proximate customer, the research project engineer, any department which receives the output of another department, or the manager who receives a report from Accounting. Recognize, however, that this notion has to be consistent with the same concept projected from the Corporate Vision statement where the satisfaction of the ultimate customer is the focus. An internal customer's needs may have to be modified to support the external customer's requirements. At what level of performance do we want to be in the future to fully satisfy or exceed the expectations of our customer?

Understand the process, the way it is now, recognize the customer's expectations (the way it should be), then apply a methodology which will be used to make improvement toward the intended expectation.

Care must be taken to record the actual process as it is and not the future state—where it could be. If necessary, the process flowchart should be reviewed for additional information which will be helpful in describing the current state.

This step of the improvement process requires that the process be analyzed carefully in order to reveal all characteristics and elements known about it. The use of value analysis is recommended in order to break the process elements down into its component parts and assign a cost to each element. By performing such a functional evaluation, one will determine what is the item, what does it cost, what does it contribute, what other options will do the job, and what might they cost? By cost, it is meant that there is an expenditure of time and/or material to accomplish the item.

A very effective way to analyze the problem is to perform a root cause analysis to determine the basic cause for the situation. A valuable tool is the fishbone or Ishikawa diagram (Figure 15), which helps to break down the analysis into neat, grouped *potential* causes. Care must be taken to place the problem statement at the ''head'' of the diagram and then major classifications for apparent causes at each of the five major bones of the fish. Subsets of theorized causes are placed on the smaller bones emanating from the major bones.

This is the most deadly stage. It is at this point that the group believes it has found the root cause(s) and can proceed to identify potential solutions. In fact, the areas highlighted are only the group's opinion (expert though that may be) and constitutes a list of theories to be investigated. Each theory must be tested to show its effect. Perhaps statistically designed experiments are needed to test combinations and to reduce the total number of tests required. The reader should review Chapter 6. Once the theory is confirmed by data and the undesirable effect can be ''turned off and on,'' there should be sufficient information available to begin *Step* 3, generating potential solutions.

Whether a team or an individual is working on the problem, this part of the process requires that some discipline be followed. It requires that creative potential solutions be listed, without much concern for the value of the solution. This is brainstorming, where all ideas are welcomed and, at this point, there are no wrong solutions. It is much too early for second guessing and critiquing of ideas. Those participating in the exercise may choose to write their proposed solutions down on an index card or piece of paper and display all solutions on the wall or board. This part of the process should not be rushed. Everyone should be urged to participate and no negative criticism is permitted. Engineers tend to be very prone to criticism because of their need for careful evaluation of alternatives. The public has a low tolerance for engineering errors, particularly since the effects can be disastrous. Engineers are not prone to give quick answers that have not been thought through. This may inhibit the free form atmosphere of a brainstorming session. A facilitator can be used to keep the interest high and stimulate the creative juices.

Step 4 is where solutions begin to take shape. The potential solutions generated in step 3 will have to be grouped into like categories and identified. The criteria for doing this might be such things as equipment, cost, training, work simplification, step elimination, and others. Each of these criteria should then be reviewed for their significance to the problem improvement and a value or weighing should be assigned. This technique will help in deciding which solutions offer more value than others, particularly if there are time or resource limitations.

The alternative solutions are evaluated by assessing their respective value to the process. This can be done in two dimensions as shown in Figure 16.

At this point, it should be noted that there is no such thing as an ultimate solution. The concept of improvement is a relative notion which allows a problem

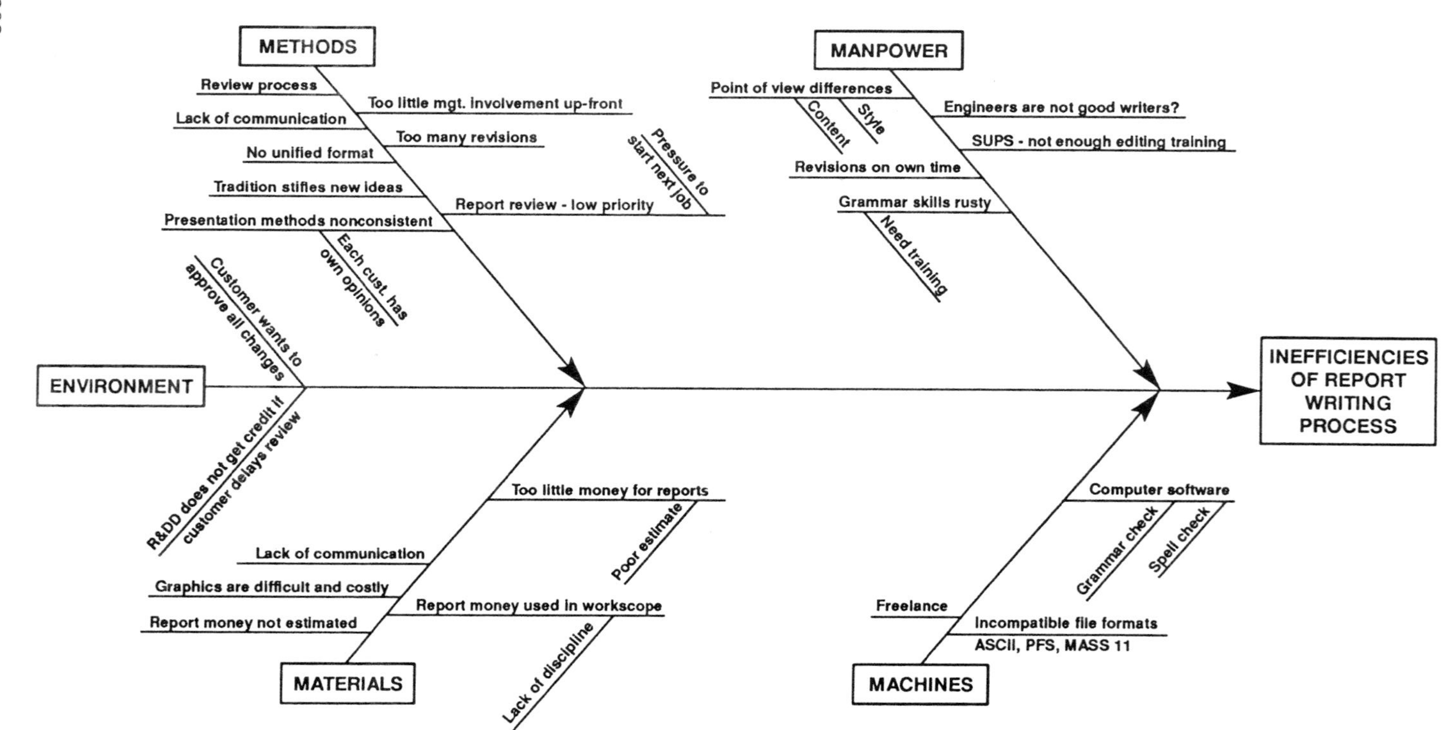

Figure 15 Fishbone diagram.

	High Impact High Cost	High Impact Low Cost	Low Impact High Cost	Low Impact Low Cost
Solution A	●			
Solution B			○	
Solution C		●		
Solution D				○

● = easy to implement
○ = difficult to implement
"ease" can be measured in time to implement or relative cost to implement

Figure 16 Two-dimensional diagram.

to be incrementally improved toward some goal of perfection. However, since the process of improvement takes some investment of time and energy, it becomes clear that the "best" solution among the candidates suggested should be the one or two which overall give the maximum improvement for the investment of time and money. Later in this chapter, the concept of R & D Process Reengineering will be addressed. That concept requires one to make a significant investment in order to restructure the entire process; however, the gain is more nearly "breakthrough" in impact. Suffice it to say that process improvement as described here can result in very significant results and should be used continuously.

To complete the Select and Plan Solution step, one needs to create a work plan expressing timing and resources requirements. Of equal importance are the impact issues. How will this solution impact others around the process? What needs to be done to smooth the transition between the "before" and "after" way of doing it? What will be the measurement criteria to validate that the solution(s) has had a positive impact on the process? The treatise on measurement earlier in this chapter should be reviewed. See also Chapter 12.

Step 5 is to implement the solution. Here is where things start to change. Management needs to adopt the change. Workers will be expected to adopt the new process and make it work. Training may be necessary. Coaching may be necessary.

The revised work plan will have to be followed closely. Deviations at this point will lead to anarchy and a destruction of the confidence the improvement process has brought to the team or individual. If there is an urge to make changes at this point, step 4 should be revisited.

Once final agreement has been reached, it is time to implement the solution. Several months, weeks, or days may have passed since the first step was taken, so there will be some feeling of accomplishment at this point. However, the proof of the effectiveness of the solution will not be recognized until the evaluation step is completed.

Step 6 will require that data be collected on the process as it has been improved. Such questions as, "Was the improvement implemented in its entirety?" and "Do the new indicators show that there was improvement made?" should be evaluated. This verification of the improvement gives legitimacy to the process which was followed. It also relies on the ability to measure the improvement which was made. At this point in the process improvement cycle, monitoring of the revised process is important to assure that it has the support of everyone involved.

With the last step, *step 7*, Institutionalize Solution, comes the stage of the process in which the greatest gains are made. Until now, the activities associated with the process change have been largely "investment." Now the "returns" begin to pay off in improved productivity, quality, time consumption, and most important, customer satisfaction. Many times the gains made in the improvement of one process can be shared with others in similar areas of the operation. Some companies (AT&T, for example) have developed electronic networks so that worldwide operations of the corporation have the ability to find out what gains are being made in other sectors of the company.

Having gone this far, teams normally disband. Before doing so, they should be asked to document the activities of their efforts in something like a final R & D report. Such items as the charter, all exhibits (fishbone diagrams, matrices, flowcharts, etc.) should be included, along with the names of team members, plans, solutions, and any recommendations. It is also very useful to document any interpersonal or behavioral reflections the team encountered while they worked together. The facilitator may be able to help prepare the final report and give an independent assessment of how well the improvement process was accepted and followed and where improvement in the improvements process is appropriate. Finally, the report should be entered into the R & D information system of the division for future retrieval and use. Key word and subject cross-references are essential for future use.

5. *Breakthrough Versus Corrective Action*

Corrective action implies that something has gone out of control, and process expectations are not being met. Fixing the problem usually gets the manager involved in a short-term corrective action solution. This application of "control" means staying on course, adherence to standard, prevention of change. What a control activity does is to merely return the process to stability as defined by performance expectations set by management. Under complete control, nothing would change—we would be in a static, quiescent world (4).

But we can be deluded by the idea that everything is under control. We can become so preoccupied with *meeting* targets or staying within process control limits that we fail to challenge the target itself. So control can be a cruel hoax, a built-in procedure for avoiding progress. This brings us to a consideration of breakthrough (4).

Breakthrough means change, a dynamic, decisive movement to new, higher levels of performance. While some improvement may be possible by better or tighter control, the concept of breakthrough and the resulting improvement of performance to meet or exceed customer expectations is the key. Breakthrough, then, is the creation of good change to enhance levels of performance, whereas control is the prevention of bad changes (4). Breakthroughs in R & D processes are intended to take the organization from one level of performance to a new and better level as a response to better satisfying our customers. If customers receive final project reports typically forty-five days after test completion, that may be acceptable because it is the industry norm. However, with an effort to improve final project report timeliness, an organization should be able to radically reduce the cycle time while maintaining or improving the quality of the report, the cost to produce it, and the overall satisfaction of the customer.

6. *Other Problem-Solving Analysis and Techniques*

This chapter has in no way touched on all or even most of the specific improvement tools currently being used. A few basic ones were described to give a general picture of the methodology for continuous improvement. The road to process improvement in R & D will wind through all levels of the organization. For it to be effective, process improvement techniques must fit the organizational needs as characterized by where the process resides in the organization structure, the source of the improvement need (whether management-directed or employee-directed), the breadth of the process (inter- or intragroup), and the scope of the process to be improved (division-wide or corporation-wide).

The choice of improvement approaches are varied: top-down team efforts (corrective action teams), individual involvement efforts, lesson-learned exercises, Area Activity Analysis (see Chapter 5), as well as bottom-up efforts (quality circles). The common element of these approaches, however, is a systematic, step-by-step analysis like the one just described. The methodology clearly works. Management needs to apply the suitable techniques in the right setting.

C. Lessons Learned

An R & D organization can overlook a very useful source of information for improvement if it ignores historical occurrences. How many situations have taken place where a problem was solved or an innovation was made but no one recognized the importance of the information for the future? The practice of recording case studies of failed and successful situations is a powerful tool which can be applied to R & D quite effectively.

The organization should make a list of the most outstanding successes and failures in recent history and select two or three of each for documenting. Case studies are not simple to prepare. The object of the study must be instructional for more employees than were originally involved in the situation. The preparation of a good "lessons learned" case requires careful deletion of names, departments, and other personal identifiers which could be embarrassing to those who were involved. This "sanitizing" forces the focus on the real issues which are important, to learn from experience without making employees look foolish. Such elements as external influences, competitive pressures, internal procedures, and people issues form a backdrop for the situation being studied. Care must be taken when preparing this portion of the case to not lead the reader to the right conclusion. Rather, the idea is to elicit a response about what the root cause might have been and what array of potential corrective actions (or preventative measures) should have been made. No two groups of people studying a case will come up with exactly the same solution. However, most groups will identify some basic faults that should have been avoided. These can be used in subsequent situations to avoid recurrence.

Examples of "good" cases are a little more difficult in that it is harder to identify all the right things which, taken in the sequence they occurred, add up to a successful project. The assumption is made that many good practices performed routinely in R & D efforts need to be recognized as such in the hope that they will continue to result in expected outcomes—no surprises. If these can be replicated, then a certain level of comfort can be achieved. Many times these elements become institutionalized in the form of standard practices. More and more, R & D organizations are adopting standard practices to set a minimum level of quality below which the researcher is prohibited from performing.

Appendix A describes the development of standard practices, quality training, and the use of case studies as "lessons learned" in an R & D organization.

II. BUSINESS PROCESS REENGINEERING FOR R & D

A. Continuous Improvement Versus Business Process Reengineering

Companies which have utilized continuous process improvement will say they have accomplished the "easy" stuff, the quick fixes, the low-risk improvements, and they have found that, for many situations, continuous improvement is appropriate. However, many other situations call for a different fix. Processes and systems can be improved only so far. To effect fundamental change, they must be reconstructed or remolded. That is why these companies have begun to reengineer their key processes—to move from improvement to total redesign (6).

What is reengineering and how does it differ from continuous improvement? First, reengineering initiatives start with a clean slate, rather than attempting to improve existing processes. Reengineering initiatives are discrete projects rather than a continuous effort—revolution versus evolution. It typically takes two

Table 1 Continuous Improvement Versus Business Process Reengineering

	Continuous improvement	Reengineering
Degree of change	Incremental	Radical
Where to begin	Existing process	Clean slate
Frequency	One-time/continuous	One-time
Time frame	Short	Long
Flow of change	Bottom-up	Top-down
Breadth of change	Narrow, within functions	Broad, cross-functional
Risk	Moderate	High
Primary enabler	Problem-solving process	Information technology
Scope of change	Cultural	Cultural/structural

to three years to fully implement a reengineering program, while continuous improvement schemes often achieve incremental benefits in a relatively short time. Reengineering initiatives tend to be higher risk but offer greater rewards. Table 1 captures the principal differences between continuous improvement and business process reengineering.

A note of caution: To avoid confusing employees, it is imperative that all improvement activities be implemented under the umbrella of a single quality program, regardless of the differences among them.

The benefits sought from reengineering programs typically have no precedent within the company. One pharmaceutical R & D organization set a "stretch" goal of reducing the time required to calculate the results of clinical trials from six months to a single day. A made-to-order capital goods company set a stretch goal of reducing the time it takes to configure and price a custom order from six weeks to several minutes—while the customer waits on the phone for both the cost and the delivery date.

Pain must be evident throughout the organization in order to justify the risk of rapid, radical change. If speed-to-market is twice as long as the competition, it clearly means one is *not doing things right*. If technological innovation and technology management are not under control, it means one is *not doing the right things*. In either case, the pain is genuine and significant, and alleviating it could mean the difference between life or death for the organization.

B. Where Does Reengineering Fit on the TQM Journey?

A four-stage quality transition model (Figure 17) was developed as a result of researching the "best" companies in the United States and Japan and how they got that way. Four basic stages of advancement in the journey toward Total Quality were identified, each distinguished by specific types of activities.

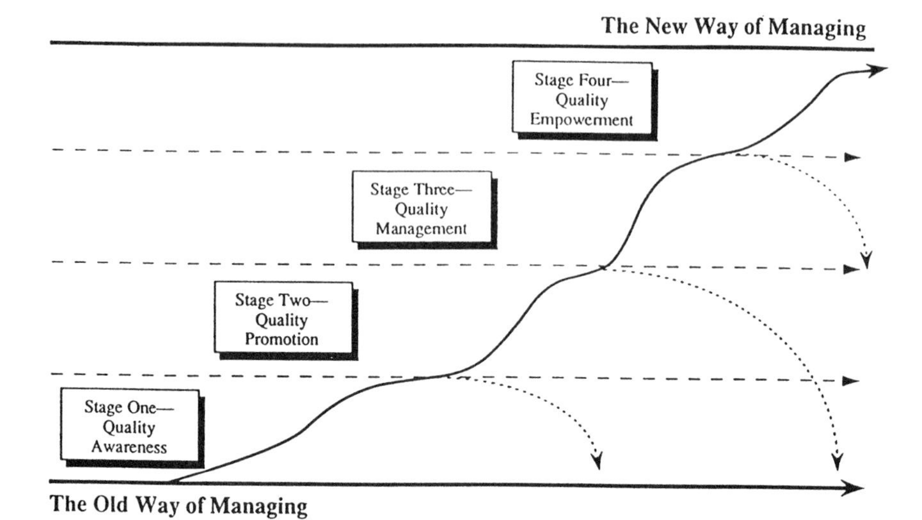

Figure 17 Four-stage quality transition model.

Stage Two is about the formation of cross-functional teams that solve problems using process improvement methodology and statistical tools. It's about continuous improvement, horizontal communication, and breaking away from hierarchical structures. This is where companies pick the low-hanging fruit.

Teams continue to spawn—until there are so many teams that they and their projects draw attention away from the business vision, or the teams fail to reap benefits because all the easy gains have been achieved. Once the teams have broken down hierarchical and departmental barriers, the organization is ready to challenge the fundamental way products are designed and developed.

At this point, the organization is ready to become a Stage Three company, to assign process owners and reengineer the business. If the organization does not take this leap, it will lose all the ground it gained in Stage Two and slip back into being a Stage One company. If the organization attempts to reengineer prior to this point, it may find itself fighting against managers who refuse to give up their pockets of power within the corporate hierarchy.

C. Management Commitment

Resistance from senior R & D management is one of the more frequent obstacles to reengineering efforts. R & D managers have to be ready for reengineering—it requires strong leadership commitment and can generate considerable angst during rollout. As the previous examples demonstrate, reengineering targets are, by their nature, extremely challenging. Hard-to-reach targets are key to promot-

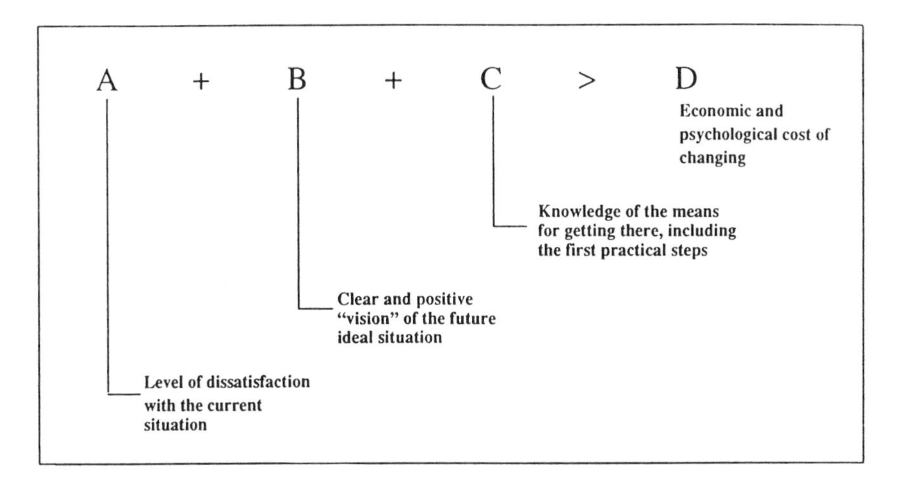

Figure 18 The Trybus model.

ing out-of-the-box thinking. However, they are alien to a continuous improvement environment and their demands are often daunting. In the face of seemingly unattainable challenges, the organization may suffer from a serious case of inertia.

In order to effect any type of strategic change within an organization, the model shown in Figure 18 must be satisfied. Named for its developer, Myron Trybus of the Massachusetts Institute of Technology, the Trybus Model says that unless the need, the vision, and the required knowledge for achieving change are greater than the cost of change to the organization, any attempt at radical redesign will be futile and will probably do more harm than good.

D. Benefits of Reengineering

Before an organization decides that reengineering simply is not worth the risk, no matter how bad things are, they should take a look at what other companies have achieved:

Feeling the need to shorten product life cycles, minimize lead times, and adopt a fresh process for launching new products, Boeing's Ballistic Systems Division implemented a program called Developmental Operations, with the goal of simplifying development practices. Design analysis was reduced from two weeks to 38 minutes (their goal is 4 minutes), with the average number of engineering changes per drawing dropping from a high of fifteen to twenty to a low of one.

The new head of R & D at Thornton Equipment Company, a large, midwestern specialty equipment manufacturer, revamped its product development process in response to declining sales, rising costs, and ailing quality. The new process included a formal ranking of development projects based on customer needs; assigning teams of design engineers, materials purchasers, and manufacturing engineers to work on major design projects; better and more formal communication between the shop floor and the design center; and the standardization of most machine tools. Within five years, the results were substantial: a 34 percent drop in average cost of design engineering for new parts; a 26 percent decline in the ratio of engineering overhead—including specification, database and quality assurance—to direct engineering; and a 40% drop in product development time.

Following an analysis of R & D performance in its industry, AT&T's Consumer Products (CP) business developed an action plan based on benchmarking best-in-class companies. CP discovered that the best-in-class companies permitted no design changes beyond product design specification, invested in and empowered strong product teams, and effectively removed support processes (such as testing) and invention from the critical path. Using these findings, CP reengineered its development process to reduce time to market by approximately half and tripled the rate of new product introductions over three years.

Marion Merrell Dow (MMD) has made significant strides in speeding the approval process for new products by reengineering its approach to R & D. This success has translated into the discovery of breakthrough drugs such as Seldane and Sabril, and in bringing new drugs to market in the United States, Europe, and Japan. MMD is reengineering its R & D process still further, to ensure that its scientists "can do what they do best." As part of this effort, MMD is taking a truly global approach, minimizing duplication at its various sites and channeling resources to the highest-priority projects. The major goals for MMD are to shorten the time needed to get a promising compound out of discovery research, more quickly conduct thorough tests for safety and efficacy, and speed up the regulatory application and review processes.

E. How It's Done

A case study will illustrate how the process works:

1. Background

- A $1 billion division of a major healthcare company
- A 400-person R & D organization in the United States and Europe

2. Process

A methodology, or approach to BPR, expressed in terms of business drivers and results is shown in Figure 19.

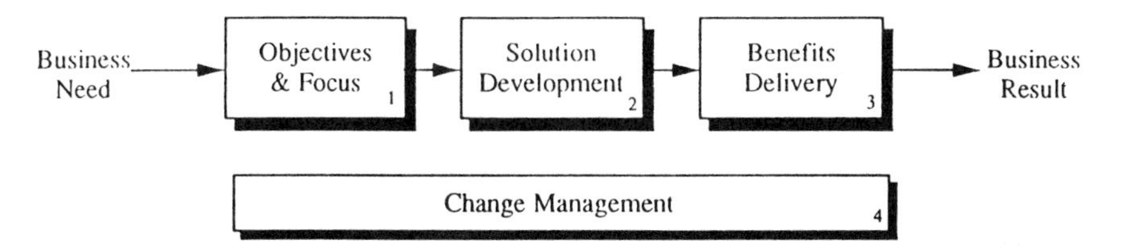

Figure 19 Approach to Business Process Reengineering.

3. Business Need, or Why Reengineer

- In response to declining sales, a new president committed to the executive group to develop technology capable of doubling sales in five years.
- The president and the VP/R & D spent half of every day for three weeks reviewing ongoing projects and concluded that the organization could not deliver as promised.
- The number of projects underway was eighty-seven, not the seventeen that had been authorized.
- R & D staff was chafing under the existing processes, and internal tension was high.
- In the first quarter a major customer, tired of waiting for the new product, took its business elsewhere.
- The president knew that if the organization did not reengineer itself, it would be reengineered by others.

4. Objectives and Focus

Guided by presidential edict, the senior R & D management team met off-site and agreed to the following targets:

- A minimum of 50% reduction in development time for new products
- Doubling the rate of new product introductions
- Reducing layers of management from six to four or less
- More doers and fewer workers
- Build on teamwork principles already in place

5. Solution Development

To redesign the organization, the team:

- Attended an intensive three-day workshop for a cross-section of seventeen R & D managers and staff
- Completely eliminated the functional management hierarchy in favor of temporary project teams

- Established the resource pool concept with groups of thirty managers, engineers, and technicians all reporting to a single resource manager
- Assigned temporary teams to specific business areas to meet business plan expectations
- Transferred all nonproject functions to a central support group.

The redesigned organization is shown in Figure 20.

6. *Process Redesign*

Process redesign was itself a four-step process as shown in Figure 21.

a. Identifying physical processes currently in place.
   It takes six months from the time we collect the last patient records at the doctor's office to the time we have a clean database.

b. Exploring logical principles behind the processes.
   - Centralized data capture.
   - Pickups from doctors' offices by messenger every eight weeks.
   - Validation of data after data entry.
   - Site visits to determine causes of data errors.

c. Exploring options for logical redesign.
   - On-line data entry and validation at doctor's office.
   - Pickups from the doctors office every week.
   - Prescreening of data at doctor's office.
   - No screening of data, just mail in every week.

d. Isolating options from sensible redesign.
   - Decentralized on-line data entry and real-time validation at host computer.
   - It now takes two weeks from last patient last visit to database release.
   - Figure 22 shows what the new product development process looks like.

7. *Motivation and Discipline*

- The group established a ''fee rate'' mechanism that effectively managed resource utilization and made possible staff rewards based on personal, measurable ''billings'' and ability to complete projects on time and under budget.
- R & D was able to control its destiny by reengineering itself.

8. *Benefits Delivery*

- The commitments for 1992 were met.
- There was a 25 percent increase in efficient utilization of resources.
- The 1993 plan included significantly more program deliverables than could be envisioned at the beginning of 1992.

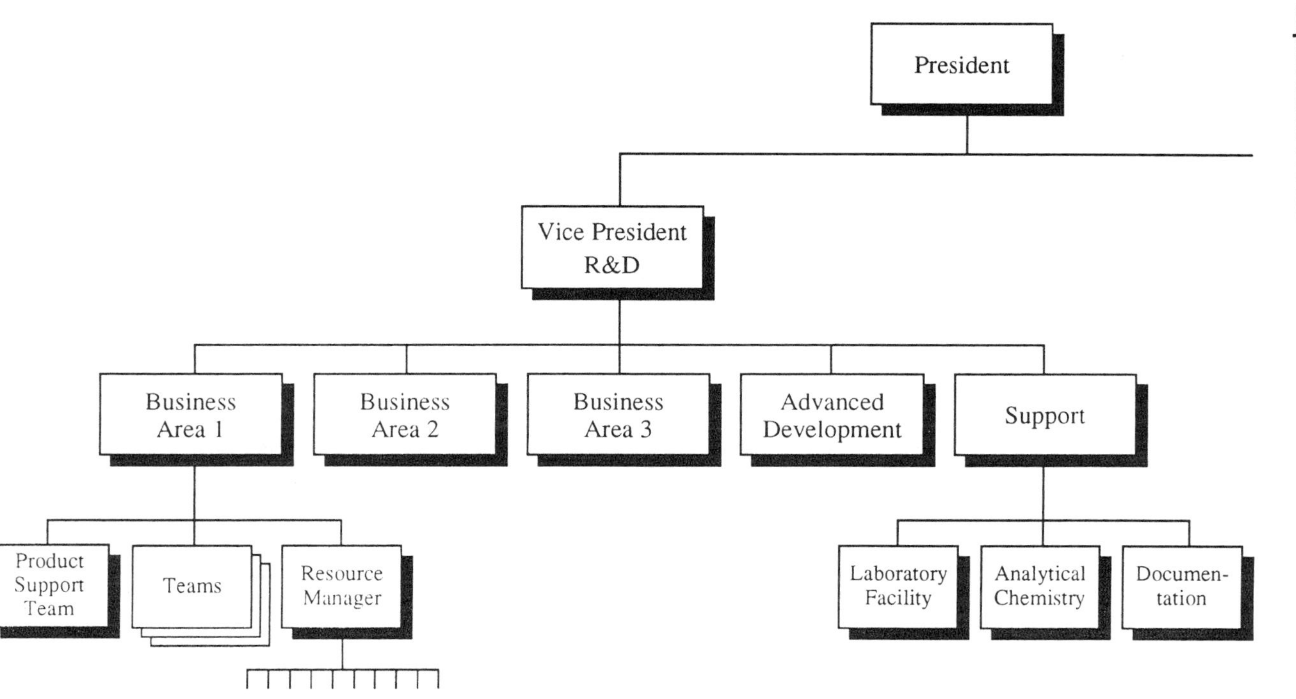

Figure 20 The redesigned organization.

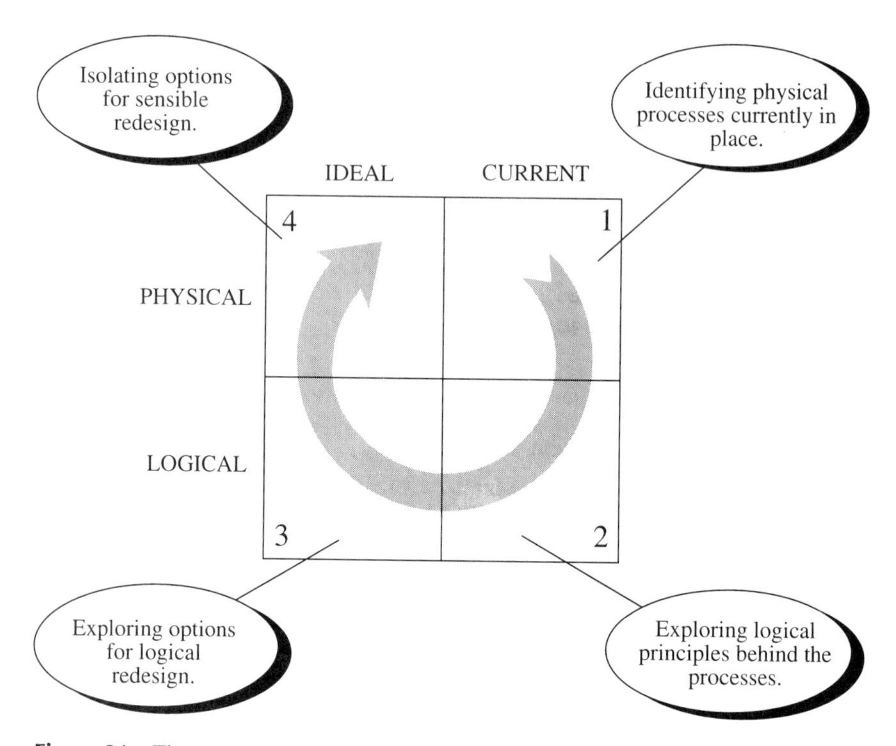

Figure 21 The process of process redesign.

9. Change Management

Figure 23 illustrates how the overall change management program operated.

10. Business Results

- New programs were contributing to revenue streams as required.
- The executive group was won over by R & D 's achievements and is increasing its investment in new technology development.

F. Some Final Words of Advice

Because reengineering is high profile and pervasive, it requires full-time commitment and unswerving dedication from senior managers and the project teams assigned to making it work. If the program derails, everyone in the organization—executives as well as staff at all levels—will know it. It takes at least one to two years to realize significant results, so be prepared with programs to bolster morale and encourage continued support for the changes.

Interestingly enough, senior managers may be most in need of bolstering and encouragement. The organization's leaders will be asked to challenge the processes they probably designed on their way up. Initially, this may not be a

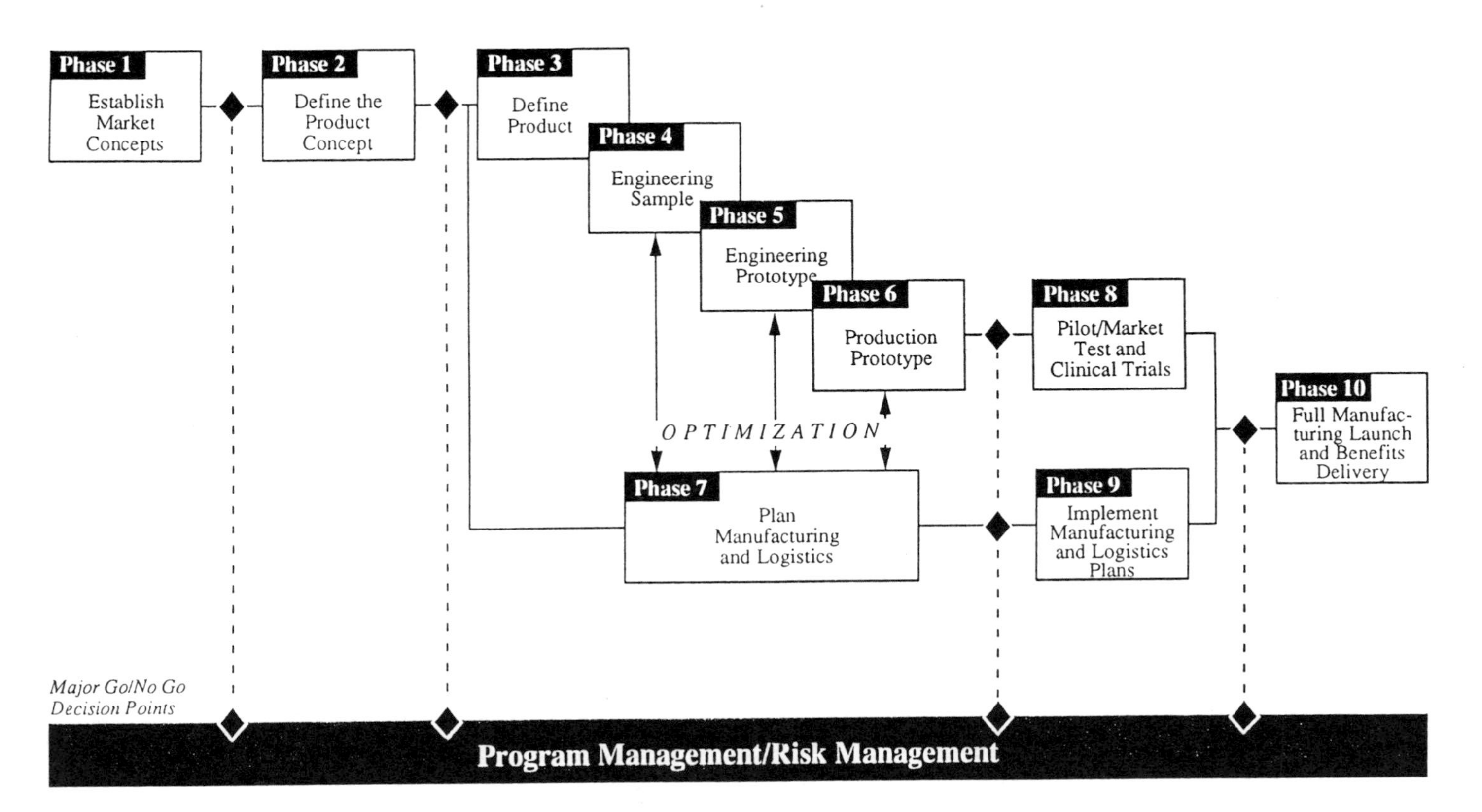

Figure 22 The redesigned product development process.

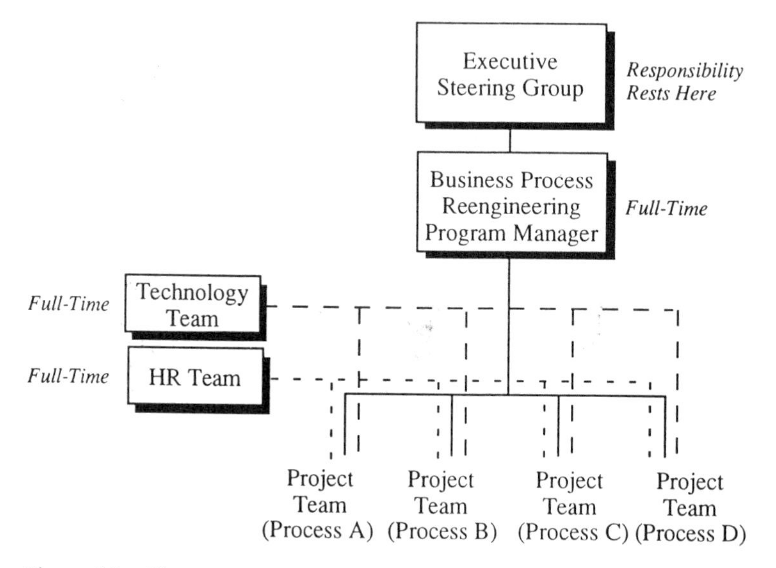

Figure 23 Change management.

problem—until they begin to see the bulwarks of their best efforts being dismantled and discarded. Leadership directives tend to start out strong, and it will be the responsibility of the project teams to sustain that level of commitment. Breaking one's own paradigms may be the most difficult cultural challenge to be faced.

The organization must be ready to reengineer! There must be a survival issue at stake, or they will not be able to secure the necessary backing from key players.

It is the responsibility of the Executive Group to define the stretch goals, derived from the business vision. If they are not the authors, they will not support the goals when the going gets really tough.

Project teams must be prepared to work closely with Marketing and Manufacturing. The business driver for BPR—the reason for anyone to go to all this trouble—is to increase customer demand for the organization's output: products or services.

Reengineering is unlike any other program or organizational change. The action will be fast and furious, so one should buckle the seat belt and enjoy the ride!

How does one know whether the organization is ready for reengineering? The survey in Appendix B should help to find out.

REFERENCES

1. H. J. Harrington, *The Improvement Process: How America's Leading Companies Can Improve Quality*, McGraw-Hill, New York, 1987.

2. David N. Griffiths, *Implementing Quality with a Customer Focus*, Quality Press, American Society for Quality Control, Milwaukee, 1990
3. A. J. Montana, ''Quality in the R & D Process,'' 1991 Symposium on Managing for Quality in Research and Development, Juran Institute, Inc., Wilton, CT, 1991
4. J. M. Juran, *Managerial Breakthrough: A New Concept of the Manager's Job*, McGraw-Hill, New York, 1964
5. J. M. Juran, ''The Non-Pareto Principle; Mea Culpa,'' *Juran on Quality Improvement Workbook*, Juran Institute, Inc., Wilton, CT, 1988
6. D. Boath and R. P. Maddams, ''Reengineering the Product Development Process,'' 1993 Management of Quality in Research and Development Symposium, Juran Institute, Inc., Wilton, CT, 1993

APPENDIX A: LESSONS LEARNED

William H. Hamilton
Consulting Engineer
Ligonier, Pennsylvania

Training for Technical Management

In 1984, Babcock & Wilcox decided to devote special energies to training technical management of the corporation in improving the quality of their technical work. A training course was developed around the idea of learning from experiences in the corporation in the past where serious technical difficulties were encountered or technical successes occurred. The course was built around a series of case studies, each one of which described an incident and then raised questions as to what could be learned from that incident. It was intended that the cases be discussed by the students in the course at length, and that the learning would come from these extensive discussions of events in the Babcock & Wilcox history.

When the course was first being designed, a group of senior engineering managers from Babcock & Wilcox came together and agreed upon a way of presenting the cases and on additional lectures to be presented by other people skilled in the engineering business and in the quality business. With a consensus of the eight people who met on that occasion, there was confidence that the course could be presented in an effective and useful way. The first session was held in New Orleans in 1984 with about twenty participants. The session featured a lecture by Dr. Juran, the quality specialist, who presented his views on American industrial management and the needs that they have to achieve quality products and services.

In preparing for the first session, it was recognized that there had to be some sort of a standard or criterion against which the practices being discussed in the various cases could be judged. Accordingly, a list of good engineering practices was developed and that list was kept in front of the students for the entire duration of the training session. In the study of each incident the list was used as a basis for judging where errors had been made in the engineering practices or where they had been used successfully. That list of engineering practices has prevailed over the years since then.

Since 1984, the course has been given eight times to over 200 managers and it has been judged as a successful way to teach senior technical people the essence of good engineering practices—to instruct the managers on the types of actions and decisions which can lead to successes or failures. In addition to the corporate presentations, the same sort of treatment has been given at several of the divisions in the corporation including the Research and Development Division. In the latter case, the session was enhanced by having a customer come to the session and present his observations on the performance of Babcock & Wilcox's research people in some contracts that he had administered. His insight into the difficulties in the projects he described were very useful for the research people.

The cases used for training were developed with a standard format:

1. A two- or three-page explanation of the incident, when it occurred, and the effects of the incident on corporate or laboratory operations (e.g., costs, delays, errors, etc.).
2. A series of questions about the incident; the questions were to be used in group deliberations where lessons to be learned were discussed and written down.
3. Attachments which describe the product and detail the history of the project, and supporting documents, such as meeting minutes, letters, and sketches.

Care has to be taken not to provide too much detail or the reader gets lost in the history. Using the case study for instructional purposes requires monitoring of the discussion periods. Technical people tend to want more and more detail about what happened, to the sacrifice of learning lessons from the incident.

Let us consider a success case and some of the aspects of that success which resulted from the case study. Consider the case of a remote-operation scheme for repairing and plugging steam generator tubes in nuclear plants. It's a robot-type machine that enters into a high-radiation area and does the mechanical job, controlled from a remote position away from the high-radiation area. Babcock & Wilcox people started the development of this machine at a small laboratory in Lynchburg, Virginia, set up a special project with a project manager, built mock-ups to test prototypes, and, although they had some difficulties in the field with the first unit and with subsequent units, they had the plan laid out and the staffing required set up so that they were able to address the problems quickly, correct them, and move on to further field operations. Indeed, the product became very popular in the nuclear field and Babcock & Wilcox produced many of the machines and did a great many repairs on steam generators.

The key to the success of this project was a small staff of dedicated and skilled people under skillful direction of a project manager who knew how to develop a new product, get it to the field, find out the field difficulties, correct them, and move on to further work. He knew how to report progress, how to maintain his sponsor's interest, and how to keep his staff focused on the forward motion required for this kind of a development project. The success of the project not only resulted in considerable profits for Babcock & Wilcox, but it also resulted in significant promotions for the individuals who had done the original work and an expansion of that business operation. The value of studying this experience was to point out to the attendees the significance of good project management, careful planning, and prompt resolution of technical problems.

Correct staffing of the project with the right people and the right number of people from the management down to the field technician came through loud and clear to those who studied the case.

One might ask if the success of such a project is a matter of quality. Success resulted from a high quality of management of the project, high quality of the engineering work done, high quality of the manufacturing work, high quality of the field operations where the units were installed in the steam generators; in other words it represented an example of total quality of the project and the success that ensues from such high-quality operations.

A problem put to the students was, "Once the development of this new piece of equipment was undertaken, a series of management decisions was made that significantly contributed to the success of the project. Name them."

A second study was written of an automatic system for retrieving parts from a storeroom to provide for shipment to customers directly or to the manufacturing floor of the factory. The project was envisioned as a way to reduce indirect costs by reducing the storeroom help required to handle the hundreds of thousands of parts every year. An outside firm was hired to develop the system and that firm did not come through in a satisfactory way. At one point, the whole factory and many customers were shut down because the parts couldn't be delivered by the automatic system. Slowly, management was able to work out the difficulties, one of which was the computer controls for the system. These were designed by the outside supplier and didn't operate very well. Management didn't recognize that this was going to be a problem and didn't staff up with people who could deal with the computer difficulties. About a year after the project was started, a young engineer was hired to help improve its operation. He was skilled in computer-controlled equipment and was able to bring the equipment into fully satisfactory operation. In this case, satisfactory meant the number of "picks" that could be made per hour was sufficient to meet customer and shop requirements, and it was also sufficient to permit significant reduction in the staff in the storeroom. That, of course, was the original intent of the project, the cost saving in the storeroom. But the case clearly illustrated the problems that can occur when the wrong people and the wrong number of people are assigned to direct a new development or a project such as this one. The engineer on the assignment was a mechanical engineer who was able to deal with the mechanical aspects of the equipment that retrieved material from the storehouse shelves, but he was not able to deal with the controls for that equipment. Frequently, the controls are overlooked in the laying out and the planning of a project, which leads to serious difficulties as the equipment is put into operation.

This case is an illustration of a quality error made by the management of the organization that was undertaking the project. One of the errors was in staffing the project with the wrong individuals. Later in the project, after much difficulty and distress from customers and the shop, the right people were brought in and the situation was corrected. But the initial error proved to be a very costly one for the company that was doing the work.

The question here is "How could the error have been avoided? Would a design review at an earlier stage have shown the need for additional staff?"

Quality is frequently thought of as it applies to the attributes of a product or to the satisfactory nature of services provided to a customer. But underlying such quality is a series of management decisions and actions which also have attributes of quality. Those decisions that are well done and carefully thought out end up with a result that is satisfactory to the customers and to the owners of the corporation. This aspect of quality management is particularly important in corporations where the products are custom designed for individual customers, such as in the case of Babcock & Wilcox. The decisions and actions of technical management are at the root of the quality of the end product.

Case Studies at the Research and Development Division

In 1988, the general manager of the Research and Development Division of Babcock & Wilcox recognized that there were occasional incidents which reflected unfavorably on the normally excellent quality of research activities of the division. Accordingly, he directed that written studies of each incident be developed. These studies would be the subject of deliberation by the Quality Team of the division and from these deliberations, changes in methods of operation could be developed. The Quality Team was made up of the senior managers who reported to the general manager of the division. This team had been meeting for several years, discussing and enacting ideas about the quality of the services provided by the Research and Development Division.

The deliberations of the Quality Team resulted in detailed evaluations of the incidents and these evaluations resulted from extended debates and conversations. But after the first few of these, it became apparent that the cases were providing an important educational result but weren't resulting in any change in the method of operation within the division. Accordingly, later deliberations of the team about case studies have focused more on what changes should be made as a result of the lessons learned from the incident. By and large, these changes have shown up in the form of changes to the Standard Practice Manual.

One case considered was a situation where the customer was highly dissatisfied with the results of a technical project. As the project progressed, there were many signals that indicated trouble brewing in the mind of the customer. Unfortunately, they were not, either individually or collectively, recognized as being signals of trouble. So when the general manager of the division wrote to the customer at the end of the project and asked for commentary, he received a very disturbing letter.

The story was written up as a case study and deliberated on by the Quality

Team, and out of that came a recognition of the need for training project managers in similar projects and in the total business of running a project in dealing with a customer, recording his interests, and making sure that the project is aimed in a satisfactory manner to meet the customer's requirements. A specific outcome of this case was a training course for project managers.

Another case was a boiler water chemistry monitoring project where the customer was highly dissatisfied with the costs incurred and with the schedule delays. The division was supplying a service of monitoring water chemistry conditions at coal-fired boilers in various parts of the country, and, while the monitoring service was viewed as satisfactory, the management of the project was severely criticized. In this instance, there was a direct customer and a customer who underwrote the program involved, so that many parties had to agree to the various parts of the program. Long delays occurred in obtaining those agreements and the communications were difficult. One outcome of this case study was the inclusion in the Standard Practice Manual of the idea of developing at the outset of a project a Functions and Responsibilities Document which stipulates the responsibilities of the various parties involved in the project, what each one will prepare or perform, where approvals will be provided, and where agreement can be reached when changes to the program are required.

In each case, the managers of the R & D division wanted to make sure that the engineers in the lab learned from their experiences—that they knew the facts, the root cause of the errors, and corrective measures that can be taken to avoid recurrence.

One difficulty that has been encountered in the overall process is the fact that the deliberations are done at a senior management level, and it is difficult to transmit the lessons to lower levels of management and to nonmanagement employees. One effort to combat this difficulty has been inviting lower-level management people to attend the deliberations of the Quality Team so that they will hear first hand what has happened. But the matter of getting the word down and the ideas behind the case studies and the lessons to be learned from them to the working-level people remains a continuing problem that requires constant attention.

Another problem in the process is making sure that the individuals who were involved in a case don't become defensive and cloud the issues during the various discussions and during the preparation of the written material. Getting the straight facts, of course, is very important, and requires special diligence on the part of the person who is preparing the written case. One technique used at Babcock & Wilcox has been to include the individuals who were involved in the project in all the deliberations. In this way, it has worked out that there is no finger pointing or accusing, but rather an objective appraisal of what happened in the incident, and, then, what corrective measures might be taken.

It may be difficult to find other instances in research and development activities where such a careful analysis of troubles has been performed and used to advantage. It has proved to be a very useful tool for the Babcock & Wilcox laboratory and might well be emulated in by others.

Once again, however, underlying the whole thought process is the importance of good management. The quality of the product or the service once again depends heavily on the manager of the project to do the work which the customer expects, at least as much as he needs, and exceeding that if possible. "There is no substitute for good management." This is a statement frequently made when an incident is being critiqued or a project manager is trying to reconstruct why his work went astray. Instituting paper systems or controls, or setting up committees to direct a program, or assigning a deputy to help a manager, or other measures or "crutches" cannot substitute for a good manager running a project effectively.

Frequently, one hears the words, "He is a good technical man but not a very good manager." The words hint at a fundamental difficulty in selecting individuals for technical management. The candidates for a position are usually from the engineering ranks. If a person has been a successful engineer—devising clever designs or analyzing designs thoroughly or putting his equipment into operation effectively and promptly—he is usually considered for more responsibility. Sometimes the complexities of supervising the work of others are beyond the capabilities of the successful engineers. So, in selecting individuals for further responsibilities, a judgment has to be made about that person's ability to stretch and to be able to handle the additional matters. That judgment is a place where errors can be made, and the result usually shows up later on in the handling of the project. The quality of the judgments made in selecting individuals for project assignments, including project management assignments, is an important element in quality of products produced or services rendered by a research and development laboratory.

APPENDIX B: REENGINEERING READINESS SURVEY FOR SENIOR R & D MANAGEMENT

David D. Boath
PA Consulting Group
Hightstown, New Jersey

Is Your Organization Ready for Change?*

Instructions for Sections A through C only:
Rank your response to each of the following questions from 1 to 5, with **1 equal to strongly disagree and 5 equal to strongly agree.**

Section A: Level of Dissatisfaction

1. Does the Executive Group recognize there is a problem? ______
2. Do they acknowledge and agree on its root causes? ______
3. Is the Executive Group committed to solving the problem? ______
4. Do they believe change is inevitable and necessary? ______
5. Is there a do or die issue? ______

Section B: Vision of the Future

1. Has the Executive Group formulated a vision for the future of the business? ______
2. Have the goals and objectives for the business been clearly defined? ______
3. Is there a clear view of the benefits to be derived from BPR? ______
4. Is the Executive Group prepared to lead the effort? ______
5. Have the concerns of all key stakeholders been included in the planning? ______

Section C: How You Will Get There

1. Is there a plan for guiding the organization through the transition period? ______

*Adapted from the Myron Trybus Change Model, which states: $A + B + C > D$, where:

A = Level of dissatisfaction with the current situation.
B = Clear and positive "vision" of the future ideal situation.
C = Knowledge of the means for getting there, including the first practical steps.
D = Economic and psychological cost of changing.

2. Does the organization have the information technology capabilities to support the reengineering process? ______

3. Are there human resource systems in place to support the reengineered organization? ______

4. Have metrics been established to gauge success along the way? ______

5. Does the existing management structure accommodate the project teams' need to exercise authority? ______

Instructions for Section D, and please note the difference:
Rank your response to each of the following questions from 1 to 5, with **1 equal to strongly agree and 5 equal to strongly disagree.**

Section D: Cost of Changing

1. Will the benefit being sought, as expressed in dollars and cents, exceed the anticipated cost of reengineering the organization by at least a factor of two? ______

2. Are employees enthusiastic about what's in it for them? ______

3. Will the new information infrastructure enhance the ability of management, staff and line personnel to perform the reengineered processes? ______

4. Will customers be served in dramatically better fashion by the reengineered organization? ______

5. Is this seen as a low risk project? ______

Scoring

For Sections A through C:
Add your number responses to all questions in Sections A, B, and C and write the total here ______.

For Section D:
Add your number responses to the questions in Section D, multiply the total by 3 and write the answer here ______.

Your R&D organization is ready to consider reengineering only if your score for Sections A, B, and C combined exceeds your score for Section D.

12

MEASURES OF EFFECTIVENESS

David D. Boath
PA Consulting Group
Hightstown, New Jersey

Mark Bodnarczuk
National Renewable Energy Laboratory
Golden, Colorado

I. USING METRICS TO DRIVE TQM IN INDUSTRIAL RESEARCH*

The first part of this chapter will discuss how industrial research organizations have successfully used metrics to improve their performance, overcoming the scientist's natural skepticism about measurement. Different types of metrics are appropriate at different stages on the journey toward total quality; three case studies illustrate the benefits that R & D organizations have achieved.

For the past ten years, extensive research has been conducted in Japan, Europe, and the United States by means of study tours to top-performing companies. Executive teams have visited companies whose operations exemplified "best practice" in their industries. Some of the research was concerned with why TQM programs fail in both research and manufacturing. An alarming 78 percent of TQM programs in the companies surveyed have *not* lived up to expectations, and chief executives are getting frustrated by the lack of progress being made.

The majority of these TQM programs plateau because they are not focused on real business issues, are not based on realistic targets, and do not have metrics

*The work described is a result of original research performed by PA Consulting Group, Inc., Hightstown, N.J.

that drive improvement. A quality program that measures its success by the increase in attendance at voluntary company meetings, for example, may be creating great vibes among employees, or perhaps simply reflecting the availability of free donuts, but what real benefit is being produced? "Touchy-feely" activities are valuable only if they produce a measurable benefit in terms of R & D performance. Teamwork and empowerment are excellent ways to create a cross-functional environment and motivate employees, but it will be the ability to measure real process improvements that will directly influence a company's capacity to keep funding this type of program until the major breakthroughs are achieved.

Certainly, a culture rooted in hypothesis and experimentation will be skeptical of any form of performance measurement, but nevertheless, R & D is based on scientific methods that *are* subject to measurement. The old adage, "if you can't measure it, you can't manage it" has never been more appropriate in R & D.

A model has been developed which defines four basic stages of advancement in the journey toward total quality. Different metrics are applicable, depending on how far an organization has proceeded on that journey.

A. There Are Initial Roadblocks to Measurement

Can measures be applied to the most complex phases of research? Certainly, those rooted in an R & D culture will often be dubious about the merits of any form of performance measurement. Scientists and engineers have comments like:

"If we did things right the first time, we would never discover anything!"
"I'll let the scientific community measure the quality of my work when I publish."
"Measurements are not relevant to what I do day-to-day—I can't influence them."
"The baseball-bat approach doesn't work well with scientists and engineers. We need to be nurtured."

Most organizations fail to implement metrics because, quite simply, scientists do not think that they are relevant. This has led to two fundamental axioms. The first is that scientists should be allowed to choose the metrics that they will be measured by. It is more important that they buy in to being measured than that the best metric is selected. After all, 30 percent of the metrics will change after the first year because they will either be too difficult to collect or not relevant. (The appendix to this chapter contains a list of the most popular measures chosen.) The second axiom is that the metrics should be linked to the critical aspects of the scientists' research—that is, the major processes that they use every day—so they will see that the measures are relevant.

B. Successful R & D Organizations Measure a Wide Variety of Factors

This section will discuss some of the metrics that successful organizations have found useful, as a guide to the reader. Each organization however, has selected metrics that are appropriate to its business, or to those aspects of R & D that it has a particular need to measure:

1. A leading pharmaceutical R & D organization has embarked upon a very successful TQM program. In one area alone—clinical research and development—it has understood its processes and implemented twenty measures relating to clinical investigation, management of clinical research assistants, data management, dossier preparation, and resource management. The benefits that it has achieved have been impressive. It has:

- Reduced the time to feed back errors on case report queries to clinical research assistants from three months to five days, and cut error rates by 60 percent.
- Reduced the time between last patient visit and database closure from five to three months.
- Reduced the number of queries from the agencies by 25 percent.
- Reduced the number of days of rework on dossiers by 40 percent.

2. The R & D division of an electronics organization applies the "Jimmy Stewart" test to measure the value of its research. Those familiar with the classic "*It's a Wonderful Life*" will remember that Jimmy Stewart was able to see the value of his life by observing with the archangel the tragedies that would have occurred had he not been born. In the same way, this organization measures its value by looking at what would *not* have happened if it had not brought its new technologies to market. This same organization also tried implementing measures at every level in its research organization by counting anything it could—that is, liaison reports, memos and papers—but also found this ineffective. It now measures "number of patents per megabuck of investment," and has seen its investment in R & D increase substantially.

3. A metals company started implementing metrics several years ago but was concerned because, although it was measuring hundreds of factors, the measures were ineffective. It concentrated, instead, on a select set of metrics appropriate for its business, namely:

Cost reduction
Increased sales from new products
Capital avoidance
Capacity expansion through the same plant
Efficient use of energy/raw materials
Adherence to environmental/safety regulations

The benefits of these focused and relevant metrics were that they legitimized R & D, improved the feedback to decision makers, thereby improving the decision-making criteria, and provided an economic and implementation focus to its efforts.

4. A major oil company measures R&D by auditing the economic value of completed developments against its original projections. Although these auditors do not make many friends, the organization has realized many benefits—for example, (1) its researchers are motivated to pursue projects with meaningful economic payback, (2) analysis of past failures means that it can stop projects earlier, and (3) the measurement system communicates the value and productivity of research, which helps to justify its R & D budget requests.

5. A major chemical company wanted to increase innovation so it decided to measure the number of patents. Most research organizations feel that this is a poor measure, but this organization went one step further by implementing a process where the patent attorneys sit down with the scientists as they design their experiments so they can look for patentable ideas. The benefit of this was that the company doubled the number of patents in 18 months.

Of course, no one set of metrics can be universally applied, since each organization is in a different business. The challenge for most R&D organizations is to select metrics that are relevant to them.

C. Different Types of Metrics Are Required at Each Stage on the TQM Journey

A majority of TQM programs plateau because they do not have metrics that drive improvement. Different types of metrics need to be applied depending on what stage of the TQM journey an organization has reached. As a result of researching the "best" companies in the United States, Europe, and Japan, a total-quality transition model has been developed. Four basic stages of advancement on the journey toward total quality are identified, each distinguished by specific types of activities (see Figure 1). This model shows that as an organization moves from the "command and control" way of managing to the "total quality" way of managing, it passes through four unique stages. The journey is not smooth because there are fundamental changes that the organization needs to make to reach the next stage. Those that do not make these fundamental shifts plateau out and, over time, fall back.

Stage 1 is called *quality awareness*. An organization at this stage will have a few quality champions trying to effect a change in culture and improve processes. The organization will plateau at this stage unless the head of the organization—in this case, the vice-president of R&D—embraces the quality concept and personally champions the change.

At Stage 2—*quality promotion*—the organization forms cross-functional

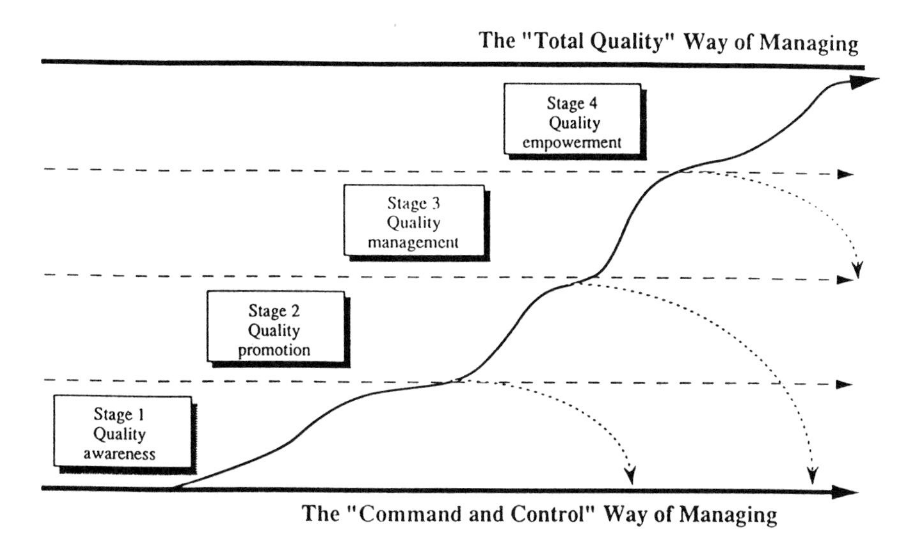

Figure 1 Total-quality transition model defines four stages of advancement toward total quality (Courtesy of PA Consulting Group, Hightstown, N.J.).

teams that solve problems using process improvement methodologies and teamwork. It is about continuous improvement, horizontal communication, and teams spawning and breaking away from hierarchical structures. Typically, in order to implement TQM, a separate quality organization is bolted onto the side of the R&D structure to drive the change. Once the teams have broken down the barriers between functional hierarchies, the organization is ready to challenge the fundamental way in which R & D is performed.

At this point, the organization is ready to become a Stage 3—*quality management*—organization. This is where the organization's vision is defined and everybody in the organization knows their role in achieving that vision. At this stage, fundamental changes are required in the way R&D is performed, and re-engineering is the methodology of choice. Winners of the Baldrige award are typically Stage 3 companies.

Stage 4 is called *quality empowerment* and is where the organization is really functioning both efficiently and effectively. TQM is self-sustaining, and the quality organization that was bolted onto the side of the organization to drive change in Stage 2 has been disbanded. Very few American companies are at this stage.

In Stages 2 and 3, different types of metrics are required to manage the improvements. These different metrics are called process metrics, strategic metrics, and culture metrics. Process metrics are most effectively employed by Stage

2 companies whose focus is on improving cross-functional processes. Strategic metrics are most appropriate for Stage 3 companies that are further along in their total quality journey. Culture metrics are useful for those organizations that are intent on driving TQM through the organization and moving between stages as quickly as possible.

D. Process Metrics Are Appropriate for Organizations Seeking to Improve Cross-Functional Processes

With process metrics, key processes must be defined before improvement objectives, with measurable targets, are set. Figure 2 depicts one illustrative R&D process. Output-control points are defined with the receivers (note that this replaces the emotional word "customers"); the organization asks whether it is satisfying the receivers and lets the receivers decide how they will measure the organization's efforts. The internal organization then develops in-process check points to ensure that the internal processes are delivering the quality required to meet the output-control-point metric.

An example outside R&D will explain the difference between an output-control point and an in-process check point. MacDonald's, the fast food chain, was concerned that the Coca Cola they served tasted different in different stores. They found this out by asking customers (output-control point). When they looked into the cause, they found that different stores put different amounts of ice in the drink, so some tasted strong and some tasted weak. They put in an in-process check point to control the amount of ice. That in-process check point is the logo M on the side of the cup. It is positioned so that the drink tastes right if the operator fills the ice up to the bottom of the M. Now, all MacDonald's operators fill the cup with ice to that level.

After the output-control point and the in-process check points have been established, cross-functional teams are created and tasked with improving processes and achieving measurable targets. In the naturally skeptical world of

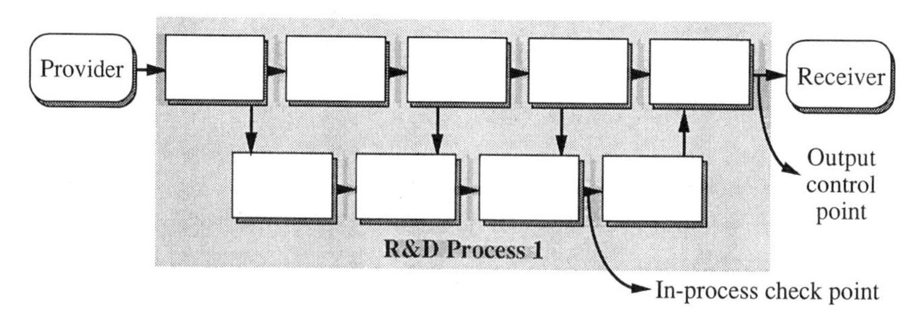

Figure 2 For each R&D process, in-process check points and an output-control point are defined (Courtesy of PA Consulting Group, Hightstown, N.J.).

R&D, demonstrable benefits are essential, so the results of the work of these first teams need to be publicized if momentum is to be maintained. A case study will illustrate how process metrics work in practice in an R&D environment.

The 450-person pharmaceutical technologies organization of a large, multinational pharmaceutical company is spread over six sites in the United States and the United Kingdom. They are responsible for five critical processes in drug development—that is, the manufacture and packaging of clinical-trial supplies, validation, testing, batch record development and stability. Each of these sites had a different way of doing these same six fundamental processes. They had recognized that they needed to standardize the processes to improve quality and efficiency. There were other forces at work that also encouraged this organization to measure its processes—primarily, a new chief executive. However, in this post-merger company, the R&D function had yet to become integrated and the task of getting agreement on one way of doing things and then measuring and further improving the processes appeared daunting.

Implementing metrics in the pharmaceutical business is particularly difficult since the industry has been high-growth and very profitable, and measurement and control have not been necessary. This should now change, because of the Clinton health care reform, but organizations are slow to change and there is still reluctance on the part of scientists to measure their own performance. The success of this project was due, in large part, to the commitment and determination of the Quality Assurance Director. The approach she took is illustrated in Figure 3 and described on the following page.

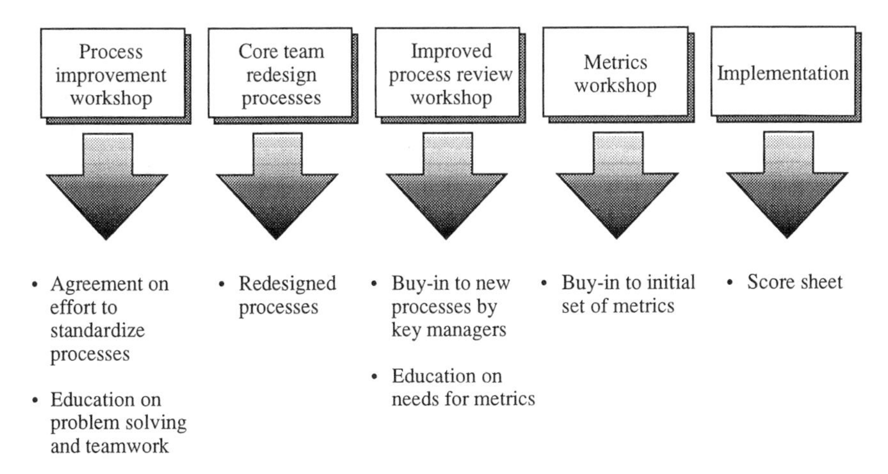

Figure 3 A multinational pharmaceutical company took a systematic approach to designing and implementing process metrics (Courtesy of PA Consulting Group, Hightstown, N.J.).

- In March 1993, a three-day process-improvement workshop was held to explain the need for common processes and to teach the group about problem-solving and team work. The workshop comprised representatives of the six processes at each of the six sites. During the workshop, the teams agreed to define common processes and to design the best system they could, rather than to pick out the best of the processes that already existed.
- A core team of seven people was selected for each process. Over the next three months, these teams redesigned the key processes.
- In September 1993, an improved process review workshop was held to review the draft processes and to reach consensus. The idea of measurement was again raised, and this time, because of the education that had been provided on the subject, the team was much more willing to accept it. This was driven somewhat by their willingness to show senior management that they had significantly improved their processes.
- In October, a one-day metrics workshop was conducted to develop preliminary measures for the processes. The group attending the workshop was also taught how to overcome the skepticism that it would encounter in the organization as it went ahead and tried to implement the metrics. Specific features included having the group apply metrics to purchasing—a process in which they had no vested interest, but with which they were familiar. This approach worked well because it helped them overcome their fear of metrics.
- The teams finally gained management buy-in to the metrics and agreed to implement them.

In a five-month period, the organization thus streamlined its processes and developed and implemented a range of metrics specific to its business. An illustration of the results is shown in Figure 4. Points A, B, and C are output-control points. At point A, for example, the measure is the percentage of on-time deliveries to the clinicians—a time measure to ensure that agreed dates of delivery to the clinical receiver are being met. At point C, the measure is the percentage of completed packaging records accepted by Quality Assurance—a quality measure to ensure that the packaging record is completed correctly.

Points D to J are in-process check points. At point D, for example, the measure is the percentage of stability reports issued on time. This is a time measure to ensure that stability reports are available for regulatory filings to meet the clinical milestones. Point I measures the percentage of master batch records requiring rework before approval by Quality Assurance—a quality measure to ensure that the documentation is correct the first time it is submitted for approval.

The benefits from this were enhanced communication and team work, reduced

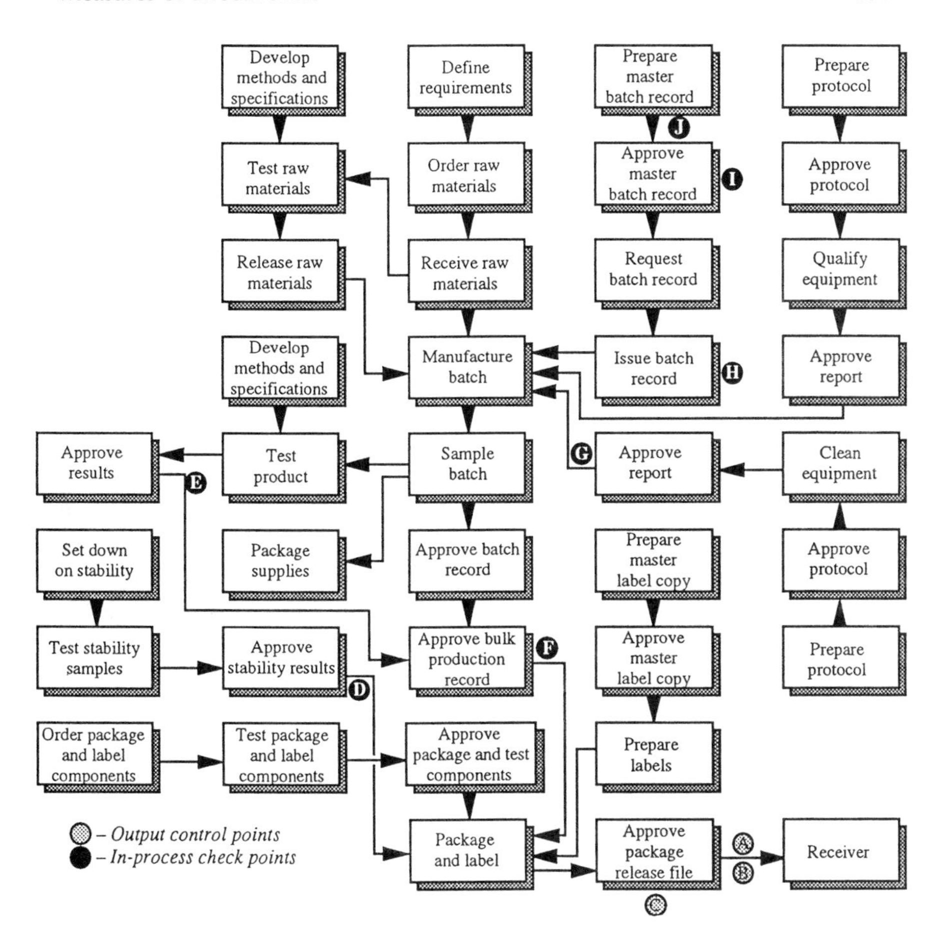

Figure 4 The organization streamlined its processes and implemented output-control points and in-process check points specific to its business (Courtesy of PA Consulting Group, Hightstown, N.J.).

complexity, and improved quality through the incorporation of the latest regulatory recommendations. The number of Standard Operating Procedures has decreased from 641 to just over 100. Fewer documents are needed in order to release bulk drug product to packaging—from fourteen to four. Approval signatures for this release have decreased from five to one. The same documentation systems and policies are in effect at all worldwide pharmaceutical technologies sites enabling greater flexibility.

E. Strategic Metrics Measure Progress Toward the Organization's Vision

Strategic metrics are used to focus everybody's efforts effectively on achieving the vision. As a first step, the vision must be quantified so that people know what to aim for. Quantifying a vision means really questioning an organization's vision statement. What does "the leading pharmaceutical company" really mean? How many new drugs need to emerge from the pipeline? What will the average development time be for these drugs?

Once the vision has been quantified, the senior management team needs to decide how they will each measure their own progress toward this vision. They then cascade the vision down through all levels of their own organizations in a series of workshops, where objectives are defined and metrics are set at each level. Figure 5 illustrates the process of the cascading of the vision and the resultant measures for an illustrative R&D hierarchy.

For example, as the figure indicates, relevant metrics at the highest level of the organization might be return on total investment in R&D, and performance versus strategic objectives. Management buy-in to this metric carries it down through the company to its relevant counterpart at the individual researcher level, where it might be translated as turnover rate, and contracted deliverables delivered. The following example will show how the focused use of strategic metrics turned around an ailing TQM program for one R&D organization.

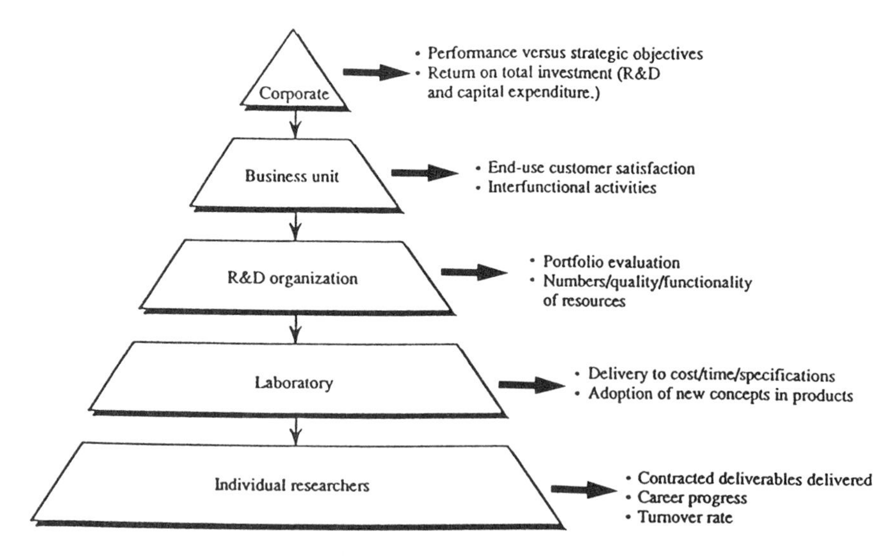

Figure 5 The development of strategic metrics should follow the cascading of the vision (Courtesy of PA Consulting Group, Hightstown, N.J.).

A certain company was a player in the telecommunications industry with 500 employees in its R&D organization, and had, for the past two years, been pursuing eight separate improvement projects across three levels of its R&D organization. Senior management was focusing on relationships with major customers, the performance appraisal system, and an environmental review. At the manager level, the expense reimbursement process and site planning were being scrutinized, as were ways to increase degrees of empowerment. Staff projects consisted of defining internal "receiver" requirements and standardizing suppliers and parts.

All of these projects appeared to have merit, but a sense of frustration permeated the organization. No one was able to see any real progress in getting new products out the door, and customer relations were still strained, as a result of missed deadlines and broken promises. Senior management was troubled by the training expenditure and was beginning to fall back into old habits, such as retaining all technical decision-making power. Critical processes that frustrated multiple levels of the organization had not been addressed, including those governing decision-making, project approval, funding, and patent application. There was no perceived improvement in development times, nor any increase in outputs from innovation. In short, the company was in a worse position than when it started its TQM journey. Before, the prevailing mood had been open and positive. By this time, management was in a state of agitation and staff were disgruntled.

Once senior management was convinced that TQM was worth another try, the cascade concept was introduced as a means of linking the different levels in the organization and getting everyone on the same improvement track. Simultaneously, total quality metrics were introduced as the appropriate drivers for total quality activities and an essential collective tool for gauging progress and results. A cascade of workshops was designed to restart the TQM program and get it focused on real improvements in R & D performance.

A two-day leadership workshop was conducted, in which senior management developed a realistic vision for the future of the company, agreed on specific and relevant objectives with meaningful measures to support the vision, and obtained a personal commitment to the required changes. Once the managers at the highest level of the organization had bought in, they led a set of workshops to get the commitment of their direct reports. Workshops to win the commitment of laboratory staff were the final step in this cascade, illustrated in Figure 6.

The results were predictable. All the activities undertaken in support of the objectives were focused on achieving real business results as a consequence of satisfying the performance-based measurements. In general terms, implementing strategic metrics created a common sense of purpose, revitalized an ailing program, and increased the sense of urgency for implementing improvements. Specifically, the initiative created a focus on pushing new products out, encouraged

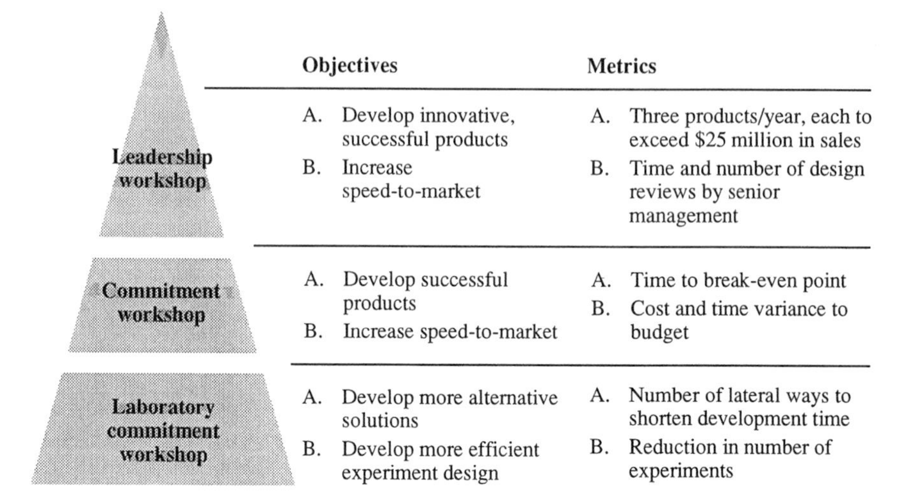

Workshop	Objectives	Metrics
Leadership workshop	A. Develop innovative, successful products B. Increase speed-to-market	A. Three products/year, each to exceed $25 million in sales B. Time and number of design reviews by senior management
Commitment workshop	A. Develop successful products B. Increase speed-to-market	A. Time to break-even point B. Cost and time variance to budget
Laboratory commitment workshop	A. Develop more alternative solutions B. Develop more efficient experiment design	A. Number of lateral ways to shorten development time B. Reduction in number of experiments

Figure 6 In a series of workshops, relevant metrics were selected to monitor the achievement of objectives at all levels of the organization (Courtesy of PA Consulting Group, Hightstown, N.J.).

the creation of four new products, reduced design review time from two months (four stages) to two weeks (three stages), reduced the time to break-even point by 15 percent, and improved time variance. Because cost variance deteriorated, as a result of more subcontracting, there is now a focus on measuring subcontracting volume.

F. Sophisticated Culture Metrics Are Useful for Organizations Pursuing TQM With a Vengeance

Culture metrics are useful for those organizations intent on pursuing TQM with a vengeance. They derive from the transition model and are used to chart an organization's progress on the TQM journey. Each of the four stages of the transition model is defined by a level of organizational performance, which research has shown to have six main dimensions:

- *Customer*—striving to find out what customers' requirements are, and continually striving to exceed their expectations.
- *Leadership*—Defining a long-term vision for the organization, ensuring that staff understand and are committed to it, and developing and maintaining an appropriate and consistent management style.
- *People*—systematically involving everyone in team work.
- *Knowledge*—making decisions based on knowledge and facts, not hunches.

- *Processes*—managing processes that result in the satisfaction of customer needs, and recognizing that those processes are inherently cross-functional.
- *Learning*—demonstrating the culture of continually striving to do better.

The levels of organizational performance associated with each of these dimensions, for each stage of the transition model, are shown in Figure 7.

Each R&D unit, either a division in a large organization, or a department in a smaller organization, decides what stage it considers itself to be at in the transition process and then conducts a survey to determine, on a scale of one to five, how well it fits the applicable descriptors. For the areas in which it scores low, the unit decides whether that is a factor that needs to be addressed as a high priority, and if so, incorporates it into its annual quality plan. A case study will illustrate how culture metrics works in practice.

A large international chemical company located in the American Midwest had embarked on a very successful TQM program. Indeed, in a matter of only a couple of years, it had made it to the end of Stage 2 of the transition model. It had a focused quality process and a senior R&D management team committed to TQM. During the last eighteen months, it had moved to Stage 3 but was encountering some fundamental impediments to progress. The most significant was that the senior management team could not decide what to do next with the limited resources it had.

The members of the senior management team locked themselves away for several hours and evaluated their progress with reference to the Figure 7 grid. To their amazement, they could not, as a group, agree on how far they had progressed, since each had a different perception of how well each item in the grid had been implemented. Because they could not agree on where they were, they could not decide on what to do next; indeed for nearly every item, half of the team thought that it was too early to do it, and half the team thought it was already done. They were able to resolve the issue by highlighting areas of weakness with reference to the grid—that is, those areas where they were only in the early part of Stage 2. Once these areas of weakness were known, it was a simple matter to set priorities for their TQM initiatives and develop a quality plan for the next year, as shown in Figure 8.

Clearly, an R&D organization has to be fairly sophisticated, in the quality sense, to get the benefits out of these culture metrics. It needs to be able to survey and evaluate itself realistically, and to have an annual quality planning process that is tied into the R&D planning process. These are traits that are typical of a Stage 3 TQM organization.

G. Summary

The first part of this chapter has discussed how industrial research organizations have successfully used metrics to improve their performance, overcoming the

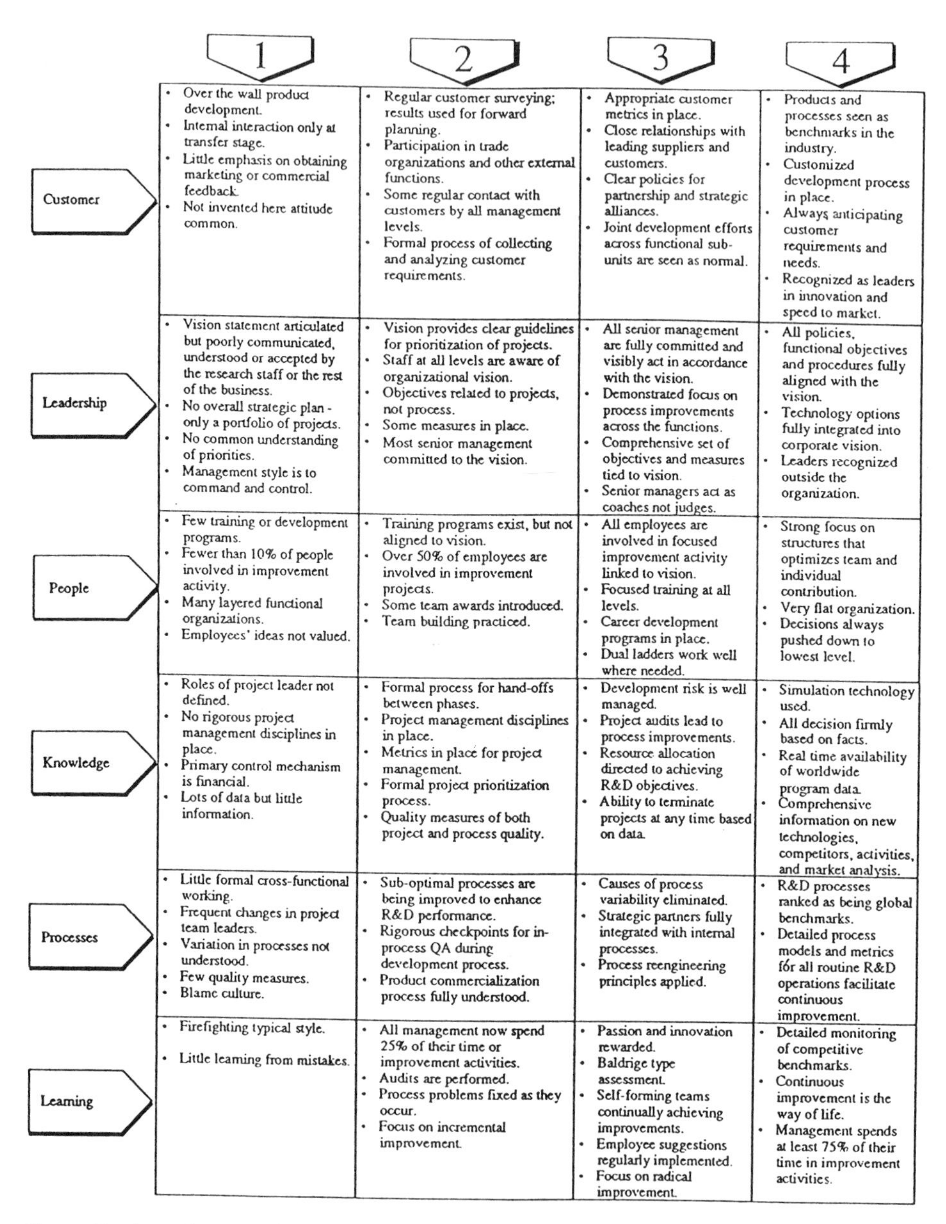

	1	2	3	4
Customer	• Over the wall product development. • Internal interaction only at transfer stage. • Little emphasis on obtaining marketing or commercial feedback. • Not invented here attitude common.	• Regular customer surveying; results used for forward planning. • Participation in trade organizations and other external functions. • Some regular contact with customers by all management levels. • Formal process of collecting and analyzing customer requirements.	• Appropriate customer metrics in place. • Close relationships with leading suppliers and customers. • Clear policies for partnership and strategic alliances. • Joint development efforts across functional sub-units are seen as normal.	• Products and processes seen as benchmarks in the industry. • Customized development process in place. • Always anticipating customer requirements and needs. • Recognized as leaders in innovation and speed to market.
Leadership	• Vision statement articulated but poorly communicated, understood or accepted by the research staff or the rest of the business. • No overall strategic plan - only a portfolio of projects. • No common understanding of priorities. • Management style is to command and control.	• Vision provides clear guidelines for prioritization of projects. • Staff at all levels are aware of organizational vision. • Objectives related to projects, not process. • Some measures in place. • Most senior management committed to the vision.	• All senior management are fully committed and visibly act in accordance with the vision. • Demonstrated focus on process improvements across the functions. • Comprehensive set of objectives and measures tied to vision. • Senior managers act as coaches not judges.	• All policies, functional objectives and procedures fully aligned with the vision. • Technology options fully integrated into corporate vision. • Leaders recognized outside the organization.
People	• Few training or development programs. • Fewer than 10% of people involved in improvement activity. • Many layered functional organizations. • Employees' ideas not valued.	• Training programs exist, but not aligned to vision. • Over 50% of employees are involved in improvement projects. • Some team awards introduced. • Team building practiced.	• All employees are involved in focused improvement activity linked to vision. • Focused training at all levels. • Career development programs in place. • Dual ladders work well where needed.	• Strong focus on structures that optimizes team and individual contribution. • Very flat organization. • Decisions always pushed down to lowest level.
Knowledge	• Roles of project leader not defined. • No rigorous project management disciplines in place. • Primary control mechanism is financial. • Lots of data but little information.	• Formal process for hand-offs between phases. • Project management disciplines in place. • Metrics in place for project management. • Formal project prioritization process. • Quality measures of both project and process quality.	• Development risk is well managed. • Project audits lead to process improvements. • Resource allocation directed to achieving R&D objectives. • Ability to terminate projects at any time based on data.	• Simulation technology used. • All decision firmly based on facts. • Real time availability of worldwide program data. • Comprehensive information on new technologies, competitors, activities, and market analysis.
Processes	• Little formal cross-functional working. • Frequent changes in project team leaders. • Variation in processes not understood. • Few quality measures. • Blame culture.	• Sub-optimal processes are being improved to enhance R&D performance. • Rigorous checkpoints for in-process QA during development process. • Product commercialization process fully understood.	• Causes of process variability eliminated. • Strategic partners fully integrated with internal processes. • Process reengineering principles applied.	• R&D processes ranked as being global benchmarks. • Detailed process models and metrics for all routine R&D operations facilitate continuous improvement.
Learning	• Firefighting typical style. • Little learning from mistakes.	• All management now spend 25% of their time or improvement activities. • Audits are performed. • Process problems fixed as they occur. • Focus on incremental improvement.	• Passion and innovation rewarded. • Baldrige type assessment. • Self-forming teams continually achieving improvements. • Employee suggestions regularly implemented. • Focus on radical improvement.	• Detailed monitoring of competitive benchmarks. • Continuous improvement is the way of life. • Management spends at least 75% of their time in improvement activities.

Figure 7 At each stage of the transition model, certain levels of organizational performance are expected (Courtesy of PA Consulting Group, Hightstown, N.J.).

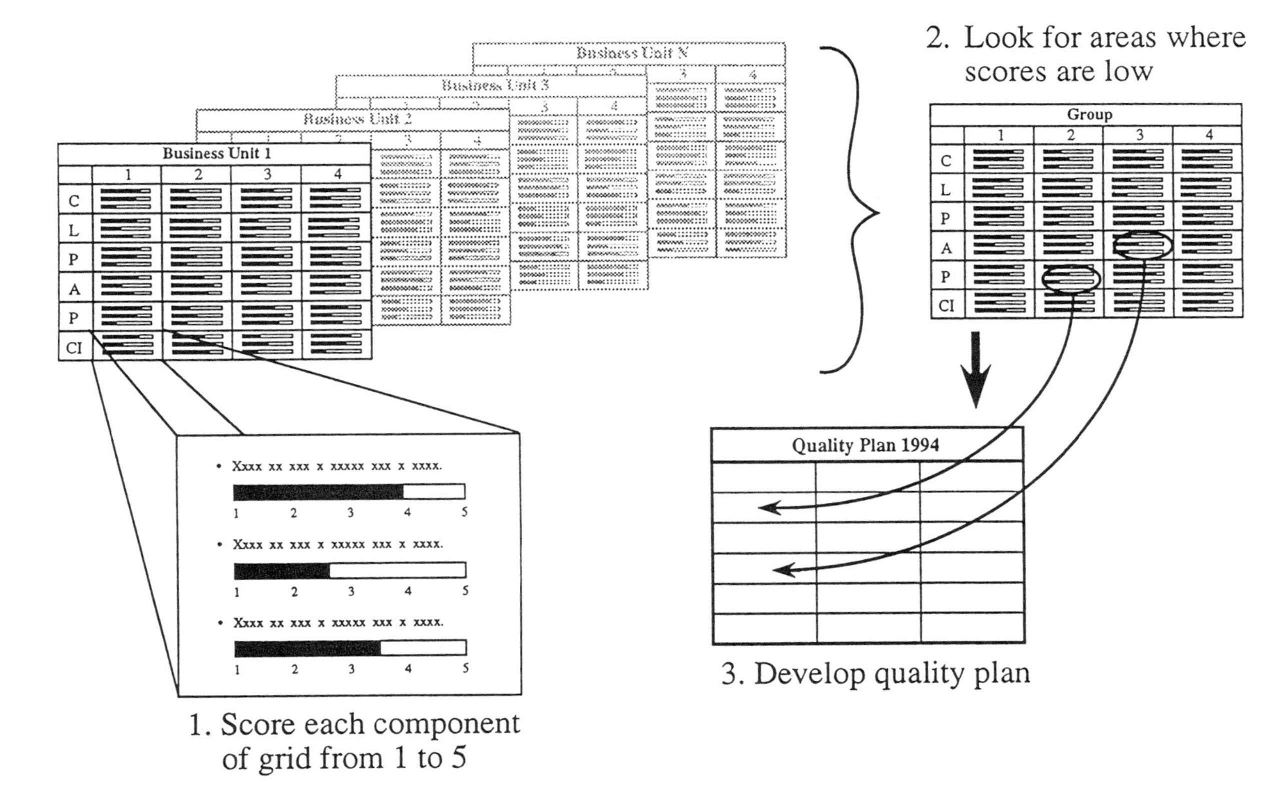

Figure 8 Once areas of weakness are known, it is simple to set priorities and devise a quality plan (Courtesy of PA Consulting Group, Hightstown, N.J.).

scientist's natural skepticism about measurement. As has been described, different types of metrics are clearly appropriate at different stages on the journey towards total quality. Hopefully, the three case studies have illustrated simple and practical approaches to implementing metrics and will encourage others to pursue the benefits that these R & D organizations have achieved.

II. DEFINING METRICS FOR EVALUATING AND IMPROVING BASIC EXPERIMENTAL SCIENCE (1)

A. Introduction

As previously discussed, industrial R & D laboratories have been surprisingly successful in developing performance objectives and metrics that convincingly show that planning, managing, and improving techniques can be value-added to the actual output of R & D organizations. This section will first discuss the more difficult case of developing analogous constructs for much of the basic experimental science performed in universities or at national laboratories, and, in order to take the most difficult case, will focus on areas like experimental high-energy physics and experimental astrophysics. Unlike most applied research and R & D in industry, the purpose of basic experimental science is producing new knowledge (usually published in professional journals) that has no immediate application to the first link (the R) of a planned R & D chain. Consequently, performance objectives and metrics are far more difficult to define. If one can successfully define metrics for evaluating and improving basic experimental science, which is the most difficult case, then defining such constructs for applied science should be much less problematic.

The remainder of the section will discuss two important issues. First, if one is going to define a value-added role for quality management principles in a basic science environment, then it is important to make the distinction between *science-as-knowledge* and *science-as-practice*. Second, defining a value-added role for quality management in a basic science environment requires that we identify believable metrics for the evaluation and improvement of basic experimental science-metrics that the scientist agrees are salient.

B. Science as Knowledge, Practice, and Map Making

One of the most important components of the application of quality management in any environment is to view the work performed by organizations as processes. In this section, some insight into the nature of the process of experimental science will be provided by discussing (and deconstructing) some modern misconceptions about the nature of science and describing a more salient account of the nature of scientific practice and the knowledge that is produced by the practitioners of basic experimental science.

1. Science as Doing, Not Just Thinking

In response to the question, "What *is* experimental science," most people (including scientists) tend to characterize science as associated to its product—knowledge—and leave out the organizational and human factors associated with laboratory life. The actual practice of experimental science is better described when we make a distinction between science-as-knowledge and science-as-practice (2,3). Although the word "science" does come from the Latin *scientia* (having knowledge), its primary meaning with reference to experimental science following the 17th century scientific revolution is the experience of *vexing* nature to understand its properties (4,5). For Galileo, the act of doing an experiment (not the voyeuristic study of Aristotelian "texts") produced a kind of knowledge that could not be obtained in any other way (6). In other words, through the practice of actually performing experiments, scientists acquire knowledge by doing and master the skills that will enable them to create physical effects in their detectors. Only when we begin to view science as both knowledge and practice can we even begin to define experimental science as a process.

To say that experimental science can be defined as a process is not to say that science as practiced in all of its disciplines is reducible to a single, systematic set of rules. One of the most widespread myths that persists about the nature and conduct of science is that scientists are governed by something called the "Scientific Method." The recalcitrance of the myth is due partly to the fact that this common, naive view of science is shared even by many scientists who not only claim they know what the scientific method is but believe that they actually use it in the practice of science. But these same scientists cannot point to a book that fully describes the scientific method as practiced in all disciplines, they cannot agree among themselves about what constitutes the scientific method, and, most importantly, when one observes the actual practices of laboratory life, they bear little or no resemblance to these rule-like procedures. Unfortunately, some quality professionals, in their zeal to support the equally naive claim that experimental science is "just like any other kind of work," have adopted the myth of the scientific method. This one-size-fits-all mentality has seriously hindered the development of the theory and practice of quality management in research environments (7).

2. Science as Map Making

A prime example of the tendency to characterize science strictly as knowledge (not practice) is the emphasis that is placed (especially by scientists) on the published scientific paper. The fact is that publications are produced after-the-fact (downstream) of the creation of physical effects in the experimental apparatus. Viewed metaphorically, all the documenting and recording activities that go into the production of a scientific paper constitute the cartographic function of science—the map making function. In the same way that explorers have to physically travel through an unexplored region of the earth (drawing maps of that region as they go), the experimental scientist uses her experimental apparatus as an extension of her physical senses to explore uncharted physical domains, drawing experimental maps of the measurements and newly created effects and phenomena she encounters. If the experimental map is drawn accurately, it enables other experimenters to move directly to the frontier of scientific knowledge and continue mapping out physical effects that lay beyond that point. It is the *practice* of science that most appropriately defines what basic experimental science is, not the experimental maps that are subsequently published in journals and traded back to funding agencies for additional resources.

Given the importance of scientific practice, it is somewhat enigmatic that scientists tend to remove references to human and social factors from their accounts of science, leaving the impression that the knowledge or "facts" contained in scientific publications *rise above* the organizational and human factors of laboratory life once the experiment is completed. This tendency to leave out social factors has been documented by numerous authors (8–10). Part of this is due to the literary convention established by the editors of scientific journals that reinforces (or helps cause) this tendency to eradicate human factors, laboratory organizational issues, social factors internal and external to the collaboration, human judgments, and interpretations of data or theories, as well as discussion of other scientists' perceptions of the skills of the experimenters (11,12). As important as these human and social factors are to describing what science is within the context of a laboratory, they are *left out* of the scientist's retrospective account. When experimental science is viewed as being constituted by both science-as-knowledge and science-as-practice, it is clear that the physical effects described in scientific publications cannot rise above the human and material setting of laboratory life once they are published as experimental maps, but are inextricably bound to the experimental apparatus and the mundane work processes of the organizational infrastructure and collaboration who performed the experiment (13,14). Modern-day experiments do not exist independent of the organizational infrastructures of the laboratories in which they are performed, and all aspects contained in the scientific publication must be ultimately reducible to work processes performed in (or associated with) the laboratory. If one can define the theoretical and experimental background against which an experiment is performed along with the measurement properties and characteristics of the experimental apparatus and the data-recording techniques (which collaborations have to define for the publication), and then add back in descriptions of the human, social, and managerial processes that constitute the laboratory organizational infrastructure and the experimental collaboration, then the factors that constitute science-as-practice begin to emerge.

C. Metrics for Basic Experimental Science

It has been commonplace for laboratories like Fermi National Accelerator Laboratory to develop strategic plans. This involves a discussion of related science that is taking place. Given the scale, cost, and complexity of high-energy physics, there is the principle of not duplicating work at other labs. This is a macro (strategic) metric and will not be discussed. Rather, attention will be focussed on the more difficult case of defining micro (process) metrics.

In this section the term "metric" is used to indicate a standard of measurement. But a distinction should be made between a direct metric (a close logical, causal, or consequential relationship to the things being measured) and an indirect metric (a collateral or circumstantial relationship to the things being measured). When possible, direct metrics are to be preferred because the data they provide more veridically characterize the things being measured. This section will describe how the metrics for evaluating science-as-knowledge and science-as-practice meet along the interface of the metric of *stability* because (as described earlier) the knowledge contained in scientific publications is inextricably bound to the experimental apparatus and the mundane realities of scientific practice. It is at the interface of stability that one sees the relationship between the in-process direct metrics defined by scientists and quality management elements. This section will describe why this type of in-process metric for science-as-practice cannot be used to evaluate science-as-knowledge until after an experiment is complete (because evaluations of science-as-knowledge are by necessity retrospective.) The section will also describe the notion of continuous improvement for both science-as-knowledge and science-as-practice.

Although his work has been plagued by various philosophical problems (15), Thomas Kuhn's notion of a paradigm is still one of the most compelling conceptual frameworks for evaluating science-as-knowledge and science-as-practice. For Kuhn, a paradigm is a matrix of "consensus standards" for a scientific community—a matrix that is constituted by a shared network of commitments to conceptual, theoretical, instrumental, and methodological ways of doing science (16). Other types of metrics are embodied in Alvin Weinberg's internal and external criteria that he claimed should be used to evaluate which types of science receive funding. His internal criteria should answer two questions: First, is the field ready for exploitation? Second, are the scientists in the field really competent? Weinburg claimed that these decisions could be made only by scientists. He identified three external criteria that could be decided by nonscientists: technological merit, social merit, and scientific merit. The criterion of scientific merit assessed the degree to which the knowledge produced by the discipline requesting funding contributed to its neighboring disciplines (17). Kuhn's "consensus standards" are articulated in the textbooks and journal publications (experimental maps) used to train new scientists in that field and scientists gains status in the community to the degree that they can articulate the parameters of the paradigm and design theoretical and experimental puzzles that heuristically probe and test it in every conceivable way. In other words, when a collaboration claims to have created physical effects that have never before been produced in the laboratory, the parameters of the effect are compared to the consensus standards of the paradigm—to everything else they know—to test the validity of the claim. These tests are most convincing when they use methodologies or apparatus that rely on laws of physics unrelated to the first experimental design.

1. Metrics for Science-as-Knowledge

If one is evaluating whether an experiment has produced "good" physics, "bad" physics, or just "meat and potato" physics, one could use Irvine's and Martin's indirect metric of counting the number of citations that a particular publication receives, but this type of indirect metric is based on the inference that if a scientific publication contains important scientific results, then it will be cited more frequently than one that does not (18,19). Whatever this type of metric tells us, it provides no direct indication of the quality of the content of the publication as well as do the following two types of metrics for science-as-knowledge:

The first type is described by Peter Galison's dual metrics of directness and stability (20). By directness, he means activities that bring experimental reasoning another rung up the ladder of causal explanation like the measurement of a background that was previously only calculated or the separate measurement of two sources of an effect previously only measured together. By stability, he means all the experimental procedures that vary some feature of the experimental setup (including changes in the test substance, the apparatus, the arrangement, or the data analysis) and leave the result basically unchanged. The metric of stability is especially important when a collaboration has controlled for all known background effects and altered numerous parameters of the experimental apparatus to control for artifacts, and the proposed effect "just won't go away." Latour (14) describes fundamentally the same notion in terms of the "trials of strength" that various effects and substances endure at the hands of experimenters. More importantly for this metric, other experimentalists can test the reality or artifactuality of the proposed effect by attempting to produce it using apparatus that rely on totally unrelated laws of physics. For example, Hacking (13) describes how interference, polarizing, phase contrast, direct transmission, and fluorescence microscopes can be used to discern the same basic microstructure using essentially unrelated aspects of light. Also, Galison (20) describes how neutral currents could be detected using bubble chambers or electronic counter detectors. The fact is that each variation introduced into the experimental design makes it more difficult to postulate an alternative causal story that will satisfy all the observations because the effect is nested within ever more complex loops of experimental demonstration (20). William Wimsatt (21) describes the theoretical equivalent, namely how our scientific theories and beliefs about the world are nested (generatively entrenched) within yet wider systems of beliefs. Theories and beliefs that are generatively entrenched are called "robust" because of their interconnectedness to the entire body of knowledge.

A second type of direct metric for science-as-knowledge is captured in the function of "crucial" experiments. While some recent philosophers of science have come to interpret the function of crucial experiments as absolute arbiters of scientific theories, this is not how it was conceived by the father of experimental science, Francis Bacon. As pointed out by Hacking (13), a more literal rendering of Bacon's Latin *Instantiae crucis* is instances of the "crossroads" or "finger posts," conjuring up the image of finger post signs that were set up where roads parted in order to indicate several directions. In other words, crucial experiments are those that indicate any new (previously unknown) direction that yields new, previously uncreated effects, phenomena, or more direct and stable measurements, not necessarily those that test competing theories. If a scientist performs an experiment that sends her down the wrong road (does not lead to new effects, etc.), this experiment may be both crucial and not problematic because it produces knowledge of ways in which nature will not be vexed, and this is valuable information to be added to the overall cartographic file on that region of nature. The scientist simply retraces her experimental steps back to the misleading finger post, selects another direction, and continues on exploring and map making.

While the principles of continuous improvement are normally associated with a manufacturing environment, designing experimental apparatus that will yield more direct and stable measurements is one crucial component of the continuous improvement of science-as-knowledge. Although it is common folklore to claim that experiments are repeated, in reality, scientists almost never repeat the same experiment. Rather, follow-up experiments are almost always attempts to improve the directness and stability of some parameter of the same phenomenon, often using improved techniques for detection, data acquisition/monitoring, or data analysis. Bodnarczuk (22) has performed a case study that shows how high-energy physics experiments are actually performed in "strings" where previous experiments and follow-up experiments meet at a transition-like interface where the first experiment is transformed into its follow-up progeny by the collaboration. Each experimental configuration in the string displays a more complex iteration (improvement) of the original apparatus design, but leaves the fundamental design of the experiment largely intact. The case study also shows how these variations in experimental design are anchored to certain continuities in the physics goals, the apparatus configuration, and the scientists who perform the experiment. These continuities allow the experimenters to (1) avoid the physics uncertainties involved in designing and building entirely new detectors, (2) continuously improve the directness and stability of the measurements, and (3) avoid the sociological uncertainties of securing resources from laboratory management or an outside funding agency. There may be similar experimental and sociological continuities in the table-top experiments that constitute research programs, with variances being due largely to the scale, cost, and complexity of the experiments.

2. *Metrics for Science-as-Practice*

When a scientist proposes a program of basic research, he or she is normally standing on the frontier of knowledge beyond which no experimental maps have yet been drawn. In U.S. Department of Energy environments, experimental proposals are normally codified into field work proposals (FWPs) which become the basis upon which funding is granted. In an FWP, the scientist makes certain knowledge claims that are calculated hypotheses of the experimental path that must be taken and the expected results that may or may not be corroborated when the experiment is actually performed. Knowledge claims define what new knowledge the scientist believes he or she will obtain by performing the experiment. But the scientist must also make certain practice claims whereby he defines planning and managing elements that can be thought of as strategic predictions of how that new knowledge will be obtained. This includes estimates of how long it will take to travel the experimental path, the types of experimental equipment that will be needed, how much engineering and technical support will be needed, and other costs and resources. If the scientist begins this experimental journey as planned and performs an experiment that turns out to be a crucial experiment (pointing in a different direction), the scientist is obliged to follow the data. Often, the practice claims in FWPs that are two or more years away are "place holders" because the knowledge claims are like moving targets that cannot be well defined. But the knowledge claims become increasingly refined and fine tuned as they move toward the performance of the current year's research, and, consequently, the practice claims can be articulated and evaluated more precisely.

In terms of defining direct metrics for science-as-practice, once the experimental design (conceived at the frontier of knowledge), procurement, installation, and overall configuration of the apparatus has settled down enough to actually perform the experiments, the crucial concern is to bring the operation of the apparatus into a "steady state" in which all possible operational parameters of the apparatus are understood and functioning as designed. But, at this stage of experimenting, the vast majority of the processes that occur in the organizational infrastructure of the laboratory in which the experiment is embedded have little or no causal efficacy on the outcome of the experiment. In addition, scientists construct organizational and experimental "protective belts" around their work to protect it from all but the most devastating laboratory perturbations—usually resource and funding related. So how does one isolate the "vital few" activities and support organizational interfaces that can actually affect the output of an experiment? These vital few cluster along the interface between science-as-knowledge and science-as-practice, that is, the stability of the experimental apparatus upon which the knowledge that will eventually be published is based. Of all possible laboratory activities, the vital few are defined as only those that exert direct causal effects on the stability of the apparatus and computing because only those can affect the outcome of the experiment itself. Experiments in laboratories should be viewed as a locus of experiment and software, and other factors, around which a collaboration gathers resources and support in ever-widening circles, with each circle exerting less and less causal efficacy on the experimental work. The vast majority of perturbations in laboratory activities never penetrate to the first few circles surrounding the experiment. Those that do are normally absorbed by the innermost protective belts.

By way of illustration, one of the "vital few" components of the metric of stability is background noise in the apparatus. If the goal of the experiment is to create new, previously undetected phenomena, then the apparatus must physically isolate these properties and, at the same time, dampen all other nonsalient effects that could present themselves as artifacts. While it is impossible to enumerate all conceivable backgrounds, Galison describes how the collaborators must demonstrate to themselves (and eventually to other scientists) that they have accounted for all known backgrounds either by constructing the apparatus to block them, or measuring and calculating them and subtracting them from the data (20). Having accounted for all known backgrounds in construction, measurement, and calculation, the vital component of the metric of stability is to reduce the variances between the parameters of the experimental design and the apparatus' actual performance, and to validate or calibrate the operational parameters of the apparatus against the values of physical phenomena that are well understood.

Other examples of the "vital few" that could affect the stability of an experiment if they are not done properly are factors like (1) visually or computationally monitoring the apparatus to ensure that components like power supplies and gases operate properly and that calibrations remain valid; (2) ensuring that the proper materials, targets, and chemicals are being used; (3) ensuring that data rates are appropriate and the data acquisition and software systems are functioning as designed and intended; (4) ensuring that the cartographic function of basic experimental science is carried out in scientific notebooks and that correct data are recorded; and (5) ensuring that the appropriate measurement uncertainty analysis is performed on experimental results that are reported in the literature. Only these types of endeavors can seriously affect the production of scientific knowledge. They are a minority of laboratory activities. They are the only components that scientists are even interested in controlling.

As stated before, knowledge claims and the associated practice claims become increasingly fine tuned as the scientist moves toward the current year's research. Consequently, the practice claims can be articulated and evaluated more precisely. One would think that, at this stage of the process, one could begin to evaluate whether or not the practice claims in the FWP are actually "cashed out" in the performance of the experiment. In other words, one ought to be able to evaluate science-as-practice using the indirect metrics that constitute the majority of the practice claims that scientists make in their FWPs (time frames, milestones, equipment, computing, and human resource costs). But there are cultural values held by the scientific community that make evaluating most of the practice claims problematic.

The direct metrics for both science-as-knowledge and science-as-practice can be defined and agreed upon only because scientists value more direct, stable measurements and crucial experiments. What a community values will define its most salient metrics because values constitute the criteria of choice (23). Previously mentioned was the enigmatic tendency for scientists to omit the mundane organizational and human factors of laboratory life from scientific accounts once an experiment is completed. While scientists are forced to make practice claims about planning and managing their experimental work in order to obtain funding, most of them would prefer not to deal with the pedestrian realities of milestones, cost estimates for equipment, computing graduate students, and especially deliverables like publications. In a scientific culture that mythologizes (and nostalgically exalts) the archetype of the solitary, irreverent, creative, rebel scientist who emerges from his laboratory only to reveal new secrets of nature, being a great theorist or experimentalist is coveted. But being a great planner or manager of experimental work is viewed (at best) as a set of necessary skills that someone else should have so the scientist can go about the business of doing "real" scientific work.

In some ways, the tendency to devalue the planning and managing aspects of science is driven by the fact that most performance claims found in FWPs or other experimental proposals do not seem to contribute meaningfully to the production of science-as-knowledge, and an evaluation of these metrics gives no indication of the quality of the basic experimental science. Once the publication has been written, no one in the scientific community asks about how much the experiment cost, whether it was completed on time, how much scientists exceeded their allotted computing time, etc. Therefore, the successful achievement of a scientist's practice claims is trivialized as a value that is not to be sought by the very best and brightest scientists. Unfortunately, this attitude is modeled by senior professors for graduate students, which ensures that the problem will be perpetuated.

Indirect metrics for science-as-practice can only be defined when the majority of scientists (and funding sponsors) agree that they need to adopt a new philosophy that does not automatically accept missed milestones, increased requests for computing, and the inability to estimate other experimental resources as an inevitable aspect of doing basic experimental science. This is Point 2 of Deming's 14-point management philosophy (24). Only when these practice claims are factored into the scientific community's evaluation of a scientist's competence as a scientist and only when funding agencies develop methods for teasing apart the difference between problems that are intrinsic to basic research and those that result from poor planning and management techniques will we be able to use the practice claims of time frames, milestones, equipment, computing, and human resource costs as indirect metrics that will enable us to improve the conduct of basic experimental science. Adopting this new philosophy will not improve the *quality* of science-as-knowledge, but it will likely improve the *quantity* of science-as-knowledge by making the practice of science more efficient and effective.

Based on the preceding discussion, there appears to be no fundamental reason why many of the continuous improvement methodologies developed by experts like Juran could not be tailored and applied to problematic practice claims. Scientists and laboratory managers can form cross-functional teams of scientists, engineers, technicians, and others to examine why milestones are missed. Following the standard method of storyboarding, teams can analyze the symptoms of missed milestones, formulate theories about why they were missed, test the theories, and identify root causes. Was the problem attributable to a limitation that nature imposed on the experiment? Was it impossible to push the technologies involved any further? Was the problem an inevitable part of the pedagogic process of obtaining knowledge by actually doing an experiment and unavoidable? Or was it a systemic problem where a spokesperson had no authority to make her collaborators come through on their commitments? Was it a lack of planning or management on the part of the principle investigator or the inability to stop introducing new parameters and changes into an experimental design that should have been frozen? Was the problem due to the lack of supervision of a graduate student by his or her senior professor or laboratory manager? Problems imposed by nature may be unavoidable, but the practice of science would certainly be improved if most of the other problems were solved.

D. Conclusion

The history of science is replete with examples of how some unknown (or improperly understood) aspect of an experimental apparatus led to problematic data results that were only realized retrospectively. The turn of the century episode with the French scientist Blondlot and N-rays (15), the incident of Fermilab experiment E-1A's "alternating currents" as described by Galison (20), and the much celebrated incident of Fleischmann and Pons' "Cold Fusion" (see Chapter 1) are all incidents where a part of the experimental apparatus was not thoroughly understood. The problem with any type of preventative approach is that, during the process, we have only indirect (or intuitive) metrics that justify the cost associated with doing them, and it is impossible to predict how many errors or problems would have occurred without this kind of up-front planning. When scientists demand that laboratory managers demonstrate the value-added of management guidelines and controls, all one can do is appeal to history—to those instances where something did go wrong due to poor planning or management. In an analogous sense, we possess the same type of indirect (intuitive, retrospective) metrics for justifying the costs of performing basic experimental science itself. Much like planning and managing, it is impossible to predict how many discoveries would have been made if we had increased our commitment to this type of esoteric research. When Congress demands that scientists demonstrate their "value-added" of basic research, all scientists can do is to appeal to history—to those instances where something was discovered and eventually developed into new technological processes or products. In a world that demands instant demonstration of the effect of an activity on the "bottom line" before that activity will be supported, the preventive approach to problems will never be popular. It will always have to justify its existence by appealing to history.

REFERENCES

1. M. Bodnarczuk, "Science as Knowledge Practice, and Map Making: The Challenge of Defining Metrics for Evaluating and Improving DOE-Funded Basic Experimental Science," NREL/TP-320–5401, National Renewable Energy Laboratory, Golden, CO, 1993
2. A. Pickering, "From Science as Knowledge to Science as Practice," *Science as Practice and Culture*, University of Chicago Press, Chicago, IL, 1992, pp. 1–26
3. D. Hull, *Science as Process: An Evolutionary Account of the Social and Conceptual Development of Science*, University of Chicago Press, Chicago, IL, 1988
4. F. Bacon, *The Works of Francis Bacon*, J. Spedding, R. Ellis, D. Heath, editors, Taggard and Thompson, Boston, 1893
5. R. Boyle, *The Works of the Honorable Robert Boyle*, London, 1772
6. G. Galilei, *Two New Sciences—Including Center of Gravity and Force of Percussion*, translated by S. Drake, University of Wisconsin Press, Madison, WI, 1974, p. 66
7. H. H. Bauer, *Scientific Literacy and the Myth of the Scientific Method*, University of Illinois Press, Chicago, IL, 1992, p. 42
8. D. J. Kevles, *The Physicists: The History of a Scientific Community in Modern America*, Harvard University Press, Cambridge, MA, 1987, p. xiv
9. A. Pickering, *Constructing Quarks: A Sociological History of Particle Physics*, University of Chicago Press, Chicago, IL, 1984, p. 5
10. B. Latour and S. Woolgar, *Laboratory Life*, Princeton University Press, Princeton, NJ, 1986, p. 28

11. H. M. Collins (ed.), ''Knowledge and Controversy: Studies of Modern Natural Science,'' special issue of *Social Studies of Science*, Vol. 11, No. 1,
12. T. J. Pinch, ''The Sun-Set: The Presentation of Certainty in Scientific Life,'' *Social Studies of Science*, Vol. 11, pp. 131–158
13. I. Hacking, *Representing and Intervening*, Harvard University Press, Cambridge, MA, 1987, p. 149
14. B. Latour, *Science in Action*, Harvard University Press, Cambridge, MA, 1987, p. 63
15. J. Horgan, ''Profile: Reluctant Revolutionary,'' *Scientific American*, May, 1991, pp. 40 and 49
16. T. Kuhn, *The Structure of Scientific Revolution*, 2nd ed., University of Chicago Press, Chicago, IL, 1970, p. 35
17. A. Weinberg, ''Criteria for Scientific Choice,'' *Minerva*, vol. 1, 1963, pp. 159–171
18. J. Irvine and B. R. Martin, ''Basic Research in the East and West: A Comparison of the Scientific Performance of High-Energy Physics Accelerators,'' *Social Studies of Science*, vol. 15, 1985, p. 300
19. S. Yearly, *Science, Technology, and Social Change*, Unwin Hyman, London, 1988, p. 88
20. P. Galison, *How Experiments End*, The University of Chicago Press, Chicago, 1987, pp. 259–260
21. W. Wimsatt, ''Robustness, Reliability, and Overdeterminism,'' M. Brewer and B. Collins (eds.), *Scientific Inquiry and the Social Sciences*, Jossey-Bass, San Francisco, 1981
22. M. Bodnarczuk, ''The Social Structure of Experimental Strings at Fermilab; A Physics and Detector Driven Model,'' Fermilab-Pub-91/63, Fermi National Accelerator Laboratory, Batavia, IL, 1990, p. 14
23. R. K. Merton, *The Sociology of Science: Theoretical and Empirical Investigations*, N. W. Storer (ed.), University of Chicago Press, Chicago, 1973
24. H. R. Neave, *The Deming Dimension*, SPC Press, Knoxville, TN, 1990, p. 293.

APPENDIX: POSSIBLE R & D METRICS

Customer

Market share
Customer satisfaction survey
Number of real-time problems solved
Reported feedback frequency or comment cards returned
Response time to customers (written or verbal)
Contracted deliverables delivered
Turnover rate of individuals/project
Number of repeat internal customers

Leadership

Number of employees who know vision or mission
Percentage of employees who know their contribution to the vision or mission
Percentage of employees who recognize senior corporate management
Employee rating of management credibility
Expected value of new technology
Commercial value of patents
Number of patents challenged (won/lost)
Broadness of coverage for patents (concept not countries)
Number of patents in use
Ratio of number of patents in use/number of patents
Shape of the R & D portfolio risk profile
Percentage of research linked to business unit and corporate strategic plans
Revenue or profit from products and processes not existing three years ago
Dollar cost reduction or capital avoidance for or to plant
Percentage improvement in plant nonfinancial performance indicators
Percentage of core technologies of world-class character
Break-even time for products and processes
Licensing income
Dollars in cost reduction or avoidance
Number or value of commercial successes
Ratio of failures due to inadequate technology/failure to implement by plant
Number or value of past R & D developments in use

People

Diversity of work force skills
Skill mix of technical staff
Number of hours devoted to technical training

Utilization, value, application of training
Percentage of R & D staff with plant experience
Time customers' (internal or external) people on teams
Number of man-years people at customer (all levels)
Number of end-customer contacts
Number or percentage of plant and R & D joint teams
Ratio of team awards or rewards/all awards or rewards
Amount of business unit involvement in performance reviews
Number of suggestions
Percentage of suggestions implemented
Ratio of managers to employees
Annual percentage of controllable terminations of above-average performers

Knowledge

Number of product or process requirement changes
Percentage of projects that have conducted a competitor analysis
R & D time or cost variance versus budget
Percentage of projects that have pre-agreed, fully defined criteria
Budget process—elapsed time, person hours
Number of pages in the monthly reports
Number of projects controlled via common project management software

Processes and Systems

Number of key processes defined
Number of key process owners with process improvement targets
Percentage of key process measurements in place
Ratio of expected value to realized value
Projects without changes in requirements
Time of customer involvement in projects
Number of technologies adapted—not developed
Percentage of revenue generated by adapted or purchased technology
Cost improvements by adapted or purchased technology
Percentage of time working on R & D
Percentage of overhead level and trends
Internal quality assurance correction rates
Number of field changes in new products or processes
Percentage of processes transferred

Learning

Percentage of time employees allocate to continuous improvement activities
Malcolm Baldrige—self audit, apply, visit, win

Number of times benchmarked by others
Number of external technical awards (Fellows)
Number of citations of publications and patents
Percentage of people involved in continuous improvement activities

INDEX